Spon's
European
Construction Costs
Handbook

Spon's European Construction Costs Handbook

Edited by

DAVIS, LANGDON & EVEREST
Chartered Quantity Surveyors

Third edition

London and New York

First edition 1992
Second edition 1995
This edition published 2000 by E & FN Spon
11 New Fetter Lane, London EC4P 4EE

E & FN Spon is an imprint of the Taylor & Francis Group

© 2000 E & FN Spon

Printed and bound in Great Britain by TJ International, Padstow, Cornwall

Publisher's Note

This book has been produced from camera-ready copy supplied by the editors.

British Library Cataloguing in Publication Data
A catalogue record for this book is available from the British Library

Library of Congress Cataloging in Publication Data
Spon's European construction costs handbook / edited by Davis, Langdon & Everest.–3rd ed.
    p. cm.
  Includes bibliographical references and index.
  1. Building–Estimates–Europe–Handbooks, manuals, etc. I. Davis, Langdon, and Everest (Firm) II. E. & F.N. Spon.

TH435 .S72538 2000   00-024967
692'.5'094–dc21

ISBN 0–419–25460–9

Contents

Part Three: Comparative Data   413

Appendices

Appendices – contd.

Preface

The second edition of the European Construction Costs Handbook covered thirty countries. That was the first time that so many of the countires of Europe had been included and the time was right to acknowledge and describe many of the newly created countries of the Former Soviet Union. The information on many of these countries including Russia is however often very difficult to access and statistical data are dubious. In this third edition only twenty countries have been included and this has enabled coverage in greater depth and detail. As the information on other countries improves and as stability of their markets develops, some of them may be included in subsequent editions.

This book is designed to be a convenient reference. Its purpose is to present coherent snapshots of the economies and construction industries of Europe in an international context with the inclusion of the USA and Japan. It is not a substitute for local knowledge and professional advice. It will, however, be extremely useful as an introduction to a country and its construction industry for clients, consultants, contractors, manufacturers of construction materials and equipment and others concerned with development, property and construction in Europe.

The book is in three parts – Part One: Construction in Europe; Part Two: Individual Countries and Part Three: Comparative Data. The twenty countries covered in the book are listed in the Contents. Two of these – the USA and Japan – are outside of Europe but are included for comparative purposes.

Part One: Construction in Europe

Part One comprises an essay entitled *The construction industry in Europe*, which provides a number of economic and construction indicators for the countries covered in this publication and makes a brief comparison of the broad regions within Europe. Basic data are included for 39 countries..

Part Two: Individual countries

After an introductory section providing general notes, explanations and definitions the twenty countries are arranged in alphabetical order. The first page of each country chapter provides key data on population, geography, the economy and construction. Subsequently data on each country are presented in a similar format under the following main headings:

- *The construction industry* outlines the structure of the industry, tendering and contract procedures, liability and insurance, and regulations and standards

- *Construction cost data* includes data on labour and materials costs, measured rates for items of construction work and unit costs (approximate estimating) for different building types
- *Exchange rates and cost and price trends* presents data on exchange rates with the £ sterling, the Euro, the Japanese Yen and the $US, and includes data on the main indices of price movements for retail prices and construction
- *Useful addresses* gives the names and addresses of authorities, professional institutions, trade associations, etc.

Part Three: Comparative data

To allow rapid comparison between countries covered in the book, Part Three brings together data from Part Two and presents them under three main headings:

- *Key macroeconomic indicators* including economic, geographic and demographic data
- *Construction output indicators* including overall construction output and output per capita
- *Construction cost data* including labour and materials costs and approximate estimating costs per square metre.

Appendices

The appendices provide a reference section with data relevant to all parts of the book.

Davis Langdon Consultancy
London
1999

Acknowledgements

The contents of this book have been gathered together from a variety of sources – individuals, organizations and publications. Construction cost data and general background information on local construction industries are based on contributions from a network of professional colleagues and associates worldwide. These include:

Austria	IL BAU, Wien, Austria
Belgium	Centre Scientifique Immobilière, Boortmeerbeek, Belgium
Republic of Cyprus	JMV QS Service, Nicosia, Cyprus
Czech Republic	URS Praha, Praha, Czech Republic
Finland	VTT Technical Research Centre, Tampere, Finland
France	Davis Langdon Economistes, Paris, France
Germany	Klaus Valet, Wendlingen, Germany
Ireland	Patterson Kempster and Shortall, Dublin, Ireland
Italy	Impregilo S.p.a., Milano, Italy
Japan	Construction Research Institute, Tokyo, Japan
Netherlands	Arcadis Bouw/infra, Utrecht, the Netherlands
Poland	PMI, Warsaw, Poland
Portugal	Engil, Lisbon, Portugal
Slovak Republic	UEOS Komercia a.s., Bratislava, Slovak Republic
Spain	Davis Langdon Edetco, Girona, Spain
Switzerland	PBK, Pfaeffikon, Switzerland
Turkey	Enka Construction & Ind. Inc., Istanbul, Turkey
UK	Davis Langdon & Everest
USA	R.S. Means Company Inc, USA

Much of the statistical data is from World Bank Development Reports, IMF International Financial Statistics, CIA Worldfactbook and Construction Statistics and official statistics. The background on individual countries is based on local sources, national yearbooks, annual reports and Economist Intelligence Unit reports.

Important sources of general and construction industry data are embassies, trade missions, statistical offices, and government departments in the UK and overseas. Information on international contracting is largely based on surveys undertaken by *Engineering News Record* and *Building* magazines. Data on exchange rates and consumer price indices come mainly from the *Financial Times*.

The research and compilation of this book were undertaken by Davis Langdon Consultancy. Special acknowledgements are due to Dr Patricia Hillebrandt.

Part One: Construction in Europe

1 The construction industry in Europe

Introduction

Since the last edition of this handbook there have been important changes in the European Union (EU). First, there has been a further enlargement from twelve to fifteen members, with the inclusion of Austria, Finland and Sweden. Other countries have applied for membership including Cyprus, the Czech Republic, Estonia, Hungary, Poland and the Slovak Republic. Secondly, in May 1998 the countries that were willing – and able – to be part of the European Monetary Union (EMU) were selected. Denmark, Sweden and the UK opted out of EMU, at least temporarily, whereas Greece failed to fulfill the Maastricht criteria for joining. Thirdly, in January 1999, The Single European Currency (the Euro) came into operation in the countries adhering to the EMU, although national currencies will continue to be used until January 2002. Monetary policy is in the hands of the European Commission and other European institutions, including the new European Central Bank. The long term scenario is for a further enlargement of the EU eastwards and a progressive convergence on many economic issues.

In recent years, economic development in the Western market economies has been beset by uncertainties on the effect of the Asian crisis, problems in Russia and other parts of the Former Soviet Union (FSU), concerns about South America and more recently the fluctuations in the Euro and the likely future of the US economy. Rarely has it been so apparent that the world economy must be seen in global terms and that the interaction of events throughout the world have impacts on individual countries often far greater than those of domestic policies.

However, overall, the countries of Western Europe have weathered the storm relatively well. Although it faltered in 1999, growth in gross domestic product (GDP) remained positive.

In Asia the initial crisis seems to be over and it is now a matter of rebuilding the financial infrastructure of the economies so that industry can once more prosper. In this, Japan is a key player and the government there is being closely watched to see whether it is prepared to take the necessary steps to place its financial policies on a sound footing and control the way in which financial institutions operate.

Russia, unfortunately is weak both politically and economically. Major improvement seems unlikely until after the Presidential elections of June 2000 and even then it is not known what the future holds. Russia and many of the other states of the FSU are now neither planned economics nor market economies. This might not matter if the roles of the two systems and their respective areas of operation were defined, but they are not. Change has been largely haphazard. A major requirement is for reform of institutions and the legal system without which a market economy cannot function efficiently.

In some of the countries formerly under the control of the FSU, progress has been better, notably in Hungary and Poland. The Czech Republic and, to a lesser extent the Slovak Republic, have experienced a setback in the last year or so, as the inadequacies of the financial management of public enterprises and of the banking system have come to light. Much of the former Yugoslavia and other countries in the Balkans, especially Albania, continue to be unstable politically and economically.

The performance of the US economy is vital to world economic activity. The economy of the USA has been expanding strongly, outperforming other areas of the world. There is concern at the future level of interest rates and the longer term impact of some of the developments in other parts of the world. Doubts are continually being expressed as to whether and how long a high growth rate can continue.

In total there are about fifty countries in the new Europe. The 39 countries covered in this section are grouped into those in the European Union, Western Europe not including the European Union, Eastern Europe in the CIS and other Eastern and Central Europe. In addition some comparative data for the USA and Japan are considered.

Construction output and GDP

The interrelationships between construction output and GDP are diverse and significant. They include the following:

- Because the new work component of construction output is investment, it has a long term effect on the level of GDP. More specifically, an increase in the level of GDP frequently requires a prior increase in investment; a decrease in the level of GDP is likely to reduce investment and the demand for construction more than proportionately.

- Because construction is a labour intensive industry, an increase in construction is likely, through the multiplier, to lead to an increase in GDP as a whole and conversely for a decline.

- The level of repair and maintenance and improvement is affected by the level of incomes and hence by GDP.

The tables that follow give some basic data on construction output and GDP for European countries for which data are available.

Table 1.1 shows GDP, overall construction output and construction output as a percentage of GDP. The common currency used to express the GDP of each country in terms of a common purchasing power is the $US. For consistency and relevant comparison the data relate to the year 1997 – the latest year for which information is available for all the countries listed. The commentary on the broad regions of Europe include some observations on changes taking place.

Table 1.1: GDP, construction output and the relationship between them, 1997

	GDP US$ billion	Construction output US$ billion	Construction output as % of GDP
European Union			
Austria (1)	214.0	29.0	13.6
Belgium	242.8	29.4	12.1
Denmark	169.9	17.0	10.0
Finland	120.0	12.6	10.5
France	1,395.6	109.7	7.9
Germany	2,099.4	222.4	10.6
Greece	120.1	14.4	12.0
Ireland	73.3	11.6	15.8
Italy	1,146.3	111.8	9.8
Luxembourg	16.4	2.4	14.6
Netherlands	360.8	39.1	10.8
Portugal	102.3	13.8	14.0
Spain	532.7	61.0	11.5
Sweden	227.9	22.6	9.9
United Kingdom	1,288.5	95.7	7.4
Total	8,095.0	793.0	9.8
Western Europe but not EU			
Republic of Cyprus	8.6	0.9	10.0
Iceland	7.2	0.7	10.0
Malta	3.5	0.4	12.0
Norway	167.6	16.5	9.9
Switzerland	255.6	25.5	10.0
Turkey	191.5	24.9	13.0
Total	634.0	68.9	10.9
Total Western Europe	8,729.0	861.9	9.9

Country	GDP US$ billion	Construction output US$ billion	Construction output as % of GDP
Eastern Europe in CIS			
Russia (2)	447.6	53.7	12.0
Ukraine	49.6	4.5	9.0
Others (3)	25.6	2.1	9.6
Total	522.9	60.7	11.6
Other Eastern and Central Europe			
Bulgaria	10.1	0.6	5.5
The Czech Republic	51.9	6.8	13.1
Estonia	4.5	1.0	22.4
Hungary	45.8	2.9	6.4
Latvia	5.4	0.7	12.6
Lithuania	8.8	1.2	14.2
Poland	135.9	13.8	10.2
Romania	35.0	4.6	13.2
The Slovak Republic	19.6	2.2	11.2
Others (4)	36.7	3.6	9.7
Total	353.7	37.4	11.6
Total Eastern and Central Europe	876.6	98.2	11.2
Total Europe	9,605.6	960.1	10.0

Notes: Percentages are based on unrounded figures.
(1) The figures for Austria have been amended in the light of the information in Euroconstruct conference proceedings, Prague, June 1999.
(2) including Russia in Asia.
(3) Armenia, Azerbaijan, Belarus, Georgia, Moldova.
(4) Croatia, Macedonia FYR, Slovenia.
Sources: 1997 data equivalent to that in key data sheets for each country in Part Two and DLC estimates based on Economist Intelligence Unit Reports.

Table 1.2 following shows population, GDP per capita and construction output per capita.

Table 1.2: Population, GDP per capita and construction output per capita, 1997

Country	Population million	GDP per capita US$ thousand	Construction output per capita US$
European Union			
Austria (1)	8.1	26.4	3,593.6
Belgium	10.2	23.9	2,888.1
Denmark	5.3	32.1	3,210.5
Finland	5.1	23.3	2,443.3
France	58.6	23.8	1,872.1
Germany	82.1	25.6	2,709.4
Greece	10.6	11.3	1,357.3
Ireland	3.6	20.3	3,210.0
Italy	57.5	19.9	1,943.3
Luxembourg	0.4	38.7	5,657.1
Netherlands	15.6	23.1	2,507.3
Portugal	9.9	10.3	1,438.4
Spain	39.3	13.5	1,551.7
Sweden	8.9	25.7	2,546.3
United Kingdom	58.8	21.9	1,626.4
Total	374.1	21.6	2,119.8
Western Europe but not EU			
Republic of Cyprus	0.7	11.6	1,162.0
Iceland	0.3	26.9	2,685.6
Malta	0.4	9.5	1,138.2
Norway	4.4	38.0	3,750.9
Switzerland	7.1	36.0	3,593.8
Turkey	63.7	3.0	390.5
Total	76.6	8.3	899.3
Total Western Europe	450.7	19.4	1,912.4
Eastern Europe in CIS			
Russia (2)	147.3	3.0	364.7
Ukraine	50.7	1.0	88.0
Others (3)	31.3	0.7	67.3
Total	229.3	2.3	264.9
Other Eastern and Central Europe			
Bulgaria	8.3	1.2	67.2
The Czech Republic	10.3	5.0	658.0
Estonia	1.5	3.1	692.7
Hungary	10.2	4.5	289.8
Latvia	2.5	2.2	276.0

Country	Population million	GDP per capita US$ thousand	Construction output per capita US$
Other Eastern & Central Europe (cont'd)			
Lithuania	3.7	2.4	334.6
Poland	38.6	3.5	358.3
Romania	22.6	1.5	205.0
The Slovak Republic	5.4	3.6	408.0
Others (4)	7.6	4.9	469.1
Total	110.6	3.2	338.6
Total Eastern and Central Europe	339.8	2.6	288.9
Total Europe	790.6	12.1	1,214.4

Notes: Percentages are based on unrounded figures.
(1) The figures for Austria have been amended in the light of information in Euroconstruct conference proceedings, Prague, June 1999.
(2) including Russia in Asia.
(3) Armenia, Azerbaijan, Belarus, Georgia, Moldova.
(4) Croatia, Macedonia FYR, Slovenia.
Sources: 1997 data equivalent to that in key data sheets for each country in Part Two and DLC estimates based on Economist Intelligence Unit Reports.

Table 1.3 below shows GDP based on a purchasing power parity basis and how that compares to exchange rates basis. Purchasing power parity is the exchange rate which would be appropriate to express an income in one country in terms of its purchasing power in another country.

Table 1.3: GDP on a purchasing power parity basis, 1997

Country	GDP on PPP basis US$ billion	GDP per capita on PPP basis US$ thousands	Construction output on PPP basis US$ billion	Construction output per capita on PPP basis US$ (1)	Ratio of exchange rate basis to PPP basis corrected to UK=100
European Union					
Austria (2)	174.1	21.6	31.1	3,847.6	118
Belgium	236.7	23.2	28.6	2,810.9	99
Denmark	122.5	23.1	12.3	2,314.9	134
Finland	102.1	19.9	10.7	2,078.6	113
France	1,320.0	22.5	103.8	1,770.7	102
Germany	1,740.0	21.1	184.4	2,245.6	116
Greece	137.4	12.9	16.5	1,553.0	84
Ireland	59.9	16.6	9.5	2,624.4	118
Italy	1,240	21.6	120.9	2,102.1	89

Country	GDP on PPP basis US$ billion	GDP per capita on PPP basis US$ thousands	Construction output on PPP basis US$ billion	Construction output per capita on PPP basis US$ (1)	Ratio of exchange rate basis to PPP basis corrected to UK=100
Luxembourg	13.5	31.8	2.0	4,642.4	117
Netherlands	343.9	22.0	37.3	2,389.6	101
Portugal	149.5	15.1	20.9	2,102.8	66
Spain	642.4	16.3	73.6	1,871.2	80
Sweden	176.2	19.9	17.4	1,968.9	125
United Kingdom	1,242	21.1	92.2	1,567.7	100
Total	7,699.8	20.6	761.0	2,034.2	101
Western Europe but not EU					
Republic of Cyprus	9.75	13.2	1.0	1,315.6	85
Iceland	5.7	21.2	0.6	2,122.7	122
Malta	4.9	13.2	0.6	1,589.2	69
Norway	120.5	27.3	11.9	2,696.1	134
Switzerland	172.4	24.3	17.2	2,424.5	143
Turkey	388.3	6.1	50.5	791.7	47
Total	701.6	9.2	81.7	1,065.9	87
Total Western Europe	8,401.3	18.6	842.7	1,869.6	100
Eastern Europe in CIS					
Russia (3)	692.0	4.7	83.0	563.7	62
Ukraine	124.9	2.5	11.2	221.7	38
Others (4)	90.7	2.9	9.4	114.7	27
Total	907.6	4.0	103.7	391.4	55
Other Eastern and Central Europe					
Bulgaria	35.6	4.3	2.0	235.9	27
The Czech Republic	111.9	10.9	14.6	1,418.2	45
Estonia	9.3	6.4	2.1	1,425.0	47
Hungary	73.2	7.2	4.7	463.4	60
Latvia	10.4	4.2	1.3	532.8	50
Lithuania	15.4	4.1	2.2	586.2	55
Poland	280.7	7.3	28.6	740.1	47
Romania	114.2	5.1	15.1	668.2	30
The Slovak Republic	46.3	8.6	5.2	966.1	41
Others (5)	44.2	5.8	4.3	564.8	80
Total	741.2	6.7	78.4	709.5	46

Country	GDP on PPP basis US$ billion	GDP per capita on PPP basis US$ thousands	Construction output on PPP basis US$ billion	Construction output per capita on PPP basis US$ (1)	Ratio of exchange rate basis to PPP prices corrected to UK=100
Total Eastern and Central Europe	1,648.8	4.9	184.7	543.4	51
Total Europe	10,050.1	12.7	1,011.9	1,279.9	92

Notes: Percentages are based on unrounded figures.
(1) For the method of calculation of construction output on a basis and its estimation see Statistical notes in Appendix.
(2) The figures for Austria have been amended in the light of information in Euroconstruct conference proceedings, Prague, June 1999.
(3) Including Russia in Asia.
(4) Armenia, Azerbaijan, Belarus, Georgia, Moldova.
(5) Croatia, Macedonia FYR, Slovenia.
Sources: 1997 data equivalent to that in key data sheets for each country in Part Two and DLC estimates based on Economist Intelligence Unit Reports. The exchange rates of mid 1997 have been used.

Construction output and GDP – the regions of Europe

Regional differences are very great in Europe, partly due to the legacy of the FSU, but also because there are very substantial differences in standards of living between countries, even in Western Europe. After these introductory comments the regions are discussed as divided in the tables for Western Europe, the EU and the rest of Western Europe; for the East, the CIS states and other Eastern and Central Europe.

In terms of size of market, Eastern and Central Europe are relatively small. Although they have 43% of the population of the whole of Europe, they had in 1997 only 10% of construction output. Their construction output per capita is less than a sixth of that of Western Europe. However, whereas, with one or two exceptions, notably Turkey, the construction industry in Western Europe is unlikely to grow very much, the potential demand for construction in Eastern and Central Europe is vast. The problem, especially for the CIS, is to assess when the demand will materialize.

In Western Europe there have been some changes since 1997. Construction output grew overall by around 1% from 1997 to 1998 and it is expected that in 1999 growth will continue, possibly at a higher rate. The share of construction in GDP decreased in 1998 as it has done over the 1990s as a whole.

The European Union

The countries with the largest share of construction output in the EU are Germany, Italy, France, UK and Spain and together they produce over 70% of total construction output of the EU. Germany is dominant in practically all aspects. However, from 1995, German construction output has been falling, partly because of the problems faced in the transition of East Germany, but also because the German economy has been in recession. Italy, France and the UK have the next highest levels of construction output and they have been increasing. Spain's output has been rising rapidly. Amongst the smaller countries of the EU are Finland and Ireland, both of which have experienced substantial increases in construction output. Denmark on the other hand has been experiencing difficulties in the construction sector.

Western Europe but not EU

Other Western European countries listed in Table 1.1 account for only about 8% of all Western Europe's output. Norway's industry has problems because the economy slowed down in 1998 and construction output fell. Output in Switzerland on the other hand recovered from previous falls. The output per capita in Switzerland and Norway is nearly ten times that of Turkey. However the potential of Turkey is great.

Eastern Europe in CIS

The countries in this group suffer from poor statistics which, together with the non-convertibility of currencies, makes the relative values of GDP and construction output difficult to establish. The statistics for some of the CIS states are not accurate to more than 50% either way and in some cases probably 100% either way. Different sources give different figures. It should be borne in mind that, because of the change from a planned economy to a market economy, the collection of data on small private entrepreneurial units may be poor and, in addition, there are no statistics on the black economy. Thus the situation may in reality be not quite so bleak as the statistics of GDP and construction output suggest.

The overwhelming fact about the CIS countries is that their GDP has fallen so that in many countries the standard of living is around half previous levels. In 1997 the trend was showing some signs of change with most of the countries actually experiencing an increase in GDP.

Table 1.4 below shows the changes in GDP from 1994 to 1997 and from 1996 to 1997.

Table 1.4: Change in GDP 1994 to 1997

	% change 1994 to 1997	% change 1996 to 1997
Armenia	16.6	3.1
Azerbaijan	−6.4	5.0
Belarus	1.5	10.0
Georgia	27.0	11.3
Moldova	−8.6	1.3
Russia	−6.8	0.8
Ukraine	−23.5	−3.2

Source: The Economist Intelligence Unit country reports.

There was improvement in Armenia and Georgia where living conditions had been very poor and where some upturn was vital. However, it is thought that GDP in Russia and the Ukraine fell further in 1998 and 1999, which would imply a fall in construction output. This is important for the whole area because Russia alone probably accounts for 80 to 90% of the total construction output of the CIS and the next most important country is the Ukraine. Most of the construction work is located in European Russia which accounts for 87% of the population of all Russia although only 25% of the area.

It is thought that, in all countries of the CIS, construction output has fallen faster than GDP. Whereas it was usual in the 1980s for the percentage of gross output of construction of GDP to be up to 20 or 25%, it is now down to around 10%.

Other Eastern and Central Europe

This group consists of the former satellites of the FSU and the Baltic states of Estonia, Latvia and Lithuania which were part of the FSU but did not join the CIS. Table 1.5 following shows changes in GDP and construction output from 1994 to 1997.

Table 1.5 below shows changes in construction output in the same period.

Table 1.5: Changes in GDP and construction output 1994 to 1997

Country	% change in GDP 1994 to 1997	% change in GDP 1996 to 1997	% change in construction output 1994 to 1997
Bulgaria	−15.3	−6.9	n.a
Czech Republic	11.7	1.0	9.2
Estonia	17.0	8.0	54.5
Hungary	7.6	4.6	8.0
Latvia	5.2	4.0	57.5
Lithuania	10.4	3.8	21.3
Poland	21.3	6.9	22.1
Romania	4.1	−6.6	−8.5
Slovak Republic	21.4	6.5	15.9

Sources: DLC estimates based on Euroconstruct conference papers, Berlin, December 1998, The Economist Intelligence Unit country reports.

Until recently Poland, Hungary the Czech Republic and the Slovak Republic were all performing well. Poland is still forging ahead at a rate well into double figures over several years (over 17% in 1998). Hungary is steadily increasing its growth of GDP; it rose by over 4% in 1998. However, both the Czech and Slovak Republics are having difficulty in sustaining growth rates. Romania and Bulgaria are probably declining both in GDP and construction output. In the former Yugoslavia, with the exception of Slovenia which is developing slowly, all countries are struggling. All three Baltic states are achieving continued growth in GDP and in construction output.

Types of work

The distribution of type of work varies considerably among countries and regions. Repair and maintenance range from 6 to 57% with an average of about 40% for Western Europe (data on maintenance for Central and Eastern Europe exist only for a few countries). The countries with the lowest GDP per capita also have the lowest proportion of repair and maintenance. Some of these differences are due to deficiencies in the collection and definition of data but great variations certainly exist. In particular, repair and maintenance has been low throughout the former planned economies.

For new work the data are more accurate. Table 1.6 shows countries sorted according to the proportion of new construction output by type of work.

Table 1.6: European countries sorted by ranges of percentages of value of output for new work, 1997

Type of work	10-20%	20-30%	30-40%	40-50%	50% and over
Residential	Czech Republic Poland Slovak Republic Sweden	Denmark Greece Norway UK	Finland Hungary Portugal Turkey	Austria Belgium France Ireland Italy Netherlands Spain	Cyprus* Germany Iceland* Switzerland
Non-residential building		Cyprus* France Germany Iceland* Portugal Netherlands Spain Switzerland Turkey	Austria Denmark Finland Greece Hungary Ireland Italy	Czech Republic Belgium Norway Sweden UK	Poland Slovak Republic
Civil engineering	Belgium Germany Iceland* Ireland	Austria Cyprus* Finland France Hungary Italy Netherlands Switzerland UK	Czech Republic Denmark Norway Poland Portugal Slovak Republic Spain	Greece Sweden Turkey	

Note: * 1996 data.
Sources: based on Euroconstruct conference papers, Berlin, December 1998 and DLC data.

In the distribution of new work, Poland and the Slovak Republic stand out with the highest proportion of non-residential building work. Turkey has the lowest share of non-residential buildings, followed by Spain and Switzerland. The share of civil engineering is very low in Belgium, Iceland, Ireland and Germany. Conversely, it is high in Sweden, followed by Turkey and Greece. In the residential sector, Germany has the highest share, followed by Switzerland. The proportions are low for Sweden, the Czech Republic, the Slovak Republic and Poland.

Characteristics and structure of the industry

In Western Europe generally, in each country, there are a few relatively large construction companies and a large number of smaller ones. Those countries which have some export business in contracting tend to have larger contractors than those serving only a domestic market. In many countries the large contractors are also involved in other businesses either related to construction – housing and property development, or building material production – or unconnected with it, for example, television.

In Europe as a whole there is little cross-border trade in contracting in relation to the size of the construction market. There is some use in Western Europe of sub contractors from the former Soviet Bloc economies, especially in Austria and Germany.

Table 1.7 below shows the distribution of large contractors in Europe in 1996 and 1997. In the top 50 contractors France is dominant, followed by Germany, the UK and Spain. In terms of sales France is still leading whereas the UK overtakes Germany in 1997. In the top 300, Italy is fourth in terms of the total number of contractors but their impact on the market is limited by their smaller output compared to other countries.

Table 1.7: European top 300 contractors, 1996 and 1997

Country	Number of contractors in top 50 1996	1997	Number of contractors in top 100 1996	1997	Number of contractors in top 300 1996	1997	Sales £ million 1996	1997
France	14	17	31	33	103	110	62,526	58,404
UK	8	9	20	22	64	63	28,908	30,599
Germany	10	10	15	12	38	36	34,174	28,080
Italy	2	1	2	4	24	23	6,792	4,764
Netherlands	5	4	11	11	12	15	10,347	11,905
Spain	6	5	9	7	13	10	10,716	9,604
Norway	1	1	2	3	6	9	7,041	7,989
Finland	0	0	2	2	9	8	2,086	2,114
Portugal	0	0	0	1	6	7	1,417	1,663
Denmark	0	0	0	0	7	6	1,326	1,345
Belgium	0	0	2	2	6	5	1,966	1,309
Sweden	4	3	5	3	5	4	9,668	8,102
Switzerland	0	0	1	0	7	4	1,362	925

Sources: Building December 1997 and December 1998.

Table 1.8 below shows the top international contractors in 1996 and 1997. This information comes from the *Engineering News Record* list of top international contractors. Here the Italians lead the field in number of contractors but France leads in value of international contracts. The UK and Germany are the other principal countries. The table also shows non-European countries. In terms of number of companies operating internationally, the USA is the most important country but Japan leads in terms of value of international contracts. The contribution of contractors from Korea and China is also significant. In the top twenty contractors 12 are from Europe – France 5, UK, Germany and Italy 2 each and Sweden 1. The remainder are USA with 4, Japan 3 and Korea 1.

Table 1.8: International contractors worldwide, 1996 and 1997

Country	Number of contractors in top 100		Number of Contractors in top 225		Value of international contracts $ million	
	1996	1997	1996	1997	1996	1997
Italy	8	7	20	15	7,432.6	6,300
Germany	9	8	14	13	13,554.5	9,432
France	8	7	10	10	16,451.1	16,533
United Kingdom	10	7	10	7	14,390.8	12,674
Spain	4	6	7	8	1,806.1	2,456
Turkey	0	2	7	9	1,094	1,194
Netherlands	3	2	4	2	3,646.1	1,481
Denmark	0	0	3	3	182.4	213.3
Sweden	2	2	2	2	3,436	4,763
Republic of Cyprus	1	1	2	2	566.3	756.3
Yugoslavia	0	1	2	1	125.6	173
Belgium	1	1	1	1	307.4	281
Greece	1	1	1	1	1268	1,318
Finland	0	1	1	2	188	219.7
Norway	0	0	1	1	30.4	47.6
Northern Ireland	0	0	1	1	102	63
Portugal	0	0	1	1	32.9	28.2
Macedonia FYR	0	0	1	1	24	41
Ireland	0	1	0	1	n.a.	251

Country	Number of contractors in top 100		Number of Contractors in top 225		Value of international contracts $ million	
	1996	1997	1996	1997	1996	1997
Other						
United States of America	16	18	48	65	22,508.3	25,142
China	4	4	27	26	4,060.5	4,079
Japan	19	14	28	19	24,255.9	12,867
Korea	8	7	12	10	6,377.5	4,922
Canada	1	1	3	7	832.4	876.7
Australia	2	3	2	3	608	990.3
Brazil	2	2	2	2	1,345	1,706
Mexico	0	0	1	2	77	272.2
Taiwan	0	0	2	2	147	155.1
India	0	0	2	1	89.2	15.8
New Zealand	1	1	1	1	1,223	811
Hong Kong	0	1	1	1	130	233
Israel	0	1	1	1	117.9	193
South Africa	0	0	1	1	170	213
Lebanon	0	0	1	1	59	85
Singapore	0	0	1	1	178	20
Philippines	0	0	1	1	18	6.9
Malaysia	0	0	1	0	24.6	n.a.
U.A.E.	0	0	1	0	94	n.a.
Argentina	0	0	1	0	11.9	n.a.

Sources: Engineering News Record 25.8.97 and 17.8.98.

Comparison with Japan and the USA

Japan and the USA are included in this volume for comparative purposes. Table 1.9 on the next page shows basic data for these two countries, Western Europe and Eastern and Central Europe.

Although Japan is a separate, relatively small but densely populated island, its GDP and construction output are high in relation to the USA and Europe. However, because prices in Japan are high on a PPP basis, the advantage of Japan diminishes considerably. On an exchange rate basis GDP in Japan is over 50% that of the USA, nearly 50% of that of Western Europe and about five times that of Eastern and Central Europe. On a PPP basis the corresponding figures are under 40% for the USA and Western Europe and less than twice that of Eastern and Central Europe. The picture for construction output varies similarly according to how it is measured. A better consensus is given by GDP per capita on a PPP basis and construction output per capita on a PPP basis. GDP per

The construction industry in Europe

Table 1.9: Comparison of Europe, Japan and USA, 1997

	Western Europe	Eastern and Central Europe	Japan	USA
Land area km^2	4,496	19,265	378	9,400
Population mm	450.7	339.8	125.7	267.9
GDP US$ bn	8,736.5	876.6	4,182.2	8,110.9
GDP on PPP basis US $ bn	8,401.3	1,648.8	3,080.0	8,110.9
Construction output US $ bn	869.7	98.2	764.8	573.1
Construction output on PPP basis US$ billion	842.7	184.7	422.8	571.1
Construction output as % of GDP US$ th	10.0	11.2	13.7	7.1
GDP per capita US $th	19.4	2.6	33.3	30.2
GDP per capita on PPP basis US$ th	18.6	4.9	24.5	30.2
Construction output per capita US$ thousand	1.9	0.3	6.1	2.1
Construction output per capita on PPP basis US$ th	1.9	0.5	3.4	2.1
Ratio of exchange rate price to UK=100 PPP prices	100	51	131	97
Land area as a % of the total	13.4	57.4	1.1	28.1
Population as a % of the total	38.1	28.7	10.6	22.6
GDP as a % of the total	39.9	4.0	19.1	37.0
Construction output as a % of the total	37.7	4.3	33.2	24.8

Sources: Tables 1.1, 1.2 and 1997 data equivalent to that in key data sheets for Japan and USA in Part Two.

capita in Japan is then about 80% of that in the USA, about 130% of that in Western Europe and five times that in Eastern and Central Europe. Because construction output as a percentage of GDP is high in Japan, construction output per capita on a PPP basis in Japan is about 160% of that in the USA, nearly 180% of that in Western Europe and nearly seven times that in Eastern and Central Europe.

The low GDP per capita of Eastern and Central Europe and the low level of construction output per capita are striking features of the table. Yet the proportion of GDP spent on construction is higher than in Western Europe and the USA. This is because both GDP and construction output are so low in absolute terms. However, in the 1980s the proportions of GDP spent on construction would have been about double that of other areas.

The USA has a remarkably low level of investment in construction as a proportion of GDP, especially as it has a high GDP per capita. The GDP per capita figure for Western Europe is lowered by the inclusion of Turkey.

Table 1.10 below shows the different percentages of types of work in Western Europe, Japan and the USA.

Table 1.10: Types of construction work as a percentage of total output, 1997

	Western Europe	Japan	USA
Residential	42.7	33.7	43.0
Non-residential building	32.8	20.1	36.3
Civil engineering	24.5	46.2	20.7

Sources: *Davis Langdon Consultancy, estimates based on Euroconstruct conference paper, Berlin December 1998, Research Institute of Construction and Economy, Japanese Ministry of Construction, Statistical Abstract of the US, US Census Bureau.*

The USA invests almost 43% of its construction output in housing, the same as in Western Europe, compared with 35% in Japan. On the other hand the expenditure on maintenance is probably much higher in Western Europe than in Japan and the USA. In Japan, routine repair and maintenance may be as low as 10%. This is largely due to the traditional 'scrap and build' approach to construction, with newbuild clearly being preferred to refurbishment. Typically Japanese housing is replaced after 30 to 40 years or less though it could last longer and may well do so in the future. In the USA, repair and maintenance is thought to fall short of West European levels – perhaps being about half. However, as in the case of all countries, the data is subject to wide margins of error due to the informal sector and statistical underrecording (see Appendix 4).

The structure of the construction industries in Western Europe, Japan and the USA is overall similar, with the usual small group of large contractors and a large number of smaller ones. However, the vast area of the USA and the separate legal and regulatory systems of individual states mean that the large local contractors operating within a limited area are also very important. Many of the largest US contractors tend to be specialists in civil engineering or process plant engineering and operate widely within and outside the USA. The Japanese large contractors all undertake building, civil engineering and industrial work. In Europe there are some large specialist firms but generally the activities of the large firms are diverse. Table 1.8 shows that the USA and Japan had 16 and 19 companies respectively in the top 100 international contractors compared with 47 for the whole of Europe. In the top ten, the USA had two and Europe seven, while Japan had one.

In Japan, construction companies undertake both design and construction with large in-house departments. In the USA, an architect is normally appointed separately, although the contractor usually prepares the detailed design. In Europe, design and build is more prevalent than in the past, but it is still usual to appoint the architect or engineer first and then to go to a contractor. All countries have a large usage of subcontractors. In Japan they often work closely and regularly with the same main contractor.

The Japanese contracting system is based on trust and mutual understanding, which in turn rests on long-term relationships between client and contractor. Litigation is rare. The potential for litigation in the USA is considerable and the costs high, so that clients have to observe a fairly rigid conformity to the contract conditions. Variations are expensive. In the European Union there is great diversity in this respect. In the UK, contractual disputes and claims are common, whereas in France, although litigation does take place, disagreement is usually settled between the parties without external intervention.

Physical measures of output

Because of the difficulties of expressing values in local currencies in terms of US$ where the currency is not convertible and because, even where it is, it does not necessarily give a true indication of real construction output, it is important to have some physical measures of output as well as monetary ones. The two used here are cement consumption and number of new dwellings constructed, both, of course, also being interesting in their own right.

Cement consumption

Table 1.11 on the next page shows cement consumption for those countries for which it is available.

Cement consumption can vary between countries because the technologies of buildings and works – for example reinforced concrete panel or brick buildings, concrete or macadam roads – vary or because of the incidence of different work types. Cement consumption per unit of output in Eastern and Central Europe was in the past probably at least double that in Western Europe. This was because of the highly concrete-intensive technology used, for example, for high rise multi-storey flats and factory buildings. It may also have been due to the low price put on construction output which would make cement, expressed in real terms, high in relation to undervalued output. Indeed Table 1.11 shows that in many cases countries which are known to have low prices of construction have high cement consumption per unit of value, and vice versa. The countries which have high apparent consumption of cement are often those where the GDP in PPP terms is higher than the money GDP and those with low consumption of cement are those where GDP in PPP terms is lower than the money GDP. Since the collapse of the command economy in Eastern and Central Europe, less cement intensive methods of construction are being used so that some of the differences in apparent cement intensity may disappear in the future.

Table 1.11: Cement consumption in total physical volume and physical volume per million $ of construction output

Country	Cement consumption (thousand tonnes)	Cement consumption (tonnes per million $ of construction output)
Western Europe		
Austria	5,100	138.5
Belgium	5,720	194.7
Cyprus*	1,014	1,179.2
Finland	1,320	105.1
France	18,730	170.7
Germany	40,000	179.8
Greece*	7,800	54.1
Ireland	2,500	215.7
Italy	33,700	301.5
Netherlands	5,500	140.6
Portugal	9,404	658.4
Spain	26,740	438.3
Switzerland	3,456	135.6
Turkey*	30,085	1,208.7
UK	13,000	135.9
Eastern and Central Europe		
Czech Republic	3,750	553.3
Poland	12,120	875.3
Russia*	35,232	655.8
Slovak Republic	1,510	687.9
Japan*	82,420	107.8
USA	80,900	141.2

Notes: * 1996 figures.
Sources: Asia construct conference paper, Hong Kong, November 1997. Statistical Abstract
of the USA, US Census Bureau. DLC estimates based on Cembureau.

It is worth noting that the cement industry is deeply regional in nature. Because the cost of transporting cement is high, customers usually buy cement from local sources. Moreover, although cement consumption is closely tied to the construction industry cycle, cement is in a sense hedged from the extreme cycles of the construction industry because it is used in almost every type of construction. While individual construction markets, such as residential building, office building and infrastructure, follow their own particular business cycle, at any given time cement is required by at least one segment of the construction industry. Therefore, figures on consumption provided at a given time are sufficiently representative of the conditions of a market. The cement industry, however, is highly subjected to seasonality effects. During the summer, always the most active period for construction, three times more cement usually is supplied than during the winter months. As a result, cement producers build up inventories of clinker during winter for grinding into cement to meet summer demand.

Dwelling completions

Table 1.12 below shows dwelling completions where the data are available. The figures for new housing completions give some indication of progress in the improvement of housing supply. However, they are not necessarily a good guide to the total amount of construction work. At one time the proportion of housing in total work was fairly stable in Eastern and Central Europe. Now, however, disruption in the economies means that there have been substantial changes in these relationships.

Table 1.12: Dwelling completions in absolute values and per 1,000 population, 1997

Country	Dwelling completions (thousand)	Dwelling completions per 1,000 population
Austria	58.0	7.2
Belgium+	39.6	3.9
Cyprus*	7.2	9.7
Finland	27.0	5.3
France+	272.0	4.6
Germany	501.0	6.1
Greece	89.6	8.4
Ireland	38.8	10.8
Italy	226.3	3.9
Netherlands	92.3	5.9
Portugal	68.6	6.9
Spain	272.0	6.9
Switzerland	34.2	4.8
Turkey	458.2	7.2
UK	178.2	3.0
Eastern and Central Europe		
Czech Republic	15.9	1.5
Poland	73.7	1.9
Russia*	474.0	3.2
Slovak Republic	7.2	1.3
Japan+	1,387.0	11.0
USA	1,400.5	5.2

Notes: + starts.   * 1996 data.

Sources: DLC estimates based on Review of Construction Markets in Europe, 1998. Asiaconstruct conference paper, Hong Kong, November 1997. US Census Bureau, The Official Statistics. National Statistical Service of Greece, Monthly Statistical Bulletin 1998.

Part Two: Individual Countries

2 Introductory notes to country sections

Introduction

Part Two gives detailed data on twenty countries of which eighteen are in Europe. Japan and the USA are included for comparative purposes. The European countries comprise:

- countries belonging to the EU. Twelve countries have been selected: Austria, Belgium, Finland, France, Germany, Greece, Ireland, Italy, the Netherlands, Portugal, Spain and the UK.
- market economies that are not part of the EU: Cyprus, Switzerland and Turkey.
- a sample of countries in transition: the Czech Republic, Poland and the Slovak Republic.

In this part of the book, twenty countries are arranged alphabetically, and each country is presented as far as possible in a similar format, under four main headings – *The construction industry*, *Construction cost data*, *Exchange rates and cost and price trends*, and *Useful addresses*. In addition, key data for each country are provided at the beginning of each country section.

These notes introduce the main sections of information and provide general notes, definitions and explanations.

Key data

The key data sheet provided at the beginning of each country section highlights some of the main indicators on population, geography, economy and construction for the particular country. It thus provides a brief statistical overview of each country. Most of these indicators are taken from official World Bank and International Monetary Fund publications and are therefore authoritative, but it takes time for international agencies to collate and publish such data, so that they can on occasion be a little dated. Official figures are used here to be consistent and to ensure that the summaries of key data in Part Three are presented on a comparable basis.

The construction industry

The main topics covered under this heading in each country section are: gross construction output, in absolute value and per capita, and its contribution to GDP at both national and local level; the characteristics and structure of the construction industry; tendering and contract procedures; and planning regulations and standards. The issues of financing of construction projects (private/public split) and liability and insurance are also covered for some countries.

Although construction is often fragmented and tends to be labour intensive with low capital investment, it is invariably the single largest industry in a country. In most countries its net output contributes between 8% and 18% to GDP and a similar percentage to direct construction employment. Indirect employment – in the construction materials industries and other related activities – can be more than double its direct contribution to the economy.

The statistical data given under each country are comparable as far as is possible. Further explanations of the data are given in Appendix 4.

Construction cost data

This section includes both construction costs incurred by contractors and the costs they charge their clients. The costs of labour and materials are input costs of construction, that is, the costs incurred by contractors. Unit rates for main work items and approximate estimating costs are output costs, that is, the costs contractors charge their clients. Problems of definition make meaningful and consistent presentation of unit rates extremely difficult. For unit rates to be useful it is essential to refer to the text of each country section in order to be clear what is included and excluded in the cost, for example, how preliminaries and profit are dealt with. Most costs are for the first quarter of 1999.

Cost of labour and materials

Typical costs for construction labour and materials are provided for most country sections. Two figures are generally given for each grade of labour. The wage rate is the basis of an employee's income – his basic weekly wage will be the number of hours worked multiplied by his wage rate per hour. The cost of labour, on the other hand, is the cost to the employer of employing that employee; it is also based on the wage rate but includes (where applicable) allowances for:

- incentive payments
- travelling time and fares
- lodging and subsistence
- public and annual holidays with pay
- training levies
- employer's liability and third party insurances
- health insurance
- payroll taxes
- other mandatory and voluntary payments.

The costs of main construction materials are given as delivered to site in quantities appropriate for a medium sized construction project, with the location of the works neither constrained nor remote in a way that would significantly affect costs. Generally tax, and particularly any value added tax, is excluded from material costs as the rate of tax levied on the materials often depends on the type of accommodation being constructed.

Unit rates

Rates for a variety of commonly occurring construction work items are provided for most countries. They are usually based on a major, if not the capital, city and always refer to the first quarter 1999. Rates include all necessary labour, materials, plant and equipment and, where appropriate, allowances for contractor's company overheads and profit, preliminary and general items associated with site overheads, and contractor's profit and attendance on specialist trades. Value added tax and other taxes are excluded. The rates are appropriate to a medium sized construction project.

A full description of each work item is presented in Appendix 3 in five languages (English, French, German, Italian and Spanish).

Approximate estimating

Approximate estimating costs per unit area (normally expressed per square metre and per square foot) are given for most countries for different building types. Notes on the method of measurement and what is or is not included in unit rates are provided in each country section. Areas are measured on all floors between the inside face of external walls and with no deduction for internal walls and columns. Generally tax, and particularly value added tax, is excluded in approximate estimating costs.

When making comparisons of construction costs between countries it is important to be clear about what is being compared. In terms of approximate estimating costs there are two main methods of comparison: first the comparison of identical buildings in each country and, second, the comparison of functionally similar buildings in each country.

In the individual country sections in this book, the approximate estimating rates given are for buildings with specifications and standards appropriate for that country. This allows for comparisons of functionally similar buildings to be made. The rate per square metre given for an office building or a warehouse, for example, in any particular country refers to the typical specification and standard of an office building or warehouse built in that country. In country sections they are presented in national currencies. A selection of approximate estimating costs are also presented in pound sterling, US dollars and Euro equivalents in Part Three, thus enabling comparisons on a common currency basis.

Exchange rates and cost and price trends

Exchange rates

Currency exchange rates are important when comparing costs between one country and another. While it is most useful to consider costs within a country in that country's currency, it is necessary, from time to time, to use a common currency in order to compare one country's costs with another. But exchange rates fluctuate dramatically and few currencies (even those considered strong) can be considered really stable. It can be risky to think in terms, for example, of one country being consistently a set percentage more or less expensive than another.

Different rates of internal inflation affect the relative values of currencies and, therefore, the rates of exchange between them. However, the reasons behind exchange rate fluctuations are complex and often political as much as economic; they include such factors as interest rates, balance of payments, trade figures and, of course, government intervention in the foreign exchange markets, and, for that matter, other actions by governments.

Graphs of exchange rates since 1990 against the US dollar, 100 Japanese yen, the ECU and the pound sterling are included for most countries. They have been calculated by using the published monthly values. They are very useful for indicating long term trends.

As far as possible, the form of the graph is kept the same, hence the vertical scale is adjusted to accommodate different currencies. It should always be checked whether marked movement in a graph is a result of erratic exchange rates or merely a result of the selected vertical scale. Currency exchange rates are defined in terms of how many units of the domestic currency are needed to purchase one unit of the four countries selected for comparisons.

If a line moves up from left to right (for example, the Spanish Peseta against the pound sterling after 1996 – see opposite) it indicates that the subject currency (the Spanish Peseta) is declining in value against the currency of the line (the Pound Sterling). The higher the line is, the more subject currency is required to purchase the line currency. If, on the other hand, a line moves down from left to right (for example, the Spanish Peseta against the Pound Sterling between 1993 and 1994 – see opposite) it indicates the subject currency strengthening against the line currency. Where there is virtually no movement at all, that is the line is horizontal, this usually indicates a currency effectively 'tied' or 'pegged' to the line currency. A number of currencies within the Exchange Rate Mechanism (ERM) of the European Monetary System (EMS) had this relationship. Things have changed with the introduction of the Euro in January 1999 in which most of the countries belonging to the European Union decided to enter the Monetary Union and adopt the new currency.

Exchange rate graph

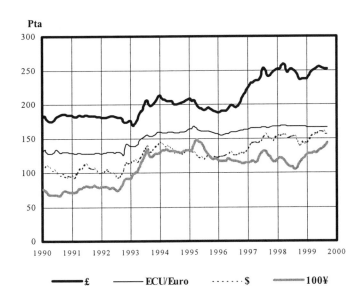

Cost and price trends

General price trends have been measured using consumer or retail price indices. These reflect price changes in a basket of goods and services weighted according to the spending patterns of a typical family. Weights are changed periodically, and new items inserted. General inflation indices usually rise and, in so doing, erode the purchasing power of a given currency unit. Other measures of inflation tend to be related to specific items. The two most commonly prepared for the construction industry are discussed below.

Building cost and tender price indices

Cost and tender price indices measure different types of inflation which occur within the construction industry. Building costs are the costs incurred by a contractor in the course of his business, the major ones being labour and materials, and are usually consistent with the movement in the general consumer price index; tender prices are the prices for which a contractor offers a client to construct a building. Tender prices include building costs but also take into account contractor's profit and on-costs that are affected by the prevailing market situation. When there is plenty of construction work, tender prices often increase at a greater rate than building costs whereas, when work is scarce, tender prices may actually fall even though building costs may still be rising.

Most countries have building cost indices – the method of compilation is generally relatively simple, basically comprising a weighted basket of the main inputs to construction. Rather fewer countries have tender price indices – their method of compilation is more complex, usually involving a detailed analysis of accepted tenders for construction work and comparing these with a common base schedule of prices.

Useful addresses

At the end of each country section, a list of useful addresses of organizations involved in the construction industry is given. This usually comprises main government and public organizations, professional and trade associations and planning, standards and research organizations.

3 Individual countries

ILBAU Ges.m.b.H.
A-9800 Spittal/Drau
Ortenburgerstr. 27
Tel.   +43 (0) 4762 620-0
Fax   +43 (0) 4762 4962
e-mail pr@bauholding.e-mail.com

A-1220 Wien
Polgarstraße 30
+43 (0) 1 21728-0
+43 (0) 1 21779

ILBAU Ges.m.b.H. is a 100% subsidiary of BAU HOLDING AG, a stock corporation listed at the Vienna Stock Exchange.

Founded in: 1835
Average number of employees in 96/97: 3,700
Sales in 96/97 business year: ATS 7,900 m
Cash flow in 96/97 business year: ATS 292 m

The company is represented throughout Austria with numerous branch offices.

Other operational units and establishments of the BAU HOLDING Group are located in Germany, Italy, Portugal, as well as in the reform countries of Hungary, the Czech Republic, Poland, the Russian Federation, Slovakia, Romania and Croatia.

Our fields of activity range from
- civil engineering and building construction
- tunnelling
- bridges
- power stations
- construction of sports facilities
- trackage
- drilling
- jointing
- special civil engineering structures
 to waste management and recycling.

Highest level of co-ordination, communication and co-operation are essential elements of our corporate identity.

Austria

All data relate to 1998 unless otherwise indicated.

Population
Population	8.1 million
Urban population (1997)	64%
Population under 15	17%
Population 65 and over	15%
Average annual growth rate (1985 to 1998)	0.6%

Geography
Land area	83,850 km^2
Agricultural area	42%
Capital city	Vienna

Economy
Monetary unit	Schilling (ASch)
Exchange rate (average first quarter 1999) to:	
the pound sterling	ASch 19.61
the US dollar	ASch 11.77
the Euro	ASch 13.76
the yen x 100	ASch 9.96
Average annual inflation (1995 to 1998)	1.3%
Inflation rate	0.8%
Gross Domestic Product (GDP)	ASch 2,641 billion
GDP PPP basis (1997)	$ (PPP) 174 billion
GDP per capita	ASch 324,900
GDP per capita PPP basis 1997	$ (PPP) 21,570
Average annual real change in GDP (1995 to 1998)	2.6%
Private consumption as a proportion of GDP	58%
Public consumption as a proportion of GDP	19%
Investment as a proportion of GDP	24%

Construction
Gross value of construction output	ASch 357 billion
Gross value of construction output per capita	ASch 44,100
Gross value of construction output as a proportion of GDP	13.5%
Average annual real change in gross value of construction output (1995 to 1998)	2.2%
Annual cement consumption	5.0 million tonnes

PPP Purchasing power parity

The construction industry

Construction output

The value of gross construction output in Austria in 1998 was ASch 357 billion or 25.7 billion ECUs (nearly the same as January 1999 Euros). This represents 13.5% of GDP. It is understood that this includes an estimate for work by other sectors, the black economy and DIY. The table below shows the breakdown of output in 1998.

Output of construction, 1998 (current prices)

Type of work	ASch billion	ECUs billion	% of total
New work			
Residential building	96.3	6.9	27
Non-residential building	67.8	4.9	19
Civil engineering	65.6	4.7	18
Total new work	229.7	16.5	64
Repair and maintenance			
Residential	66.3	4.8	19
Non-residential	44.2	3.2	12
Civil engineering	16.4	1.2	5
Total repair and maintenance	127.0	9.1	36
Total	**356.7**	**25.7**	**100**

Note: the value of the Euro in January 1999 was very similar to that of the ECU in 1998.
Source: based on Czerny, M., 'Austria' in Euroconstruct conference proceedings, Prague, June 1999.

The output of the construction industry has grown in the years 1995 to 1998 and growth was 3% from 1997 to 1998. Renovation work has been especially buoyant with increases in residential work of over 15% in 1998.

New residential construction is declining. The pent-up demand of some years ago has been substantially met. In 1998 the total value of new residential construction fell, as did the number of completions of both houses and flats. The level of building permits granted has also declined signalling that the fall is unlikely to be reversed in the near future. In 1998 the estimated proportion of building permits for flats was 70% of the total. There has been a decline in subsidized housing. Tax incentives for the renovation of old building stock cease at the end of 1999, which may cause an easing of the boom in residential renovation.

The outlook for other sectors is more positive. Non-residential building is being buoyed up by office construction though other markets are less strong. Civil engineering may well increase. In the past years there has been restricted investment in the public sector in order that Austria meet the Maastricht criteria and there is a need for infrastructure, especially for transport projects. Amongst other influences, the proposed enlargement of the European Union eastwards has created a need for expansion and improvement of the road network in eastern Austria. Austria is in many ways at a focal point for trade between Eastern and Central and Western Europe.

The geographical distribution of construction work in 1995 compared with that of population is shown below. It is noteworthy that the amount of building and civil engineering work in Vienna was nearly double that which would be expected on the basis of population alone. It accounted for over a third of total output. Nearly another third of output occurs in Niederösterreich and Oberösterreich (Lower and Upper Austria). Most other regions receive less than would be expected on the basis of population.

Population and construction output, 1995 regional distribution

Regions	Population (%)	Building and civil engineering work (%)
Burgenland	3.5	2.3
Kärnten	7.0	8.0
Niederösterreich	18.9	16.5
Oberösterreich	17.1	12.8
Salzburg	6.2	5.3
Steiermark	15.2	9.5
Tirol	8.1	9.0
Vorarlberg	4.3	1.1
Wien	19.7	35.5
Total	**100.0**	**100.0**

Sources: Statistisches Jahrbuch für die Republik Österreich 1995; Fachverband der
Bauindustrie Österreichs, Kenndaten der Öösterreichischen Bauindustrie 1995.

Clients and finance

At one time the amount of work financed directly or indirectly by government was about 60% of construction output but this is now changing and the emphasis is switching to private financing. Major public projects are increasingly managed by private organizations with private funding.

However, Government offers incentives for investment at federal, provincial and local level. These are aimed primarily at employment generation, especially in areas in need of special assistance. In general, central Government assistance takes the form of loans to supplement available commercial credits. The *Österreichische Kommunal Kredit AG*, for example, grants loans for up to 35 years at a low rate of interest for the development of industrial sites. Provincial authorities have similar schemes. Local authorities provide help often in the form of accommodation or sites at low rents or even free of charge.

Contractual arrangements

In the case of large contracts to be awarded by central or local Government agencies, public advertisement and open tendering is generally the method of obtaining competitive tenders. Private sector work is usually awarded as a result of selective tendering, tenders typically being invited from between three and five firms. In the public sector, contracts are invariably awarded to the best tenderer which is usually the lowest. In the private sector, particularly where design is at a less advanced stage, the capability and suitability of the tenderer may, on occasion, be considered more important than his tender bid alone.

Standard conditions of contract exist but are revised as appropriate for public works contracts – thus, in effect, every Government department has its own conditions of contract. Similarly, where an architect is appointed to handle the project management of private sector work, contract conditions, though derived from published conditions, are produced on a 'one-off' basis. Contract documents consist of general and special conditions of contract, drawings and bills of quantities. Specifications and quantities of work are often approximate and subject to remeasurement by the architect. Standard methods of measurement are defined in standards published by *Österreichisches Normungsinstitut*. The successful contractor is selected by the architect and the building owner. In Austria, architects are usually also employed for project management but they have no direct contractual relation to the contractors. The contractual relation is between the client and the contractor.

During construction, interim payments are made as detailed in the contract and, although standardised in ON A2060 at 7%, a retention of 5% to 10% of the monies due is usually held. Advance payments are seldom made. On completion, usually the employer retains 5% during a defects liability period. In ON A2060 this retention is fixed at 3%. The period is typically two years long, though retention may be released earlier if a letter of warranty is received from a bank. The final account is prepared by the contractor and checked by the employer, taking into account remeasured quantities. Variations are valued using the tender information as a basis.

In the event of a contractual dispute arising, a settlement will be reached as laid down in the contract. An arbitrator may be nominated by both parties to settle a dispute or it may ultimately be sent to the courts.

Development control and standards

Building regulations covering the satisfactory construction and maintenance of buildings – including requirements for energy conservation – are laid down and administered by the individual federal provinces, though they tend to follow a common pattern. The conditions laid down are usually in the form of functional requirements. The regulations for civil engineering are different from those for building but the two are complementary. For many types of civil works, e.g. road building, power stations, military works, mining, etc. the regulations are under the control of the Austrian Federal Government.

The Austrian Standards Institute attributes the classification 'Ö-NORM' or "ON" to the standards that it establishes. The standards are, in principle, recommendations but they may be declared legally binding or as 'deemed to satisfy' either in whole or in part through reference or inclusion in laws or statutory regulations. The Standards Institute may also recommend foreign or international standards.

Construction cost data

Cost of labour

The figures below are typical of labour costs in the Vienna area as at the first quarter 1999. The wage rate is the basis of an employee's income, while the cost of labour indicates the cost to a contractor of employing that employee. The difference between the two covers a variety of mandatory and voluntary contributions - a list of items which could be included is given in section 2.

	Wage rate (per hour) ASch	Cost of labour (per hour) ASch	Number of hours worked per year
Site operatives			
Mason/bricklayer	117	303	1,850
Plumber	–	440	1,850
Electrician	–	480	1,850
Structural steel erector	–	380	1,850
HVAC installer	–	440	1,850
Semi-skilled worker	113	110	1,850
Unskilled labourer	100	258	1,850

	Wage rate (per hour) ASch	Cost of labour (per hour) ASch	Number of hours worked per year
Equipment operator	116	300	1,850
Watchman/security	92	237	1,850
Site supervision			
General foreman *	31,000	45,000	2,100
Trades foreman	133	343	1,850
Clerk of works	129	334	1,850
Contractors' personnel *			
Site manager	38,000	55,100	2,100
Resident engineer	29,000	42,500	2,100
Resident surveyor	29,000	42,500	2,100
Junior engineer	23,000	33,400	2,100
Junior surveyor	23,000	33,400	2,100
Planner	26,000	37,200	2,100
Consultants' personnel *			
Senior architect	38,000	55,100	2,100
Senior engineer	29,000	42,500	2,100
Senior surveyor	29,000	42,500	2,100
Qualified architect	23,000	33,400	2,100
Qualified engineer	23,000	33,400	2,100
Qualified surveyor	23,000	33,400	2,100

* monthly average salary/labour cost.

Cost of materials

The figures that follow are the costs of main construction materials, delivered to site in the Vienna area, as incurred by contractors in the first quarter 1999. These assume that the materials would be in quantities required for a medium sized construction project and that the location of the works would be neither constrained nor remote. All the costs in this section exclude value added tax VAT which is at 20%.

	Unit	Cost ASch
Cement and aggregates		
Ordinary portland cement in 50kg bags	tonne	1,500
Coarse aggregates for concrete	m³	160
Fine aggregates for concrete	m³	180
Ready mixed concrete (B225)	m³	750
Ready mixed concrete (B400)	m³	880

	Unit	Cost ASch
Steel		
Mild steel reinforcement	tonne	5,000
High tensile steel reinforcement	tonne	8,000
Structural steel sections	tonne	8,000
Bricks and blocks		
Common bricks (25 x 12 x 6.5cm)	1,000	4,400
Good quality facing bricks(25 x 12 x 6.5cm) KLINKER	1,000	11,000
Hollow concrete blocks (25 x 25 x 22cm) HLZ	1,000	23,500
Solid concrete blocks (25 x 12 x 6.5cm)	1,000	5,500
Precast concrete cladding units with exposed aggregate finish	m^2	150
Timber and insulation		
Softwood for carpentry	m^3	1,900
Softwood for joinery	m^3	4,000
Hardwood for joinery	m^3	8,000
Exterior quality plywood (4mm)	m^2	120
Plywood for interior joinery (4mm)	m^2	105
Softwood strip flooring (10mm)	m^2	150
Chipboard sheet flooring (6mm)	m^2	225
100mm thick quilt insulation EPS	m^2	80
100mm thick rigid slab insulation XPS	m^2	200
Softwood internal door with frames and ironmongery	no.	6,000
Glass and ceramics		
Float glass (4mm)	m^2	150
Sealed double glazing units	m^2	450
Good quality ceramic wall tiles (15 x 15cm)	m^2	210
Plaster and paint		
Plaster in 50kg bags	tonne	1,600
Plasterboard 12.5mm thick	m^2	35
Emulsion paint in 5 litre tins	litre	5
Gloss oil paint in 5 litre tins	litre	20
Tiles and paviors		
Clay floor tiles (15 x 15 x 0.4cm)	m^2	260
Vinyl floor tiles (25 x 25 x 0.1cm)	m^2	110
Precast concrete paving slabs (50 x 50 x 5cm)	m^2	180
Clay roof tiles	1,000	7,000
Precast concrete roof tiles	1,000	5,000
Drainage		
WC suite complete	each	3,100
Lavatory basin complete	each	1,500
100mm diameter clay drain pipes	m	120
150mm diameter cast iron drain pipes	m	450

Unit rates

The descriptions of work items below are generally shortened versions of standard descriptions listed in five languages (English, French, Italian, German and Spanish) in Appendix 3.

Where an item has a two digit reference number (e.g. 05 or 33), this relates to the full description against that number in Appendix 3 where an item has an alphabetic suffix (e.g. 12A or 34B) this indicates that the standard description has been modified. Where a modification is major the complete modified description is included here and the standard description should be ignored.

Where a modification is minor (e.g. the insertion of a named hardwood) the shortened description has been modified here but, in general, the full description in Appendix 3.

The unit rates below are for main work items on a typical construction project in the Vienna area in the first quarter of 1999. The rates include labour, materials and equipment. No allowance has been made to cover contractor's overheads and profit and preliminary and general items or contractor's profit and attendance on specialist trades. All the rates in this section exclude VAT which is at 20%.

		Unit	Rate ASch
Excavation			
01	Mechanical excavation of foundation trenches	m³	120
02	Hardcore filling in making up levels	m²	25
03	Earthwork support	m²	350
Concrete work			
04	Plain insitu concrete in strip foundations in trenches	m³	1,200
05	Reinforced insitu concrete in beds	m³	1,350
06	Reinforced insitu concrete in walls	m³	1,600
07	Reinforced insitu concrete in suspended floor or roof slabs	m³	1,500
08	Reinforced insitu concrete in columns	m³	1,900
09	Reinforced insitu concrete in isolated beams	m³	1,850
10	Precast concrete slab	m²	290
Formwork			
11	Softwood formwork to concrete walls	m²	290
12	Softwood formwork to concrete columns	m²	520
13	Softwood formwork to horizontal soffits of slabs	m²	310
Reinforcement			
14	Reinforcement in concrete walls	tonne	12,000
15	Reinforcement in suspended concrete slabs	tonne	11,000
16	Fabric reinforcement in concrete beds	m²	55

	Unit	Rate ASch

Steelwork

17	Fabricate, supply and erect steel framed structure	tonne	45,000

Brickwork and Blockwork

18	Precast lightweight aggregate hollow concrete block walls	m^2	600
19	Solid clay or concrete common bricks 100mm thick	m^2	620
20	Solid perforated sand lime bricks	m^2	630
21	Facing bricks, flush pointed	m^2	900
22	Concrete interlocking roof tiles 430 x 380mm	m^2	570
23	Plain clay roof tiles 260 x 160mm	m^2	620
24	Fibre cement roof slates 600 x 300mm	m^2	480
25	Sawn softwood roof boarding	m^2	200
26	Particle board roof coverings with t & g joints	m^2	960
27	3 layers glass-fibre based bitumen felt roof covering	m^2	280
28	Bitumen based mastic asphalt roof covering in 2 layers	m^2	410
29	Glass-fibre mat roof insulation 160mm thick	m^2	200
30	Rigid sheet resin-bonded glass-fibre roof insulation	m^2	560
31	0.8mm troughed galvanised steel roof cladding	m^2	450

Woodwork and metalwork

34	Single glazed casement window in hardwood	each	3,100
35	Two panel door with panels open for glass in hardwood	each	18,000
36	Solid core half hour fire resisting hardwood flush door	each	6,000
37	Aluminum double glazed window, size 1200 x 1200mm	each	6,200
38	Aluminum double glazed door set, hardwood frame	each	8,300
40	Framed structural steelwork in universal joist sections	tonne	35,000
41	Structural steelwork lattice roof trusses.	tonne	36,000

Plumbing

42	UPVC half round eaves gutter	m	200
43	UPVC rainwater pipes with pushfit joints	m	200
44	Light gauge copper cold water tubing	m	135
45	High pressure UPVC pipes for cold water	m	180
46	Low pressure UPVC pipes for cold water	m	155
47	UPVC soil and vent pipes with ring seal joints	m	200
48	White vitreous china WC suite	each	4,000
49	White vitreous china lavatory basin	each	3,700
50	Glazed fireclay shower tray	each	4,000
51	Stainless steel single bowl sink and double drainer	each	4,000

Electrical work

52	PVC insulated and copper sheathed cable	m	180
53	Socket outlet with PVC insulated copper cable	each	120
54	Flush mounted I way light switch with insulated copper cable	each	160

Finishings

55	2 coats gypsum based plaster on brick walls 13mm thick	m^2	400
56	White glazed tiles on plaster walls	m^2	650
57	Red clay quarry tiles on concrete floors	m^2	610

		Unit	Rate ASch
58	Floor screed. Cement and sand screed	m²	180
59	Thermoplastic floor tiles on screed 2.5mm thick	m²	220
60	Suspended ceiling system. Mineral fibre tiles, concealed grid		
	(625 x 625 x 15mm)	m²	330
Glazing			
61	Glazing to wood	m²	320
Painting			
62	Emulsion on plaster walls	m²	42
63	Oil paint on timber	m²	60

Approximate estimating

The building costs per unit area given below are averages incurred by building clients for typical buildings in the Vienna area as at the first quarter 1999. They are based upon the total floor area of all storeys, measured between external walls and without deduction for internal walls.

Approximate estimating costs generally include mechanical and electrical installations but exclude furniture, loose or special equipment, and external works; they also exclude fees for professional services. The costs shown are for specifications and standards appropriate to Austria and this should be borne in mind when attempting comparisons with similarly described building types in other countries. A discussion of this issue is included in section 2. Comparative data for countries covered in this publication, including construction cost data, are presented in Part Three.

Approximate estimating costs must be treated with caution; they cannot provide more than a rough guide to the probable cost of building. All the rates in this section exclude VAT which is at 20%.

	Cost m²	Cost ft²
	ASch	ASch
Industrial buildings		
Factories for letting	4,500	418
Factories for owner occupation (light industrial use)	5,800	539
Factories for owner occupation (heavy industrial use)	6,200	576
Factory/office (high tech) for letting – shell and core only	7,000	650
Factory/office (high tech) for letting – ground floor shell		
and first floor offices	7,400	687
Factory/office (high tech) for owner occupation		
(controlled environment)	10,000	929

	Cost m² ASch	Cost ft² ASch
High tech laboratory workshop centres (air conditioned)	16,000	1,490
Warehouses, low bay (6 to 8m high) for letting	4,000	372
Warehouses, low bay for owner occupation	5,900	548
Warehouses, high bay for owner occupation	5,900	548
Cold stores / refrigerated stores	12,000	1,110
Administrative and commercial buildings		
Civic offices, non air conditioned	14,000	1,300
Civic offices, fully air conditioned	15,000	1,390
Offices for letting, 5 to 10 storeys, non air con.	9,500	883
Offices for letting, 5 to 10 storeys, air con.	10,500	975
Offices for letting high rise, air con.	14,000	1,300
Offices for owner occupation, 5 to 10 storeys, air con	17,000	1,580
Offices for owner occupation, 5 to 10 storeys, non air con	18,000	1,670
Offices for owner occupation, high rise, air con	20,000	1,860
Prestige/headquarters office, 5-10 storeys, air con.	24,000	2,230
Prestige/headquarters office, high rise, air con.	26,000	2,420
Residential buildings		
Social/economic single family housing (multiple units)	18,000	1,670
Private/mass market single family housing (multiple units)	15,000	1,390
Purpose designed single family housing		
2 storey detached	22,000	2,040
Social/economic apartment housing , low rise	7,500	697
Social/economic apartment housing , high rise	7,800	725
Private sector apartment building	12,000	1,110
Private sector apartment building (luxury)	22,000	2,040
Student/nurses halls of residence	8,200	762
Home for the elderly (shared accommodation)	12,000	1,110
Home for the elderly (self contained)	11,500	1,070
Hotel, 5 star, city centre	21,000	1,950
Hotel 3 star, city/provincial	17,000	1,580
Motel	14,000	1,300

Regional variations

The approximate estimating costs are based on costs in Vienna which are about the average for Austria. Costs may vary from about 10% higher in Salzberg and Innsbruck to 20% less in southern Styria.

Value added tax (VAT)

The standard rate of VAT is currently 20%, chargeable on general building work and materials.

Exchange rates and cost and price trends

The combined effect of exchange rates and inflation on prices within a country and on price comparisons between countries is discussed in section 2.

Exchange rates

The graph below plots the movement of the Austrian Schilling against sterling, the ECU/Euro, the US dollar and 100 Japanese yen since 1990. The values used for the graph are quarterly and the method of calculating these is described and general guidance on the interpretation of the graph provided in section 2. The average exchange rate in the first quarter of 1999 was ASch 19.61 to the pound sterling, ASch 11.77 to the US dollar and ASch 13.76 to the Euro.

The Austrian Schilling against sterling, the ECU/Euro, the US dollar and 100 Japanese yen

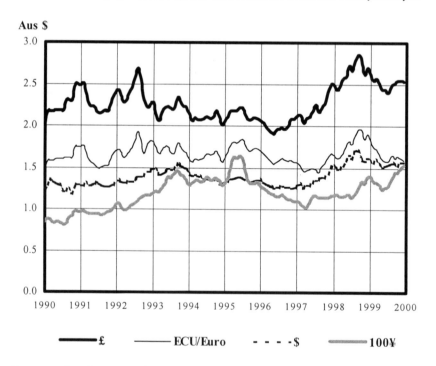

Cost and price trends

The following tables present indices for consumer prices, residential building and civil engineering costs and building labour costs in Austria since 1990. The indices have been rebased to 1990=100. The annual change is the percentage change from the average index of the previous years. Notes on inflation are included in section 2.

Consumer price, residential building cost and civil engineering cost indices

Year	Consumer prices annual average	change %	Residential building costs annual average	change %	Civil engineering costs annual average	change %
1990	100.0	3.3	100.0	4.1	100.0	3.5
1991	103.3	3.3	101.1	1.1	104.3	4.3
1992	107.5	4.1	106.4	5.3	106.0	1.6
1993	111.4	3.6	111.9	5.1	107.7	1.6
1994	114.7	3.0	114.8	2.6	110.3	2.4
1995	117.3	2.3	118.6	3.3	112.9	2.4
1996	119.4	1.8	120.4	1.5	116.3	3.0
1997	120.8	1.2	123.7	2.8	120.2	3.3
1998	121.8	0.8	–	–	–	–

Sources: International Monetary Fund, Yearbook 1998 and monthly supplements 1999.
Österreisches Statistiches Zentralamt, 1997.

Building labour cost index

Year	annual average	change %
1990	100.0	7.9
1991	106.5	6.5
1992	112.6	5.7
1993	119.7	6.3
1994	123.8	3.4
1995	127.7	3.2
1996	131.0	2.6
1997	135.0	3.1

Sources: Österreisches Statistiches Zentralamt, 1997.

It will be seen that cost of building has risen roughly in line with consumer prices. Building labour costs have risen faster then consumer prices. The higher increases in 1991 to 1993 may partly account for the steeper rise in residential building cost in 1992 and 1993. It is estimated that 1998 construction costs increased at a low rate compared with 1997.

Useful addresses

Government and public organizations

Bundesministerium für Wirtschaftliche Angelegenheiten

Federal Ministry of Economic Affairs
Amtsgebäude, Stubenring 1
1010 Wien

Tel: (43-1) 71100
Fax: (43-1)7110055

Bundesministerium für Wissenschaft u Verkehr

Federal Ministry of Public Economy
and Transport
Minoritenpl. 5
1010 Wien

Tel: (43-1) 53120

Statistisches Zentralamt

Central Statistics Office
Zollamtsstrasse 2b,
A-1033 Wien

Tel: (43-1) 71128
Fax: (43-1)7156829

Österreichisches Normungsinstitut

Austrian Standard Institute
Heinestrasse 38
1020 Wien

Tel: (43-1) 213000
Fax: (43-1) 21300818

Professional and trade associations

Fachverband der Osterreichischen Bauindustrie

Association of the Austrian
Construction Industry
Karlsgstrasse
1040 Wien

Tel: (43-1) 50415510
Fax: (43-1) 5041555

Bundeskammer der Architekten und Ingenieurkonsulenten

National Union of Architects and
Engineers
Karlsgasse 9
1040 Wien

Tel: (43-1) 5055807
Fax: (43-1) 5053211

Verband Österreichischer Ingenieure (VOI)

Association of Austrian Engineers
Eschenbachg 9
1010 Wien

Tel: (43-1) 5874198

Verband Österreichischer Beton und Fertigteilwerke

Leopold-Hasnerstrasse 36/1/4
4020 Linz/Donau

Tel: (43-732) 656062
Fax: (43-732)666683

Other organizations

Technische Universität Wien

Karlsplatz 13
A-1040 Wien

Tel: (43-1) 58801
Fax: (43-1) 58801

Technische Universität Graz

Rechbauerstrasse 12
A-8020 Graz

Tel: (43-316) 8730

Belgium

All data relate to 1998 unless otherwise indicated.

Population
Population	10.2 million
Urban population (1997)	97%
Population under 15	17%
Population 65 and over	17%
Average annual growth rate (1985 to 1998)	0.2%

Geography
Land area	30,510 km^2
Agricultural area	45%
Capital city	Brussels

Economy
Monetary unit	Belgian Franc (BFr)
Exchange rate (average first quarter 1999) to:	
the pound sterling	BFr 57.53
the US dollar	BFr 34.53
the Euro	BFr 40.34
the yen x 100	BFr 29.21
Average annual inflation (1995 to 1998)	1.6%
Inflation rate	1%
Gross Domestic Product (GDP)	BFr 8,968 billion
GDP PPP basis (1997)	$ (PPP) 236 billion
GDP per capita	BFr 881,900
GDP per capita PPP basis (1997)	$ (PPP) 23,200
Average annual real change in GDP (1995 to 1998)	2.4%
Private consumption as a proportion of GDP	63%
Public consumption as a proportion of GDP	15%
Investment as a proportion of GDP	17%

Construction
Gross value of construction output	BFr 1,089 billion
Gross value of construction output per capita	BFr 107,000
Gross value of construction output as a proportion of GDP	12.1%
Average annual real change in gross value of construction output (1995 to 1998)	2.7%
Annual cement consumption	5.7 million tonnes

PPP Purchasing power parity

The construction industry

Construction output

The estimated value of the gross output of the construction sector in Belgium in 1998 was BFr 1,089 billion equivalent to 26.8 billion ECUs (nearly the same as January 1999 Euros). This represents 12.1% of GDP. Information on work by other sectors, the black economy or DIY is not available. The table below shows the breakdown of output in 1998.

Estimated output of construction sector, 1998 (current prices)

Type of work	BFr billion	ECUs billion	% of total
New work			
Residential building	312	7.7	29
Non-residential building			
Office and commercial	163	4.0	15
Industrial	41	1.0	4
Other	73	1.8	7
Total	277	6.8	25
Civil engineering	102	2.5	9
Total new work	691	17.0	63
Repair and maintenance			
Residential	204	5.0	19
Non-residential	122	3.0	11
Civil engineering	72	1.8	7
Total repair and maintenance	398	9.8	37
Total	**1,089**	**26.8**	**100**

Note: the value of the Euro in January 1999 was very similar to that of the ECU in 1998.
Source: based on Binotto, C, 'Belgium' in Euroconstruct conference proceedings,
     Prague, June 1999.

The rate of growth of the construction sector in 1998 at 3.6% was lower than that of 1997 at 6.8% but still considerable and spread fairly evenly over all sectors. The generally buoyant levels of output are a result of a favourable economic situation including a growth in real disposable income and a fall in unemployment. Lower interest rates have also helped.

The new residential sector grew by almost 5% in 1997 and in 1998 continued to grow at a lower rate. The favourable economic situation will probably keep the level of construction up but in the longer term a fall is likely following a decrease in the permits for construction of new dwellings. One or two family dwellings accounted for around two thirds of housing starts in 1998, the other third being flats.

The private non-residential sector increased in 1998 but at a lower rate than in the 1997 boom when office buildings rose by about 24%. This high rate is not being maintained. Industrial building in 1998 continued the modest growth of previous years. Public sector non-residential construction marked time in 1998 with only slight increases including in school and hospital buildings.

Civil engineering investment in transport, power and water were high in 1998 and the potential demand is great. With the greater role of the private sector, especially in the freeing of the telecommunications sector from constraints, investment is likely to continue. Local authorities account for over half public investment largely in civil engineering projects. With local elections in 2000 investment may increase. However, the proportion of civil engineering construction in Belgium is low, at less than 10% compared to other countries in Europe.

Renovation is likely to continue at a satisfactory level in line with the economy in general. The financial position of households is favourable to residential repair and maintenance expenditure.

The geographical distribution of work is shown in the table below. Brussels is not as dominant as might be expected, especially in view of its importance in the EU. Most of the work is undertaken in the Flemish region.

Construction output 1996 regional distribution

Regions	Building and civil engineering work (%)
Bruxelles	15.1
Flemish Region	
Antwerpen	20.4
Limburg	9.0
Oost-Vlaanderen	15.6
Vlaams-Brabant	6.3
West-Vlaanderen	12.3
Total	63.6
Wallon Region	
Brabant Wallon	2.3
Hainault	7.5
Liege	9.1
Namur	2.3
Total	21.3
Total	**100.0**

Source: Confederation de la Construction, Chiffres-Cles de la Construction 1997.

Characteristics and structure of the industry

The title of engineer is protected by law. Some engineers are employed in state organizations; others are employed in engineering *bureaux* or operate as freelance consultants. The choice of a *bureau* for a particular project is influenced by political factors, and by regional location (partly because of language).

The title of architect is also protected. It is obligatory, according to laws of February 1939 and 26 June 1963, to employ an architect for the design, execution and control of works for which planning permission is necessary. The *Ordre des Architectes* (the professional organization for architects) is responsible for maintaining a register of qualified architects. Some architects working within architectural practices do not need to register. The size of practices is generally small. Pay is poor compared with other professions. The architect (as it is defined in the Code of Architects) designs the project, prepares and directs its physical realization and ensures that the work is done according to the best possible standards. Traditionally the country is dominated by one-family houses which are characterised by a high degree of individual design, therefore the number of architects (around 10,000) is quite large.

Contractors are required by law to be registered. Belgium is a small country and has no very large companies. In 1996 there were around 2,200 construction companies in the industry, employing around 18,000 people. Only 200 companies had more than 100 employees. Most of Belgian companies work on a regional scale. In the last few years, the number of Belgian companies investing in foreign companies increased substantially.

In 1997 Belgium had two firms in the top 100 European firms and six in the top 300 according to *Building* magazine's list of top European contractors. These are shown in the table below. 'Besix' is the only Belgian contractor listed in *Engineering News Record's* list of the 'Top 225 International Contractors' for 1997, at rank 73.

Major Belgium contractors, 1997

Major contractors	Place in Building's Top 300 European Contractors'
CFE	76
BESIX	95
SOFICOM	136
CIB	206
SCREG BELGIUM	292

Source: Building December 1998.

Clients and finance

Around 10% of houses are financed by housing associations, for example by the *Société Nationale de Logement*. Low cost housing, designed and operated by communal low cost housing companies, is financed partly by the regions. However, housing in Belgium is largely dominated by individual units: private single family houses built by individuals and housing units built by real estate developers.

Since regionalisation, the communities have assumed responsibility for education, culture, health care and social services. Many of the services of the federal government-controlled organization – the *Régie des Bâtiments* – in the management and finance of new projects have been dispersed among the various regions.

Selection of design consultants

Public sector design work is often undertaken by in-house design teams. Alternatively, if an external designer is employed, he is selected on the basis of his reputation through a wide-ranging interview process. It is therefore important for architects to be well known. Competitions are rare.

In the private sector the process for selection of designers is varied but is normally based on reputation and past association with the client.

Contractual arrangements

Belgian contracting methods are very similar to those adopted in the Netherlands and the construction industries in the two countries share many similarities.

Open tendering is the method of contractor selection used for Government work and the Government is bound to pay damages to the lowest tenderer if his offer is not accepted. Contractors cover the risks involved in open tendering by a well established system of price adjustment before tenders are submitted; tenders are adjusted to the arithmetical average and allowance is also made for the cost of tendering. There can also be a considerable amount of post-tender negotiation. It is normal to employ a general contractor who may sublet work to specialist trades; however, this does not relieve him of his obligations to the client and he is directly responsible for the trades work contracts. There is also normally a fluctuation clause that involves adjustments by a pre-agreed formula based upon monthly price indices published by the Government Statistical Office.

Selective tendering is most common in private work but negotiated tendering is on the increase. Another method, mainly for private work, is where contractors submit unit prices for labour, materials and plant with a percentage addition for on-costs, and valuations are based on an assessment of resources used.

Private construction generally follows the General Contract Conditions issued by the *Fédération Royale des Sociétés d'Architectes* along with the *Confédération Nationale de la Construction*, which is made up of civil engineering, building and decorating contractors. The three most frequently used contracts, defined according to the method of establishing the cost of the works, are the following:

- *Marché à Forfait Absolu* (the fixed lump sum contract) with no provision for variations
- *Marché à Forfait Relatif* which includes a variation and fluctuation clause and a remeasurement contract
- *Marché à Borderaux de Prix* based upon a schedule of rates.

The forms are often adapted to particular circumstances. There are also Government standard forms of contract for public works and dwellings built with government subsidies.

Liability and insurance

Both the contractor and the architect are jointly responsible. Particularly relevant is the fact that the contractor is responsible for all aspects of the construction work and for informing the architect of any anomalies or discrepancies that may exist. Since each project is unique, the special contract conditions generally provide a detailed description of the respective responsibilities of the parties.

Insurance is obligatory, although the amount of the insurance is not stipulated. Moreover, it is advisable that the client or his representative (e.g. the architect) ensures that, at least for larger projects, all the parties are insured. The client will often take out a blanket insurance to cover all parties. Insurance is considerably cheaper than in France and probably adds about 1%, or a little more, to the cost of a project. In addition, it is necessary to be insured for injury and accidents.

Provision is made in the contract conditions for phased payments at an instalment rate of two per cent (work costing less than 500,000 Belgian Francs) and one per cent (works costing more than 500,000 Belgian Francs). Moreover, a retention of five per cent is withheld until final completion. The architect is considered responsible for the decision of the date at which the instalments are due. In order to be certified to full value, works must be fully completed in accordance with the contract, unless it has been differently agreed in the special contract conditions. Materials and other building elements which have not been used will not be included in any certificate, even if they have been accepted by the architect and delivered on site. Invoices are payable after approval by the architect within two weeks. In case this period is exceeded, due interest is contractually set at a premium over the bank rate paid at the Banque National de Belgique. This rate increases after 101 days of delay. In case of delay of more than 30 days after the approval, the contractor is entitled to suspend his work, given that he notifies his intention to do so at least one week before the suspension, and this does not harm his right to interest or an indemnity. Where contentions arise because the employer will not make a payment for defective work, a designated expert will value the work and the employer will be obliged to place the extra amount with a mutually agreed third party, or in default of an agreement, a designated third party. The special contract conditions will specify in each case the jurisdiction or arbitral institution which will be held responsible for any litigation or arbitration which might arise among the employer, the architect and the contractor, or any pair of them. In any case, clause 35 provides that any arbitration agreement comes outside the jurisdiction of the Belgian courts in relation to the resolution of contention which arise within the scope of the arbitration segment.

Development control and standards

The planning system is largely controlled by the three regional Governments. The basic rules and regulations for obtaining planning consent are well formulated and, if they are adhered to, planning permission will normally be given.

A planning application must be signed by a registered architect. Moreover, the *Ordre des Architectes* receives a copy of every building application to check that the architect is in fact registered. The decision on granting a planning application is taken by the mayor. There are, however, frequently specific local peculiarities and early consultation with the mayor is advisable.

Building control in Belgium is also exercised through local authorities, each of which is authorized to issue by-laws on planning, and on safety and hygiene which may affect design and equipment. There are no national laws covering technical requirements for building construction, though the local authority may request submission of technical calculations and drawings. There is conflict between the various levels of authority, especially in Brussels, so that permission is often delayed by up to a year.

There are technical controllers, (*bureau de contrôle*) who are employed by the client alone. The quality of their advice is high.

The *Institut de Normalisation* centralises regulations produced by several institutions and produces a large number of standards for construction. Standards produced by the Institute are not, in general, mandatory unless referred to by a body with legal authority.

There are several inspecting organizations in Belgium including the Safety of Building in Belgium (SECO) and AIB. The Inspectorate for SECO carries out the technical inspection of plans, calculations, construction, and of products used in construction. It operates its own (SECO) approval scheme for certain products; it also operates two schemes in conjunction with other organisations. One of these schemes is the SECO-CSTC (*Le Centre Scientifique et Technique de la Construction* : the Scientific and Technical Centre for Building) scheme which is administered jointly by SECO and CSTC; the other is the BENOR (*l'Institut Belge de Normalisation*) approval scheme which is sponsored by the Belgian Standards Institution. SECO tests products in its own laboratories; SECO, AIB and other such organizations' inspectors will then make periodic visits to manufacturers' works to check the production of approved products, which must be subjected to quality control by the manufacturer.

For new products or products for which there are no Belgian standards, *Agrément* certificates issued by the National Housing Institute are widely accepted.

Research and development

The *Centre Scientifique et Technique de la Construction* is a non-profit making organisation financed by government grants and a levy on the building materials. It provides an information service and publishes research reports and information notes. The notes contain recommendations and rules for good practice.

Construction cost data

Cost of labour

The figures in the following tables are typical of labour costs in the Brussels area as at the first quarter 1999. The wage rate is the basis of an employee's income, while the cost of labour indicates the cost to a contractor of employing that employee. The difference between the two covers a variety of mandatory and voluntary contributions – a list of items which could be included is given in section 2.

	Cost of labour (per hour) BFr	Number of hours worked per year
Site operatives		
Mason/bricklayer	1,230	1,824
Carpenter	1,240	1,824
Plumber	1,220	1,824
Electrician	1,220	1,824
Structural steel erector	1,230	1,824
HVAC installer	1,260	1,824
Semi-skilled worker	1,110	1,824
Unskilled labourer	1,010	1,824
Site supervision		
Watchman/security	1,040	1,824
Site Supervision	1,230	1,824
General foreman	1,330	1,824
Contractors' personnel		
Site manager	1,330	
Junior engineer	3,140	1,880
Planner	2,500	1,880
Consultants' personnel		
Senior architect	3,150	1,880
Senior engineer	3,140	1,880
Qualified architect	3,300	1,880
Qualified engineer	3,330	1,880

Cost of materials

The figures that follow are the costs of main construction materials, delivered to site in the Brussels area, as incurred by contractors in the first quarter of 1999. These assume that the materials would be in quantities as required for a medium sized construction project and that the location of the works would be neither constrained nor remote. All the costs in this section exclude value added tax (VAT) which is at 6%.

	Unit	Cost BFr
Cement and aggregates		
Coarse aggregates for concrete 7/20	m^3	378
Fine aggregates for concrete 2/7	m^3	230
Ready mixed concrete (mix 16N/mm^2)	m^3	2,540

	Unit	Cost BFr
Steel		
Mild steel reinforcement (12mm)	tonne	11,700
High tensile steel reinforcement (12mm)	tonne	19,200
Structural steel sections	tonne	16,500
Bricks and blocks		
Common bricks (230 x 140 x 90mm)	1,000	9,600
Good quality facing bricks	1,000	8,100
Hollow concrete blocks (390 x 190 x 190mm)	1,000	32,200
Solid concrete blocks (320 x 190 x 190mm)	1,000	33,000
Timber and insulation		
Softwood internal door with frames and ironmongery	each	8,500
Glass and ceramics		
Float glass (8mm thick)	m^2	1,470
Sealed double glazing units (4 x 12 x 4mm)	m^2	1,640
Good quality ceramic wall tiles (20 x 20mm)	m^2	1,200
Plaster and paint		
Plaster in 25kg bags	tonne	4,500
Plasterboard (9.5mm thick)	m^2	81
Emulsion paint in 5 litre tins	litre	350
Gloss oil paint in 5 litre tins	litre	350
Tiles and paviors		
Clay floor tiles (20x20cm)	m^2	1,000
Clay roof tiles	m^2	400
Precast concrete roof tiles	m^2	200
Drainage		
WC suite complete	each	20,000
Lavatory basin complete	each	15,000
100mm diameter clay drain pipes	m	410
100mm diameter cast iron drain pipes	m	2,060

Unit rates

The descriptions of work items following are generally shortened versions of standard descriptions listed in five languages (English, French, Italian, German and Spanish) in Appendix 3.

Where an item has a two digit reference number (e.g. 05 or 33), this relates to the full description against that number in Appendix 3. Where an item has an alphabetic suffix (e.g. 12A or 34B) this indicates that the standard description has been modified. Where a modification is major the complete modified description is included here and the standard description should be ignored.

Where a modification is minor (e.g. the insertion of a named hardwood) the shortened description has been modified here but, in general, the full description in Appendix 3 prevails.

The unit rates below are for main work items on a typical construction project in the Brussels area in the first quarter of 1999. The rates include labour, materials and equipment and allowances for contractor's overheads and profit and preliminary and general items. Allowances have also been made to cover contractor's profit and attendance on specialist trades. All the rates in this section exclude VAT which is at 21%.

		Unit	Rate BFr
Excavation			
01	Mechanical excavation of foundation trenches	m³	490
02	Hardcore filling making up levels	m³	1,240
03	Earthwork support	m	188
Concrete work			
04	Plain insitu concrete in strip foundations in trenches 20N/m²	m³	12,800
05	Reinforced insitu concrete in beds 20N/m²	m³	4,520
06	Reinforced insitu concrete in walls	m³	10,300
07	Reinforced insitu concrete in suspended floor or roof slabs 20N/m²	m³	19,000
08	Reinforced insitu concrete in columns 20N/m²	m³	33,400
09	Reinforced insitu concrete in isolated beams 20N/m²	m³	26,800
Formwork			
11	Softwood formwork to concrete walls	m²	1,210
12	Softwood or metal formwork to concrete columns	m²	1,220
Reinforcement			
14	Reinforcement in concrete walls (16mm)	kg	77
16	Fabric reinforcement in concrete beds	m²	95
Brickwork and blockwork			
18	Precast lightweight aggregate hollow concrete block walls (100m thick)	m²	1,220
19	Solid (perforated) concrete bricks	m²	1,030
20	Solid (perforated) sand lime bricks	m²	1,100
21	Facing bricks	m²	2,250
Roofing			
22	Concrete interlocking roof tiles 430 x 380mm	m²	766
23	Plain clay roof tiles 260 x 160mm	m²	1,020
24	Fibre cement roof slates 600 x 300mm	m²	1,210

		Unit	Rate BFr
27	2 layers glass-fibre based bitumen felt roof covering include chippings	m²	1,780
28	Bitumen based mastic asphalt roof covering	m²	1,180
29	Glass-fibre mat roof insulation 120mm thick	m²	467
30	Rigid sheet resin-bonded loadbearing glass-fibre insulation	m²	600
31	Troughed galvanised steel roof cladding	m²	853

Woodwork and metalwork

33	Preservative treated sawn softwood (52 x 155mm)	m	268
34	Single glazed casement window in Merahti hardwood, Size 600 x 900mm	each	9,080
35	Two panel glazed door in hardwood, size 850 x 2000mm	each	22,400
36	Solid core half hour fire resisting hardwood internal flush doors, size 800 x 2000mm	each	22,300
37	Aluminium double glazed window, size 1200 x 1200mm	m²	14,200
39	Hardwood Oak skirtings 20 x 100mm	each	4,250

Plumbing

42	UPVC half round eaves gutter (110mm)	m	163
43	Copper rainwater pipes (100mm)	m	1,290
44	Light gauge copper cold water tubing (15mm)	m	141
45	High pressure plastic pipes for cold water supply (12.5mm)	m	447
46	Low pressure plastic pipes for cold water distribution (16mm)	m	111
48	White vitreous china WC suite	each	9,490
49	White vitreous china lavatory basin	each	7,010
50	White glazed fireclay shower tray	each	10,800
51	Stainless steel single bowl sink and double drainer	each	9,640

Electrical work

52	PVC insulated and copper sheathed cable	m	213
53	13 amp flush mounted, unswitched socket outlet	each	657
54	Flush mounted, 1 way light switch	each	742

Finishings

55	2 coats gypsum based plaster on brick walls	m²	466
56	White glazed tiles on plaster walls	m²	1,210
57	Red clay quarry tiles on concrete floor	m²	2,150
58	Cement and sand screed to concrete floors 50mm thick	m²	850
59	Thermoplastic floor tiles on screed	m²	4,550
60	Mineral fibre tiles on concealed suspension system (600 x 600 x 15mm)	m²	1,050

Glazing

61	Glazing to wood 4mm	m²	1,500

Painting

62	Emulsion on plaster walls	m²	555
63	Oil paint on timber	m²	452

Approximate estimating

The building costs per unit area given below are averages for the Brussels area incurred by building clients for typical buildings as at the first quarter 1999. They are based upon the total floor area of all storeys, measured between external walls and without deduction for internal walls.

Approximate estimating costs generally include mechanical and electrical installations but exclude furniture, loose or special equipment, and external works; they also exclude fees for professional services. The costs shown are for specifications and standards appropriate to Belgium and this should be borne in mind when attempting comparisons with similarly described building types in other countries. A discussion of this issue is included in section 2. Comparative data for countries covered in this publication, including construction cost data, are presented in Part Three.

Approximate estimating costs must be treated with caution; they cannot provide more than a rough guide to the probable cost of building. All the rates in this section exclude VAT which is at 21%.

	Cost m^2 BFr	Cost ft^2 BFr
Industrial buildings		
Factories for letting	20,000	1,860
Factories for owner occupation	20,000	1,860
Factories for owner occupation(heavy use)	20,000	1,860
Factory/office for letting (shell and core)	25,000	2,320
Factory/office for letting (ground floor and first floor offices)	25,000	2,320
Factory/office for owner occupation (controlled environment)	25,000	2,320
High-tech laboratory workshop centres	25,000	2,320
Warehouses low bay for letting (no heating)	15,000	1,390
Warehouses low bay for owner occupation	15,000	1,390
Warehouses high bay for owner occupation	15,000	1,390
Administrative and commercial buildings		
Civic offices, non air conditioned	31,000	2,880
Civic offices, fully air conditioned	31,000	2,880
Offices for letting, 5-10 storeys, non air con.	37,000	3,440
Offices for letting, 5-10 storeys, air con.	37,000	3,440
Office for letting, high rise air conditioned	37,000	3,440
Office for owner occupation, 5-10 storeys, air conditioned	37,000	3,440
Office for owner occupation, 5-10 storeys, non air conditioned	37,000	3,440
Office for owner occupation, high rise, air conditioned	46,000	4,270
Prestige/headquarters office, 5-10 storeys	46,000	4,270
Prestige/headquarters office, high rise	46,000	4,270

	Cost m² BFr	Cost ft² BFr
Health and education buildings		
General hospitals	35,000	3,250
Teaching hospitals	35,000	3,250
Private hospitals	35,000	3,250
Health centres	33,000	3,070
Nursery schools	33,000	3,070
Primary/junior schools	33,000	3,070
Secondary/middle schools	33,000	3,070
University art buildings	31,000	2,880
University science buildings	31,000	2,880
Management training centres	31,000	2,880
Recreation and arts buildings		
Theatres over 500 seats, seating and stage	40,000	3,720
Theatres less than 500 seats, seating and stage	40,000	3,720
Concert halls including seating and stage	40,000	3,720
Sports halls, changing and social facilities	40,000	3,720
Swimming pools, changing and social facilities (international standard)	40,000	3,720
Swimming pools, changing and social facilities (schools standard)	40,000	3,720
National museums full air conditioning	48,000	4,460
Local museums air conditioning	48,000	4,460
City centre/central libraries	48,000	4,460
Branch local libraries	48,000	4,460
Residential buildings		
Social/economic single family housing	24,000	2,230
Private single family, 2 storey, multiple units	28,000	2,600
Purpose designed, single family, 2 storey, single unit	30,000	2,790
Social/economic apartment housing, low rise	25,000	2,320
Social/economic apartment housing, high rise	30,000	2,790
Private sector apartment building (standard)	32,000	2,970
Private sector apartment building (luxury)	40,000	3,720
Student/nurses halls of residence	25,000	2,320
Homes for the elderly (shared)	32,000	2,970
Homes for the elderly (self contained with shared communal facilities)	32,000	2,970

Regional variations

The approximate estimating costs are based on average costs in Brussels. These costs may vary by up to 10% either side depending on the region.

Value added tax (VAT)

The standard rate of VAT is currently 21%, chargeable on general building work and 6% on building materials.

Exchange rates and cost and price trends

The combined effect of exchange rates and inflation on prices within a country and on price comparisons between countries is discussed in section 2.

Exchange rates

The graph below plots the movement of the Belgian franc against sterling, the ECU/Euro, the US dollar and 100 Japanese yen since 1990. The values used for the graph are quarterly and the method of calculating these is described and general guidance on the interpretation of the graph provided in section 2. The average exchange rate in the first quarter of 1999 was BFr 57.53 to the pound sterling, BFr 34.5 to the US dollar and BFr 40.3 to the Euro.

Belgian franc against sterling, the ECU/Euro, the US dollar and 100 Japanese yen

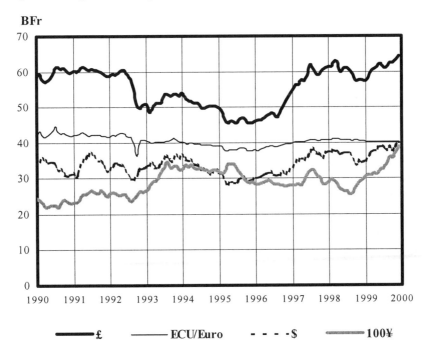

Cost and price trends

The table below presents the consumer price and input price indices in Belgium since 1990. The indices have been rebased to 1990=100. Notes on inflation and the distinction between costs and prices are included in section 2.

The input cost index is based on the costs of the labour and materials inputs to the construction sector. It aims to cover the whole field of construction: new construction and renovation of residential buildings (both public and private); non-residential buildings; and civil engineering. The materials selected for the index accounted for 68% of the sector's total inputs in 1980.

Consumer price and input costs indices

	Consumer prices		Input costs	
	annual	change	annual	change
Year	average	%	average	%
1990	100.0	3.4	100.0	40.9
1991	103.2	3.2	94.8	−6.2
1992	105.7	2.4	93.5	−1.4
1993	108.6	2.7	92.7	−1.1
1994	111.2	2.4	95.1	2.6
1995	112.8	1.4	97.5	2.5
1996	115.2	2.1	97.5	0.0
1997	117.0	1.6	99.7	2.3
1998	118.2	1.0	99.1	−0.6

Sources: International Monetary Fund, Yearbook 1998
and monthly supplements 1999.
Institut National de statisques/National Institut voor de Statistiek.

Useful addresses

Government and public organizations

**Centre Scientifique et Technique d
e la Construction (CSTC)**
Rue de la Violette 21–23
1000 Bruxelles
Tel: (32-2) 5026690
Fax: (32-2) 5028180

Le Ministère des Travaux Publics

Flemish Community
Quartier Arcade – Blok F (6e étage)
1010 Bruxelles
Tel: (32-2) 2276411
Fax: (32-2) 2276455

French Community
Rue Mazy 25–27
5100 Jambes
Tel: (32-81) 331211
Fax: (32-81) 331299

Brussels Region
Boulevard du Régent 21–23
1000 Bruxelles
Tel: (32-2) 5063311
Fax: (32-2) 5135080

Professional and trade associations

Confédération Nationale de la Construction (CNC)
Rue du Lombard 34–42
B-1000 Bruxelles
Tel: (32-2) 5104723
Fax: (32-2) 5133004

Fédération Royale des Sociétés d'Architectes de Belgique
21 Rue Ernest Allard
1000 Bruxelles
Tel: (32-2) 5123452
Fax: (32-2) 5028204

Société Royale Belge des Ingénieurs et des Industriels
Hotel Ravenstein
Rue Ravenstein 3
1000 Bruxelles
Tel: (32-2) 5115856
Fax: (32-2) 5145795

Groupement des Producteurs de Matériaux de Construction
Rue César Franck 46
1050 Bruxelles
Tel: (32-2) 6455211
Fax: (32-2) 6400670

Belgian International Construction Industry (BICI)
Avenue Grand Champ 148
1150 Bruxelles
Tel: (32-2) 7716108
Fax: (32-2) 7713093

Association Royale Belge des Ingénieurs – Conseils et Experts
Avenue des Gaulois 32
1040 – Brussels
Tel: (32-2) 7361198

National Union of Metal, Timber and Building Companies (NACEBO)
Spastraat 8
1040 Bruxelles
Tel: (32-2) 2380605
Fax: (32-2) 239354

Association des Entrepreneurs Belges de Travaux de Génie Civil (ABEB)
Avenue Grand Champ 148
1150 Bruxelles
Tel: (32-2) 7710044
Fax: (32-2) 7713093

National Federation of Builders Merchants' Professional Unions (FE-MA)
Rue de la Limite 23
1030 Bruxelles
Tel: (32-2) 2180473
Fax: (32-2) 2175472

Other organizations

Institut Belge de Normalisation
Avenue de la Brabanconne 29
B-1040 Bruxelles
Tel: (32-2) 7349205
Fax: (32-2) 7334264

Belgian Quality Centre (BCQ)
Rue des Drapiers 21
1050 Bruxelles
Tel: (32-2) 5102311
Fax: (32-2) 5102301

Association Belge des Experts (ABE)
Residence Aurore 1
Avenue Frans Van Kalnen 1-104
1070 Bruxelles
Tel-Fax: (32-2) 5207813

Republic of Cyprus

All data relate to 1998 unless otherwise indicated.

Population
Population	0.75 million
Urban population (1995)	70%
Population under 15	25%
Population 65 and over	10%
Average annual growth rate (1985 to 1998)	0.2%

Geography
Land area	5,895 km^2
Agricultural area	17%
Capital city	Nicosia

Economy
Monetary unit	Cyprus Pound (C£)
Exchange rate (average first quarter 1999) to:	
the pound sterling	C£ 0.83
the US dollar	C£ 0.49
the Euro	C£ 0.58
the yen x 100	C£ 0.42
Average annual inflation (1985 to 1998)	3%
Inflation rate	2.3%
Gross Domestic Product (GDP)	C£ 4.65 billion
GDP PPP basis (1997)	$ (PPP) 9.75 billion
GDP per capita	C£ 6,200
GDP per capita PPP basis (1997)	$ (PPP) 13,175
Average annual real change in GDP (1995 to 1998)	4.9%
Private consumption as a proportion of GDP	61%
Public consumption as a proportion of GDP	18%
Investment as a proportion of GDP	20%

Construction
Gross value of construction output	C£ 0.6 billion
Gross value of construction output per capita	C£ 800
Gross value of construction output as a percentage of GDP	12%
Average annual real change in gross value of construction output (1995 to 1998)	2.6%
Annual cement consumption (1997)	1.0 million tonnes

PPP Purchasing power parity

The construction industry

Introduction

This section deals with the Republic of Cyprus, which occupies the southern part of the island of Cyprus and comprises almost two thirds of the land area and around 80% of the population. The remainder of the island is occupied by Turkey. The Republic has applied for membership of the European Union and in March 1998 discussions for accession to the Union commenced. The Union does not recognize the Turkish occupation of the north of the island.

Construction output

It is estimated that the gross value of construction output in the Republic of Cyprus in 1998 was C£0.6 billion or 1.0 billion ECUs (the same as January 1999 Euros). This represents around 12% of GDP. The estimated breakdown by main types of work for 1996 is shown below. 1997 construction output was stationary when compared to 1996 levels but there was an increase in 1998.

Output of construction, 1996 (current prices)

Type of work	C£ billion	ECUs billion	% of total
New work			
Residential building	0.2	0.35	52
Non-residential building	0.1	0.17	22
Civil engineering	0.1	0.17	26
Total	**0.4**	**0.69**	**100**

Source: Department of Statistics and Research, Ministry of Finance.

Of the civil engineering work the largest component was roads and bridges, followed by telecommunications and electrical transmission lines, dams and irrigation works and land development and water supply.

In the past ten years, the main construction activity in the Republic consisted of infrastructure projects (roads and water supply), tourism and housing. There are further plans for major infrastructure upgrading. Residential building was mainly driven by tourism and the need for rehousing the refugees after the Turkish invasion but now both these needs have largely been catered for.

The geographical distribution of construction work in 1995 is shown below.

Building permits, 1995 regional distribution

Districts	% of total
Lefkosia	34.7
Ammochostos	9.2
Larnaka	13.7
Lemesos	27.7
Pafos	14.7
Total	**100.0**

Sources: Statistical Abstract, 1996.

Characteristics and structure of the industry

In the Republic of Cyprus in 1997 there were 20 construction firms with over 50 employees. There are two Cypriot firms in the *Engineering News Record's* 'Top 225 International Contractors'. They are as follows:

Principal contractors in Cyprus, 1997

Contractor	Place in ENR *list*
Joannou and Paraskevaides (Overseas) Ltd	44
Charilaos Apostolides and Co Ltd	187

Source: Engineering News Record 17.08.1998.

The Cyprus Civil Engineers and Architects Regulation Board is charged, by the 'Architects and Civil Engineers Law' of 1962, with the duty of maintaining suitable professional standards.

Contractual arrangements

In the Republic of Cyprus tenders for Government contracts are usually invited by public advertisement. Contractors must hold the appropriate licence for the class of work involved to be eligible to participate. In the private sector, most contracts are awarded as a result of competitive tenders received from a short list of firms selected by the client's consultants.

The following forms of contract are used in the Republic of Cyprus:

- the standard form of contract issued by the Association of Architects and Civil Engineers and the Association of Building Contractors. This is the form most widely used on all types of construction projects and is based on a 1947 RIBA form. An updated version was introduced in January 1992
- the Government standard form. This is a form based on the FIDIC (3rd edition) with minor changes
- FIDIC (3rd Edition).

The tender documentation under the standard forms of contract (public and private) comprises:

- drawings
- specifications or schedules of works
- bills of quantities.

Contracts under the standard forms can fluctuate for increases in the cost of labour and materials and involve the use of adjustment formulae based on quarterly published indices.

The contractor and nominated subcontractors receive a 10% advance payment based on the contract value and proportionate deductions are made from subsequent payments. The client retains 5% of the contract sum until practical completion. The contractor and subcontractors provide the client with Performance and Advance Payment Guarantees. The general contractor carries out the work with his own labour and subcontractors and with specific subcontractors nominated by the client's consultants.

Development control and standards

Development is controlled by the Town Planning Act which was fully implemented in 1990 and planning permits are issued by the Planning Authority. In addition, building permits are subject to the control of the local district office or municipality and the application for a permit must be signed by an architect or engineer registered with the Cyprus Architects Association or the Cyprus Civil Engineers and Architects Association. The only existing law on standards in the Republic of Cyprus is the *Streets and Building Regulations Law* which lays down certain standards, many of which are necessary for the proper functioning of buildings.

Examination of the submission is undertaken by trained building inspectors who also inspect the building work at stages throughout the operations.

The only control on quality of materials is that imposed by the manufacturers.

Construction cost data

Cost of labour

The figures below are typical of labour costs in the Nicosia area as at the first quarter 1999. The wage rate is the basis of an employee's income, while the cost of labour indicates the cost to a contractor of employing that employee. The difference between the two covers a variety of mandatory and voluntary contributions – a list of items which could be included is given in section 2.

	Wage rate (per hour) C£	Cost of labour (per hour) C£	Number of hours worked per year
Site operatives			
Mason/bricklayer	4.30	6.00	1,730
Carpenter	4.30	6.00	1,730
Plumber	4.30	6.00	1,730
Electrician	5.40	7.50	1,730
Structural steel erector	5.50	7.75	1,730
HVAC installer	4.30	6.00	1,730
Semi-skilled worker	3.75	5.25	1,730
Unskilled labourer	3.50	4.90	1,730
Equipment operator	4.50	6.25	1,730
Watchman/security	3.50	4.90	1,730
Site supervision			
General foreman	5.75	8.00	1,730
Trades foreman	5.00	7.00	1,730
Clerk of works	5.75	8.00	1,730
Contractors' personnel			
Site manager	9.00	15.50	1,730
Resident engineer	7.65	10.70	1,730
Resident surveyor	7.00	9.80	1,730
Junior engineer	4.80	6.70	1,730
Junior surveyor	4.80	6.70	1,730
Planner	5.50	7.75	1,730
Consultant Personnel			
Senior architect	–	20.00	–
Senior engineer	–	20.00	–
Senior surveyor	–	20.00	–
Qualified architect	–	14.00	–
Qualified engineer	–	14.00	–
Qualified surveyor	–	14.00	–

Cost of materials

The figures that follow are the costs of main construction materials, delivered to site in the Nicosia area, as incurred by contractors in the first quarter of 1999. These assume that the materials would be in quantities as required for a medium sized construction project, and that the location of the works would be neither constrained nor remote. All the costs in this section exclude value added tax (VAT) which is at 8%.

	Unit	Cost C£
Cement and aggregates		
Ordinary portland cement in 50kg bags	tonne	33.00
Coarse aggregates for concrete	m³	4.85
Fine aggregates for concrete	m³	6.10
Ready mixed concrete (mix 1:2:4)	m³	21.50
Ready mixed concrete (mix 1:1½:4)	m³	24.00
Steel		
Mild steel reinforcement	tonne	190.00
High tensile steel reinforcement	tonne	190.00
Structural steel sections	tonne	350.00
Bricks and blocks		
Common bricks (300 x 200 x 100 mm)	1,000	170.00
Good quality facing bricks (200 x 50 x 50 mm)	1,000	225.00
Hollow concrete blocks	1,000	300.00
Solid concrete blocks	1,000	310.00
Precast concrete cladding units with exposed aggregate finish	m²	26.00
Timber and insulation		
Softwood sections for carpentry	m³	240.00
Softwood for joinery	m³	260.00
Hardwood for joinery	m³	600.00
Exterior quality plywood (18 mm)	m²	6.50
Plywood for interior joinery (18 mm)	m²	5.50
Softwood strip flooring (12 mm)	m²	8.25
Chipboard sheet flooring (12 mm)	m²	1.75
100mm thick quilt insulation	m²	3.50
100mm thick rigid slab insulation	m²	5.00
Softwood internal door complete with frames and ironmongery	each	130.00
Glass and ceramics		
Float glass (6mm thick)	m²	6.50
Sealed double glazing units (5-12-5mm)	m²	18.50
Good quality ceramic wall tiles (100 x 100mm)	m²	9.00

	Unit	Cost C£
Plaster and paint		
Plaster in 50kg bags	tonne	82.00
Plasterboard (12.5mm thick)	m^2	1.50
Emulsion paint in 5 litre tins	litre	2.00
Gloss oil paint in 5 litre tins	litre	2.75
Tiles and paviors		
Clay floor tiles (200 x 100 x 20mm)	m^2	11.00
Vinyl floor tiles (400 x 400 x 20mm)	m^2	6.00
Precast concrete paving slabs (400 x 400 x 35mm)	m^2	2.75
Clay roof tiles	1,000	380.00
Precast terrazzo floor tiles (400 x 400 x 25mm)	m^2	6.00
Drainage		
WC suite complete	each	125.00
Lavatory basin complete	each	97.00
100mm diameter clay drain pipes	m	6.50
150mm diameter cast iron drain pipes	m	13.40

Unit rates

The descriptions of work items overleaf are generally shortened versions of standard descriptions listed in five languages (English, French, Italian, German and Spanish) in Appendix 3.

Where an item has a two digit reference number (e.g. 05 or 33), this relates to the full description against that number in Appendix 3. Where an item has an alphabetic suffix (e.g. 12A or 34B) this indicates that the standard description has been modified. Where a modification is major the complete modified description is included here and the standard description should be ignored. Where a modification is minor (e.g. the insertion of a named hardwood) the shortened description has been modified here but, in general, the full description in Appendix 3 prevails.

The unit rates overleaf are for main work items on a typical construction project in the Nicosia area in the first quarter of 1999. The rates include labour, materials and equipment and allowances for contractor's overheads and profit (12%), preliminary and general items (2%) and contractor's profit and attendance on specialist trades (5%). All the rates in this section exclude VAT which is at 8%.

		Unit	Rate C£
Excavation			
01	Mechanical excavation of foundation trenches	m³	3.50
02	Hardcore filling making up levels	m²	2.55
03	Earthwork support	m²	1.00
Concrete work			
04	Plain insitu concrete in strip foundations in trenches	m³	29.50
05	Reinforced insitu concrete in beds	m³	32.00
06	Reinforced insitu concrete in walls	m³	30.50
07	Reinforced insitu concrete in suspended floor or roof slabs	m³	30.50
08	Reinforced insitu concrete in columns	m³	34.00
09	Reinforced insitu concrete in isolated beams	m³	32.50
10	Precast concrete slab	m²	118.00
Formwork			
11	Softwood formwork to concrete walls	m²	4.50
12	Softwood or metal formwork to concrete columns	m²	6.50
13	Softwood or metal formwork to horizontal soffits of slabs	m²	5.00
Reinforcement			
14	Reinforcement in concrete walls	tonne	330.00
15	Reinforcement in suspended concrete slabs	tonne	330.00
16	Fabric reinforcement in concrete beds	m²	2.20
Steelwork			
17	Fabricate, supply and erect steel framed structure	tonne	1,160.00
Brickwork and blockwork			
18	Precast lightweight aggregate hollow concrete block walls (100mm thick)	m²	7.50
19	Solid (perforated) concrete bricks (100mm thick)	m²	7.00
21	Facing bricks	m²	42.00
Roofing			
22	Concrete interlocking roof tiles (430 x 380mm)	m²	16.50
25	Sawn softwood roof boarding	m²	8.50
26	Particle board roof coverings	m²	19.00
27	3 layers glass-fibre based bitumen felt roof covering include chippings	m²	8.00
28	Bitumen based mastic asphalt roof covering	m²	9.75
29	Glass-fibre mat roof insulation (160mm thick)	m²	5.00
30	Rigid sheet loadbearing roof insulation (75mm thick)	m²	8.00
31	Troughed galvanised steel roof cladding	m²	16.50
Woodwork and metalwork			
32	Preservative treated sawn softwood (50 x 100mm)	m	1.00
33	Preservative treated sawn softwood (50 x 150mm)	m	1.30
34	Single glazed casement window in hardwood, (650 x 900mm)	each	80.00

		Unit	Rate C£
35	Two panel glazed door in hardwood, (850 x 2000mm)	each	230.00
36	Solid core half hour fire resisting hardwood internal		
	fush doors, size 800 x 2000mm with frame	each	145.00
37	Aluminium double glazed window, size 1200 x 1200mm	each	115.00
38	Aluminium double glazed door, (850 x 2100mm)	each	175.00
39	Hardwood skirtings (Mahogany) 25 x 100mm	m	3.00
40	Framed structural steelwork in universal joist sections	tonne	1,150.00
41	Structural steelwork lattice roof trusses	tonne	1,270.00

Plumbing

42	UPVC half round eaves gutter (110mm)	m	7.50
43	UPVC rainwater pipes (100mm)	m	4.50
44	Light gauge copper cold water tubing (15mm)	m	2.50
45	High pressure plastic pipes for cold water supply (15mm)	m	1.25
46	Low pressure plastic pipes for cold water distribution	m	1.20
47	UPVC soil and vent pipes	m	7.50
48	White vitreous china WC suite	each	140.00
49	White vitreous china lavatory basin	each	110.00
50	White glazed fireclay shower tray	each	70.00
51	Stainless steel single bowl sink and double drainer	each	62.00

Electrical work

52	PVC insulated and PVC sheathed copper cable core and earth	m	1.20
53	13 amp unswitched socket outlet	each	16.00
54	Flush mounted 20 amp, 1 way light switch	each	28.00

Finishings

55	2 coats gypsum based plaster on brick walls	m^2	6.00
56	White glazed tiles on plaster walls	m^2	12.50
57	Red clay quarry tiles on concrete floor	m^2	17.80
58	Cement and sand screed to concrete floors	m^2	6.00
59	Thermoplastic floor tiles on screed	m^2	12.50
60	Mineral fibre tiles on concealed suspension system	m^2	10.50

Glazing

61	Glazing to wood, 4mm glass	m^2	7.50

Painting

62	Emulsion on plaster walls	m^2	1.40
63	Oil paint on timber	m^2	4.10

Approximate estimating

The building costs per unit area given below are averages incurred by building clients for typical buildings in the Nicosia area as at the first quarter 1999. They are based upon the total floor area of all storeys, measured between external walls and without deduction for internal walls.

Approximate estimating costs generally include mechanical and electrical installations and external works but exclude furniture and loose or special equipment; they also exclude fees for professional services. The costs shown are for specifications and standards appropriate to the Republic of Cyprus and this should be borne in mind when attempting comparisons with similarly described building types in other countries. A discussion of this issue is included in section 2. Comparative data for countries covered in this publication, including construction cost data, are presented in Part Three.

Approximate estimating costs must be treated with caution; they cannot provide more than a rough guide to the probable cost of building. All the rates in this section exclude VAT which is at 8%.

	Cost m² C£	Cost ft² C£
Industrial buildings		
Factories for letting (include lighting, power and heating)	110	10
Factories for owner occupation (light industrial use)	135	13
Factories for owner occupation (heavy industrial use)	160	15
Factory/office (high-tech) for letting (shell and core only)	150	14
Factory/office (high-tech) for letting (ground floor shell, first floor offices)	170	16
Factory/office (high-tech) for owner occupation (controlled environment, fully furnished)	210	20
High tech laboratory (air conditioned)	325	30
Warehouses, low bay (6 to 8m high) for letting (no heating)	95	9
Warehouses, low bay for owner occupation (including heating)	105	10
Warehouses, high bay for owner occupation (including heating)	137	13
Cold stores/refrigerated stores	230	21
Administrative and commercial buildings		
Civic offices, non air conditioned	270	25
Civic offices, fully air conditioned	335	31
Offices for letting, 5 to 10 storeys, non air conditioned	290	27
Offices for letting, 5 to 10 storeys, air conditioned	330	31
Offices for letting, high rise, air conditioned	320	30
Offices for owner occupation, 5 to 10 storeys, non air conditioned	365	34
Offices for owner occupation, 5 to 10 storeys, air conditioned	400	37
Offices for owner occupation, high rise, air conditioned	385	36

	Cost m² C£	Cost ft² C£
Prestige/headquarters office, 5 to 10 storeys, air conditioned	510	47
Prestige/headquarters office, high rise, air conditioned	500	47
Health and education buildings		
General hospitals (482 beds) (per bed)	74,500	6,920
Teaching hospital (482 beds) (per bed)	78,000	7,250
Private hospitals	470	44
Health centres	390	36
Nursery schools	290	27
Primary/junior schools	300	28
Secondary/middle schools	325	30
University (arts) buildings	420	39
University (science) buildings	490	46
Management training centres	420	39
Recreation and arts buildings		
Theatres over 500 seats including seating and stage equipment	520	48
Theatres under 500 seats including seating and stage equipment	500	47
Concert halls including seating and stage equipment	500	47
Sports halls including changing and social facilities	290	27
Swimming pools (international standard) including changing facilities	330	31
Swimming pools (schools standard) including changing facilities	230	21
National museums including fully air conditioning	400	37
Local museums including air conditioning	365	44
City centre/central libraries	360	33
Branch/local libraries	310	29
Residential buildings		
Social/economic single family housing (multiple units)	230	21
Private/mass market single family housing 2 storey detached/ semidetached (multiple units)	290	27
Purpose designed single family housing 2 storey detached (single unit)	400	37
Social/economic apartment housing, low rise (no lifts)	240	22
Social/economic apartment housing, high rise (with lifts)	270	25
Private sector apartment building (standard specification)	290	27
Private sector apartment building (luxury)	335	31
Student/nurses halls of residence	285	27
Homes for the elderly (shared accommodation)	280	26
Hotel, 5 star, city centre, (per bed)	27,000	2,510
Hotel, 3 star, city/provincial (per bed)	17,000	1,580
Motel	410	38

Regional variations

There are no significant variations within Cyprus except for work in mountainous areas which may be 5% to 10 % higher.

Value added tax (VAT)

The standard rate of value added tax (VAT) is currently 8%, chargeable on general building work and materials.

Exchange rates and cost and price trends

The combined effect of exchange rates and inflation on prices within a country and on price comparisons between countries is discussed in section 2.

Exchange rates

The graph on the next page plots the movement of the Cyprus pound against sterling, the ECU/Euro, the US dollar and 100 Japanese yen since 1990. The values used for the graph are quarterly and the method of calculating these is described and general guidance on the interpretation of the graph provided in section 2. The average exchange rate in the first quarter of 1999 was C£0.83 to the pound sterling, C£0.49 to the US dollar and C£0.58 to the Euro.

Cost and price trends

The tables on the next two pages present the indices for consumer prices, labour costs, building materials and residential and non-residential prices in the Republic of Cyprus since 1990. All indices have been rebased to 1990=100. The annual change is the percentage change between the last quarters of consecutive years. Notes on inflation are included in section 2. The labour cost index and the building materials index are compiled by the Ministry of Finance, Department of Statistics and Research in the Republic of Cyprus.

The Cyprus pound against sterling, the ECU/Euro, the US dollar and 100 Japanese yen

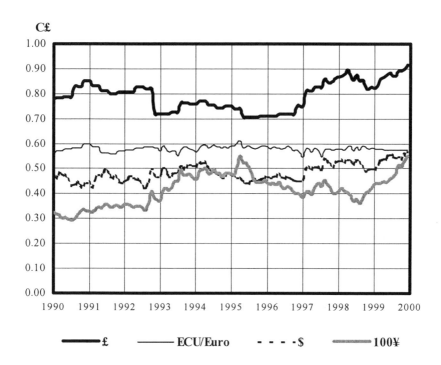

| | £ | ECU/Euro | - - - -$ | 100¥ |

Consumer price, labour cost and building materials price indices

	Consumer price		Labour costs		Building material prices	
	annual	change	annual	change	annual	change
Year	average	%	average	%	average	%
1990	100.0	4.5	100.0	8.1	100.0	4.4
1991	105.0	5.0	108.5	8.5	103.5	3.5
1992	111.9	6.6	119.9	10.5	106.5	2.9
1993	117.3	4.8	133.6	11.4	108.6	2.0
1994	122.8	4.7	143.0	7.0	110.6	1.8
1995	126.0	2.6	155.1	8.5	114.4	3.4
1996	129.8	3.0	165.2	6.5	116.7	2.0
1997	134.5	3.6	176.6	6.9	119.1	2.1
1998	137.5	2.3	–	–	–	–

Sources: International Monetary Fund, Yearbook 1998 and monthly supplements 1999.
Cyprus' Ministry of Finance.

Residential and non-residential building cost indices

Year	Residential building cost		Non-residential building cost	
	annual average	change %	annual average	change %
1990	100.0	5.8	100.0	5.6
1991	105.7	5.7	105.5	5.5
1992	111.4	5.4	111.4	5.6
1993	117.3	5.3	115.8	3.9
1994	121.1	3.2	120.1	3.7
1995	125.9	4.0	125.7	4.7
1996	129.9	3.2	129.7	3.2
1997	134.5	3.5	134.3	3.5

Sources: Cyprus' Ministry of Finance.

Useful addresses

Government and public organizations

Cyprus Organization of Standards and Control of Quality

Ministry of Commerce and Industry 6 A. Araouzos Street

Nicosia

Tel: (357-2) 867190

Fax: (357-2) 375120

Department of Statistics and Research, Ministry of Finance

13 Byron Avenue

Nicosia

Tel: (357-2) 309300

Fax: (357-2) 374830

Ministry of Communication and Works

Dem. Severis Ave

Nicosia

Tel: (357-2) 303235

Fax: (357-2) 465462

Planning Bureau

Apellis Street

Ay. Omoloyitae

Nicosia

Tel: (357-2) 302078

Fax: (357-2) 666810

Trade and professional associations

Cyprus Architects Association
P.O. Box 25565
Nicosia
Tel: (357-2) 672887
Fax: (357-2) 672512

Cyprus Civil Engineers and Architects Association
12 Komninis Street
Nicosia
Tel: (357-2) 751221
Fax: (357-2) 769371

Cyprus Chamber of Commerce and Industry
38 Gr. Dhigenis Ave
PO Box 1455
1509 – Nicosia
Tel: (357-2) 669500
Fax: (357-2) 667433

Technical Chamber of Cyprus
P.O. Box 21826
Nicosia
Tel: (357-2) 672822
Fax: (357-2) 676840

ÚRS Praha, a.s.,
Pražská 18, 102 00 Prague 10
Czech Republic

Engineering And Advisory Organization With The Construction Tradition Over 35 Years,

Member of EUROCONSTRUCT, EDIBUILD, and ICIS
Registered In PHARE Program in Brussels

OFFERS
AND PROVIDES THE FOLLOWING SERVICES:

- Marketing studies and strategy, research in the market of construction work, and production of building materials, advisory and expert activity for implementation of new products and new activities in the Czech market

- Analyses of Building industry and its individual sectors, the processing of development strategies and prognostic studies

- Creation and publishing of valuation data, construction work price lists, material price lists, price indicators and software

- Advisory for valuation of property and valuation of enterprises

* * *

Connection:

Phone: (**420 – 2) 6721 9111
Fax: (**420 – 2) 7175 1175, 7175 0074
E-mail: URS@urspraha.cz
Internet: WWW.URSPRAHA.CZ

Managing Director
Ing. František Glazar
Contact person:
Ing. Zdenek Kadlec, CSc.
Phone: (**420 – 2) 71751116

Czech Republic

All data relate to 1998 unless otherwise indicated.

Population

Population	10.3 million
Urban population (1997)	66%
Population under 15	18%
Population 65 and over	13%
Average annual growth rate (1995 to 1998)	0%

Geography

Land area	78,703 km^2
Agricultural area	54%
Capital city	Prague

Economy

Monetary unit	Koruna (Kcs)
Exchange rate (average first quarter 1999) to:	
the pound sterling	Kcs 50.16
the US dollar	Kcs 29.97
the Euro	Kcs 35.31
the yen x 100	Kcs 25.36
Average annual inflation (1995 to 1998)	9.3%
Inflation rate	10.7%
Gross Domestic Product (GDP)	Kcs 1,743 billion
GDP PPP basis (1997)	$ (PPP) 112 billion
GDP per capita	Kcs 169,200
GDP per capita PPP basis (1997)	$ (PPP) 10,860
Average annual real change in GDP (1995 to 1998)	0.7%
Private consumption as a proportion of GDP	51%
Public consumption as a proportion of GDP	20%
Investment as a proportion of GDP	31%

Construction

Gross value of construction output	Kcs 167 billion
Gross value of construction output per capita	Kcs 16,200
Gross value of construction output as a proportion of GDP	9.6%
Average annual real change in gross value of construction output (1995 to 1998)	−1.3%
Annual cement consumption	3.9 million tonnes

PPP Purchasing power parity

The construction industry

Construction output

The value of gross construction output of the construction sector in the Czech Republic in 1998 was about Kcs 167 billion equivalent to 4.7 billion ECUs (nearly the same as January 1999 Euros). However, on a purchasing power parity basis it was probably about double. Construction output was about 9.6% of GDP in 1998. The breakdown of construction output in 1998 is shown below.

Output of construction sector, 1998 (current prices)

Type of work	Kcs billion	ECUs billion	% of total
New work			
Residential building	19.6	0.6	12
Non-residential building			
Office and commercial	25.6	0.7	15
Industrial	10.5	0.3	6
Other*	7.8	0.2	5
Total	43.9	1.2	26
Civil engineering	49.7	1.4	30
Total new work	113.2	3.2	68
Repair and maintenance			
Residential	7.2	0.2	4
Non-residential	30.5	0.9	18
Civil engineering	15.7	0.4	9
Total repair and maintenance	53.4	1.5	32
Total	**166.6**	**4.7**	**100**

Note: the value of the Euro in January 1999 was very similar to that of the ECU in 1998.
     * schools, universities and hospitals only.
Source   based on Hezký, J 'Czech Republic' in Euroconstruct conference proceedings, Prague, June 1999.

In addition to the output of the construction sector, it is estimated that another Kcs 69 billion, or 2 billion ECUs, of construction work is undertaken by sectors other than construction, by the black economy and DIY. Much of this work would consist of renovation work so that the share of construction renovation in the total work including other sectors and the informal sector is probably much higher. It has in any case been increasing in the construction sector partly due to a fall in new work but also to absolute increases in repair and maintenance in housing and civil engineering.

Construction output has been decreasing from 1996 to 1998, with the fall in the output of the sector in 1998 being 9% due to difficulties in the economy as a whole and especially in the restructuring of enterprises. In 1998 unemployment increased, GDP fell and consumer prices rose. The largest fall in construction output took place in the non-residential sector, especially in industrial buildings. This decrease in industrial investment in 1998 is particularly serious since it followed falls in 1997 and 1996. Reasons include the worsening of the balance sheets of some industrial and service sectors, difficult access to funding and increasing insolvency along with a decrease in foreign direct investment. Recovery is not expected to be quick.

Contraction also occurred in the public non-residential sector, due to a decrease in spending for education and health care. The tightening of the State budget, privatization and a decreasing trend of the birth rate are among the major causes of this trend. In 1998 about 44% of non residential building construction was public, compared with about 57% in 1993. Housing as a percentage of total construction output is amongst the lowest in Europe and, although the number of dwellings completed rose from 1995 to 1998, the level in 1998 is only just over half the levels of 1991. The housing problem is acute and is an obstacle to growth in the economy because it hampers mobility. Various schemes to increase housing such as loans and tax advantages have not so far had great success.

Around 50% of completed dwellings in 1998 were family homes, the remainder being flats. The proportion of family homes has grown steadily since 1991 when it was around 25%. There has been an increase in modernization of dwellings.

Civil engineering was driving the construction sector in the Czech Republic in the early and mid-nineties. Its share of construction sector output increased from 29% to 39% from 1993 to 1998. It was affected by tighter fiscal policies in 1998 and these controls are likely to continue. Private finance in civil engineering projects a minimal.

The geographical distribution of construction output compared with that of population is shown below.

Construction output 1996 regional distribution

Regions	Population (%)	Construction output (%)
Prague	11.7	31.3
Central Bohemia	10.7	4.9
South Bohemia	6.8	7.5
West Bohemia	8.3	5.7
North Bohemia	11.4	7.7
East Bohemia	12.0	9.0
South Moravia	19.9	21.4
North Moravia	19.1	12.5
Total	**100.0**	**100.0**

Sources: Czech Agency for Foreign Investment.
   Construction in the Czech Republic 1997, Association of Building Entrepreneurs of
   the Czech Republic.

The situation in 1998 is probably not very different so far as the dominance of the whole Moravia region is concerned. In 1998 the two Moravias together had 33% of construction orders. The city of Prague may be relatively less busy as it had 16% of all orders in 1998, but the Central Bohemian region as a whole, which includes Prague, has 29% of total orders.

Characteristics and structure of the industry

The structure of the construction contracting industry in the present Czech Republic has in the past been dominated by the State sector, as in most planned socialist economies. However, in the former Czechoslovakia construction firms were generally larger and their number fewer than in other Eastern European countries.

In the five years to 1998 the volume of work undertaken by small firms decreased. The work of the very large firms, employing over 1000, increased as also did that of the medium firms with from 23 to 200 persons. By 1999 nearly all firms had been privatized. Foreign participation in construction had increased. Firms with mixed ownership obtained a quarter of construction output in 1998. About 70% of the work by large firms was publicly funded.

The structure of construction firms in the Czech Republic is now very similar to that in any market economy.

In the Czech Republic in 1996 there were 157,736 construction firms registered by the Czech Statistical Office. The largest construction firms (in alphabetical order) are as follows:

Principal contractors in the Czech Republic, 1996

Contractor	Head office
Armabeton	Praha
Dopravní stavby IES	Olomouc
Dopravní stavby	Uherské Hradišt
Ekoingstav	Brno
Ingstav	Opava
IPS	Praha
Metrostav	Praha
Pozemní stavby	Zlín
Prⴰmstav	Praha
Prmyslové stavitelství	Brno
PSG	Zlín
SS	Praha
Strabag Bohemia	Budjovice
Subterra	Praha
Víktovické staveb	Ostrava
Vodní staveb	Praha
Vodní stveb Bohemia	Sez. Ústí
Vojenské staveb	Praha
VOKD	Ostrava
Zakládání staveb	Praha
elezniní stavitelství	Brno
elezniní stavitelství	Praha

Sources: Construction in the Czech Republic 1997, Association of Building Entrepreneurs of the Czech Republic.

Design offices in the government sector were in the past large organizations of several hundred persons, sometimes connected to large contracting organizations, but often independent. In addition, there were some architects working on their own account on design projects. Now this old structure has been broken down and many more architects are establishing their own practices. The qualifications of designers are not controlled but there is a central register of those authorized to undertake design work. There are two types of architect: one is trained only in the beaux arts tradition, while the other is part engineer and part architect with the title of engineer/architect. Both engineers and architects have good reputations and Czech designers have won design awards at international exhibitions.

Construction cost data

Cost of labour

The figures below are typical of labour costs on a national average basis at the first quarter 1999. The wage rate is the basis of an employee's income, while the cost of labour indicates the cost to a contractor of employing that employee. The difference between the two covers a variety of mandatory and voluntary contributions - a list of items which could be included is given in section 2.

	Wage rate (per hour) Kcs	Cost of labour (per hour) Kcs	Number of hours worked per year
Site operatives			
Mason/bricklayer	58	90	2,150
Carpenter	57	88	2,150
Plumber	57	89	2,150
Electrician	60	94	2,150
Structural steel erector	71	112	2,150
HVAC installer	57	89	2,150
Semi-skilled worker	49	77	2,150
Unskilled labourer	40	62	2,150
Equipment operator	59	92	2,150
Site supervision *			
General foreman	18,700	29,200	2,150
Trades foreman	19,400	30,200	2,150
Watchman/security	13,500	21,000	2,150
Contractors' personnel *			
Site manager	18,700	29,200	2,150
Resident engineer	18,700	29,200	2,150
Resident surveyor	18,700	29,200	2,150
Junior engineer	18,700	29,200	2,150
Junior surveyor	18,700	29,200	2,150
Planner	19,400	30,200	2,150
Consultants' personnel *			
Senior architect	19,400	30,200	2,150
Senior engineer	19,400	30,200	2,150
Senior surveyor	19,400	30,200	2,150
Qualified architect	19,400	30,200	2,150
Qualified surveyor	19,400	30,200	2,150
Qualified engineer	19,400	30,200	2,150

* monthly average salary/labour cost.

Cost of materials

The figures that follow are national average costs of main construction materials, delivered to site, as incurred by contractors in the first quarter of 1999. These assume that the materials would be in quantities as required for a medium sized construction project and that the location of the works would be neither constrained nor remote. All the costs in this section exclude value added tax (VAT) which is at 22%.

	Unit	Cost Kcs
Cement and aggregates		
Ordinary portland cement in CEM III/B, 5 LA BULK	tonne	2,240
Coarse aggregates for concrete	m³	163
Fine aggregates for concrete	m³	203
Ready mixed concrete (mix OI)	m³	1,210
Ready mixed concrete (mix III)	m³	1,790
Steel		
Mild steel reinforcement	tonne	17,900
High tensile steel reinforcement	tonne	25,900
Structural steel sections	tonne	17,300
Bricks and blocks		
Common bricks (29 x 14 x 6.5 cm)	1,000	4,710
Good quality facing bricks (29 x 15 x 6.5 cm)	1,000	5,480
Hollow concrete blocks (38 x 30 x 22 cm)	1,000	27,000
Solid concrete blocks (29 x 14 x 6.5 cm)	1,000	9,000
Precast concrete cladding units with exposed aggregate finish (190 x 20 x 282 cm)	each	2,780
Timber and insulation		
Softwood sections for carpentry	m³	4,000
Softwood for joinery	m³	4,500
Hardwood for joinery	m³	6,000
Exterior quality plywood (12mm)	m²	232
Plywood for interior joinery (12mm)	m²	208
Softwood strip flooring	m²	90
Chipboard sheet flooring (22mm)	m²	94
100mm thick rigid slab insulation	m²	115
Soft wood internal door with frames and ironmongery	each	855
Glass and ceramics		
Float glass (4mm thick)	m²	135
Sealed double glazing units (2 x 4)	m²	550
Good quality ceramic wall tiles	m²	224

	Unit	Cost Kcs
Plaster and paint		
Plaster in bags, 50 kg	tonne	2,400
Plasterboard (18mm thick)	m^2	108
Emulsion paint in 5 litre tins	litre	57
Gloss oil paint in 5 litre tins	litre	70
Tiles and paviors		
Clay floor tiles (300 x 300 x 9mm)	m^2	280
Vinyl floor tiles (400 x 400 x 3mm)	m^2	254
Precast concrete paving slabs	m^2	137
Clay roof tiles	1,000	15,000
Precast concrete roof tiles	1,000	17,000
Drainage		
WC suite complete	each	2,000
Lavatory basin complete	each	700
100mm diameter clay drain pipes	m	91
150mm diameter cast iron drain pipes	m	730

Unit rates

The descriptions of work items opposite are generally shortened versions of standard descriptions listed in five languages (English, French, Italian, German and Spanish) in Appendix 3.

Where an item has a two digit reference number (e.g. 05 or 33), this relates to the full description against that number in Appendix 3.   Where an item has an alphabetic suffix (e.g. 12A or 34B) this indicates that the standard description has been modified.   Where a modification is major the complete modified description is included here and the standard description should be ignored.

Where a modification is minor (e.g. the insertion of a named hardwood) the shortened description has been modified here but, in general, the full description in Appendix 3 prevails.

The unit rates following are for main work items on a typical construction project on a national average basis at the first quarter of 1999.   The rates include all necessary labour, materials and equipment and, where appropriate, allowances for contractor's overheads and profit, preliminary and general items and contractor's profit and attendance on specialist trades.

		Unit	Rate Kcs
Excavation			
01	Mechanical excavation of foundation trenches	m^3	97
02	Hardcore filling making up levels	m^2	56
03	Earthwork support	m^2	54
Concrete work			
04	Plain insitu concrete in strip foundations in trenches	m^3	1,860
05	Reinforced insitu concrete in beds	m^3	1,910
06	Reinforced insitu concrete in walls	m^3	2,070
07	Reinforced insitu concrete in suspended floor or roof slabs	m^3	2,050
08	Reinforced insitu concrete in columns	m^3	2,470
09	Reinforced insitu concrete in isolated beams	m^3	2,020
10	Precast concrete slab	m^3	1,620
Formwork			
11	Softwood or metal formwork to concrete walls	m^2	187
12	Softwood or metal formwork to concrete columns	m^2	284
13	Softwood or metal formwork to horizontal soffits of slabs	m^2	194
Reinforcement			
14	Reinforcement in concrete walls	tonne	22,000
15	Reinforcement in suspended concrete slabs	tonne	19,500
16	Fabric reinforcement in concrete beds	m^2	19,000
Steelwork			
17	Fabricate, supply and erect steel framed structure	tonne	13,000
Brickwork and blockwork			
18	Precast lightweight aggregate hollow concrete block walls	m^2	296
19	Solid (perforated) concrete bricks	m^2	249
20	Solid (perforated) sand lime bricks	m^2	252
21	Facing bricks	m^2	1,030
Roofing			
22	Concrete interlocking roof tiles 430 x 380mm	m^2	343
23	Plain clay roof tiles 260 x 160mm	m^2	326
25	Sawn softwood roof boarding	m^2	44
26	Particle board roof covering	m^2	411
27	3 layers glass-fibre based bitumen felt roof covering	m^2	425
28	Bitumen based mastic asphalt roof covering	m^2	360
29	Glass-fibre mat roof insulation 160mm thick	m^2	530
30	Rigid sheet loadbearing roof insulation 75mm thick	m^2	420
31	Troughed galvanised steel roof cladding	m^2	380
Woodwork and metalwork			
32	Preservative treated sawn softwood 50 x 100mm	m^3	4,000
33	Preservative treated sawn softwood 50 x 150mm	m^3	4,000

		Unit	Rate Kcs
34	Single glazed casement window in hardwood, size 650 x 900mm	each	3,900
35	Two panel glazed door in hardwood, size 850 x 2000mm	each	4,500
36	Solid core half hour fire resisting hardwood internal flush doors, size 800 x 2000mm	each	3,100
37	Aluminium double glazed window, size 1200 x 1200mm	each	6,600
38	Aluminium double glazed door, size 850 x 2100mm	each	7,200
39	Hardwood skirtings	m	30
40	Framed structural steelwork in universal joist sections	tonne	13,000
41	Structural steelwork lattice roof trusses	tonne	13,000

Plumbing

42	UPVC half round eaves gutter	m	250
43	UPVC rainwater pipes	m	300
44	Light gauge copper cold water tubing	m	290
45	High pressure plastic pipes for cold water supply	m	140
46	Low pressure plastic pipes for cold water distribution	m	60
47	UPVC soil and vent pipes	m	260
48	White vitreous china WC suite	each	2,300
49	White vitreous china lavatory basin	each	2,000
50	Glazed fireclay shower tray	each	2,200
51	Stainless steel single bowl sink and double drainer	each	1,700

Electrical work

52	PVC insulated and copper sheathed cable	m	35
53	13 amp unswitched socket outlet	each	260
54	Flush mounted 20 amp, 1 way light switch	each	280

Finishings

55	2 coats gypsum based plaster on brick walls	m^2	207
56	White glazed tiles on plaster walls	m^2	450
57	Red clay quarry tiles on concrete floor	m^2	276
58	Cement and sand screed to concrete floors	m^2	151
59	Thermoplastic floor tiles on screed	m^2	230
60	Mineral fibre tiles on concealed suspension system	m^2	670

Glazing

61	Glazing to wood	m^2	165

Painting

62	Emulsion on plaster walls	m^2	115
63	Oil paint on timber	m^2	106

Approximate estimating

The building costs per unit area given below are averages incurred by building clients for typical buildings on a national average basis at the first quarter 1999. They are based upon the total volume of all storeys, measured between external walls and without deduction for internal walls.

Approximate estimating costs generally include mechanical and electrical installations and external works but exclude furniture and loose or special equipment; they also exclude fees for professional services. The costs shown are for specifications and standards appropriate to the Czech Republic and this should be borne in mind when attempting comparisons with similarly described building types in other countries. A discussion of this issue is included in section 2. Comparative data for countries covered in this publication, including construction cost data, are presented in Part Three.

Approximate estimating costs must be treated with caution; they cannot provide more than a rough guide to the probable cost of building. Note that rates given are per m^3.

	Cost Kcs per m^3	Cost Kcs per Ft^3
Industrial buildings		
Factories for letting	2,840	264
Factories for owner occupation (light industrial use)	2,840	264
Factories for owner occupation (heavy industrial use)	2,840	264
Factory/office (high-tech) for letting (shell and core only)	2,840	264
Factory/office (high-tech) for letting (ground floor shell, first floor offices)	2,840	264
High tech laboratory workshop centres	2,840	264
Factory/office (high-tech) for owner occupation (controlled environment, fully furnished)	2,840	264
Warehouses, low bay (6 to 8m high) for letting (no heating)	2,050	191
Warehouses, low bay for owner occupation (including heating)	2,050	191
Warehouses, high bay for owner occupation (including heating)	2,050	191
Cold stores/refrigerated stores	2,050	191
Administrative and commercial buildings		
Civic offices, non air conditioned	4,350	404
Civic offices, fully air conditioned	4,350	404
Offices for letting, 5 to 10 storeys, non air conditioned	4,350	404
Offices for letting, 5 to 10 storeys, air conditioned	4,350	404
Offices for letting, high rise, air conditioned	4,350	404
Offices for owner occupation, 5 to 10 storeys, non air conditioned	4,350	404
Offices for owner occupation, 5 to 10 storeys, air conditioned	4,350	404

	Cost Kcs per m³	Cost Kcs per Ft³
Offices for owner occupation, high rise, air conditioned	4,350	404
Prestige/headquarters office, 5 to 10 storeys, air conditioned	4,350	404
Prestige/headquarters office, high rise, air conditioned	4,350	404
Health and education buildings		
General hospitals	4,740	440
Teaching hospitals	4,740	440
Private hospitals	4,740	440
Health centres	4,740	440
Nursery schools	3,720	345
Primary/junior schools	3,720	345
Secondary/middle schools	3,720	345
University (arts) buildings	3,720	345
University (science) buildings	3,720	345
Management training centres	3,720	345
Recreation and Arts Buildings		
Theatres over 500 seats including seating and stage equipment	5,210	484
Theatres under 500 seats including seating and stage equipment	5,210	484
Concert halls including seating and stage equipment	5,210	484
Sports halls including changing and social facilities	3,740	348
Swimming pools including changing and social facilities (international)	3,740	348
Residential buildings		
Social/economic single family housing (multiple units)	3,760	349
Private/mass market single family housing 2 storey detached/semidetached (multiple units)	3,760	349
Purpose designed single family housing 2 storey detached (single unit)	3,760	349
Social/economic apartment housing, low rise (no lifts)	3,350	311
Social/economic apartment housing, high rise (with lifts)	3,350	311
Private sector apartment building (standard specification)	3,350	311
Student/nurses halls of residence	4,240	394
Homes for the elderly (shared accommodation)	4,240	394
Homes for the elderly (self contained with shared communal facilities)	4,240	394
Hotel, 5 star, city centre	4,860	452
Hotel, 3 star, city/provincial	4,860	452
Motel	4,860	452

Exchange rates and cost and price trends

The combined effect of exchange rates and inflation on prices within a country and on price comparisons between countries is discussed in section 2.

Exchange rates

The graph following plots the movement of the Czech koruna against sterling, the ECU/Euro, the US dollar and 100 Japanese yen since 1990. The values used for the graph are quarterly and the method of calculating these is described and general guidance on the interpretation of the graph provided in section 2. The average exchange rate in the first quarter of 1999 was Kcs 50.2 to the pound sterling, Kcs 30.0 to the US dollar and Kcs 35.3 to the Euro.

The Czech koruna against sterling, the ECU/Euro, the US dollar and 100 Japanese yen

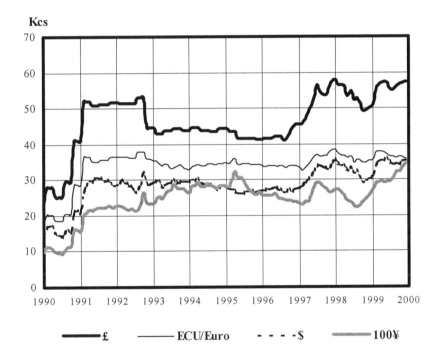

Cost and price trends

The table overleaf presents the indices for construction costs and consumer prices in the Czech Republic since 1993, rebased to 1990=100. The annual change is the percentage change between the average index of consecutive years. Notes on inflation are included in section 2.

Construction costs and consumer price indices

Year	Consumer prices		Construction costs	
	annual average	change %	annual average	change %
1993	210.4	20.4	153.4	18.8
1994	231.6	10.1	111.9	−27.1
1995	252.6	9.1	111.5	-0.4
1996	299.1	18.4	112.5	0.9
1997	330.6	10.5	111.9	−0.5
1998	366.0	10.7	–	–

Sources: Cenovézprávy Úrs, Úrs Praha a.s.
International Monetary Fund, Yearbook 1998 and supplements 1999.

While the consumer price index has risen steadily the index of construction costs has behaved erratically with a large fall in 1994 and little change at least until 1998.

Useful addresses

Government and public organizations

Ministry of Industry and Trade of the Czech Republic

Na Františku 32
11015 Praha 1

Tel: (420-2) 24851111

Fax: (420-2) 24853216

Czech Statistical Office

Sokolovska 142
186 04 Praha 8

Tel: (420-2) 66042451

Fax: (420-2) 66310429

Trade and professional associations

Czech Association of Structural Engineers

Legerova 52
120 00 Praha 2

Tel: (420-2) 24912812

Housing Association of the Czech Republic

Podolská 50
147 00 Praha 4

Tel: (420-2) 61214247

Fax: (420-2) 61225560

Union of Building Entrepreneurs

Národní 10
10 00 Praha 1

Tel: (420-2) 24912341

Fax: (420-2) 293062

Czech Chamber of Authorized Engineers

Legerova 52
120 00 Praha 2

Tel: (420-2) 24913397

Fax: (420-2) 24913397

Association of Building Materials Traders

Národní 10
110 00 Praha 1

Tel: (420-2) 24951409

Fax: (420-2) 293062

Association of Producers of Concrete Building Components

Národní 10
110 00 Praha 1

Tel: (420-2) 24912341

Fax: (420-2) 293062

Economic Chamber of the Czech Republic

Argentinska 38
170 05 Praha 7

Tel: (420-2) 66794845

Fax: (420-2) 804894

Other organizations

Building Information Agency

Ostrovní 8
111 21 Praha 1

Tel: (420-2) 24914093

Fax: (420-2) 297120

Úrs Praha

Pra'aská 18
102 00 Praha 10

Tel: (420-2) 67219111

Fax: (420-2) 71751175

VTT BUILDING TECHNOLOGY
Construction
Facility Management

Expert services in construction and facility management

VTT Building Technology is one of the main eleven research units of the Technical Research Centre of Finland (VTT).

Our main goal is to provide our clients with information and knowledge on construction technology and market covering all the possible aspects.

We specialize in the R&D of the procedures and processes, markets and technologies of construction and facility management. We focus on both economic and technical aspects.

Our activities are in the first place determined by the needs of our clients and the commissions we execute for them.

Our versatile know-how and links to broad domestic and international networks enable us to acquire the most up-to-date information and to develop solutions to enhance clients' competitiveness.

AREAS OF INVOLVEMENT AND EXPERTISE

Development of the processes and procedures of construction and facility management

Construction and property management sector and development

Facility management technologies

PRODUCTS AND RESEARCH SERVICES

Tools of design and production control

Business information for planning strategies and operations

Continuosly compiled market reports on construction branch as a whole and on its main sectors to be used in strategic decision making.

Surveys on construction markets in Russia, the Baltic countries, Poland and some parts in Asia

Tools of facility management

A member of EUROCONSTRUCT organization.

Contacts

Research Manager: Pekka Pajakkala, tel: int. + 358 3 316 3404

VTT BUILDING TECHNOLOGY
- P.O. Box (Tekniikankatu 1),
  FIN-33101 TAMPERE
  Tel: int + 358 3 316 3111
  Fax: + 358 3 316 3497
- P.O. Box 18021 (Kaitoväylä 1),
  FIN-90571 OULU
  Tel: int + 358 8 551 2111
  Fax: + 358 8 551 2090
- P.O. Box 1801 (Kivimiehentie 4),
  FIN-02044 VTT
  Tel: int + 358 9 4561
  Fax: + 358 9 456 6251

Construction Management and Concurrent Engineering
Pekka Huovila,
tel: int. + 358 9 456 5903
Veijo Nykänen,
tel: int. + 358 3 316 3415

Business Intelligence in Construction and Real Estate
Erkki Lehtinen,
tel: int. + 358 3 316 3420
Facility Management
Kauko Tulla,
tel: int. + 358 8 551 2012

For further information please contact: http:\\www.vtt.fi/rte/

Finland

All data relate to 1998 unless otherwise indicated.

Population

Population	5.1 million
Urban population (1997)	64%
Population under 15	19%
Population 65 and over	14%
Average annual growth rate (1985 to 1998)	0.4%

Geography

Land area	337,030 km^2
Agricultural area	8%
Capital city	Helsinki

Economy

Monetary unit	Markka (Fmk)
Exchange rate (average first quarter 1999) to:	
the pound sterling	Fmk 8.48
the US dollar	Fmk 5.09
the Euro	Fmk 5.95
the yen x 100	Fmk 4.30
Average annual inflation (1995 to 1998)	1.1%
Inflation rate	1.4%
Gross Domestic Product (GDP)	Fmk 680 billion
GDP PPP basis (1997)	$ (PPP) 102.1 billion
GDP per capita	Fmk 132,000
GDP per capita PPP basis (1997)	$ (PPP) 19,860
Average annual real change in GDP (1995 to 1998)	4.8%
Private consumption as a proportion of GDP	53%
Public consumption as a proportion of GDP	21%
Investment as a proportion of GDP	17%

Construction

Gross value of construction output	Fmk 77.2 billion
Gross value of construction output per capita	Fmk 15,100
Gross value of construction output as a proportion of GDP	11.4%
Average annual real change in gross value of construction output (1995 to 1998)	4.8%
Annual cement consumption	1.5 million tonnes

PPP Purchasing power parity

The construction industry

Construction output

The value of gross total construction output in Finland in 1998 was Fmk 77.2 billion or $12.9 billion ECUs (nearly the same as January 1999 Euros). This includes estimates for work by other sectors. The black economy and DIY and represents 11.4% of GDP. The breakdown by type of work in 1998 is shown below:

Total output of construction, 1998 (current prices)

Type of work	Fmk billion	ECU's billion	% of total
New work			
Residential building	15.7	2.6	20
Non-residential building			
Office and commercial	4.7	0.8	6
Industrial	4.1	0.7	5
Other	10.9	1.8	14
Total	19.7	3.3	26
Civil engineering	11.3	1.9	15
Total new work	46.7	7.8	61
Repair and maintenance			
Residential	14.1	2.4	18
Non-residential	10.4	1.7	13
Civil engineering	6.0	1.0	8
Total repair and maintenance	30.5	5.1	39
Total	**77.2**	**12.9**	**100**

Note: the value of the Euro in January 1999 was very similar to that of the ECU in 1998.
Source: based on Lehtinen, E., 'Finland' in Euroconstruct conference proceedings, Prague, June 1999.

Finland has suffered severely from recession in the early 1990s with construction in 1994 standing at about half its peak level of the late 1980s. From 1995 output started to increase again and grew to 1998 with increases of 11% in each of the last two years of that period.

Residential output contracted during the years of the recession but recovered substantially in 1997 with an increase of 35% and with a continued rise in 1998 of 12%. Higher household confidence and easier access to borrowing are the major engines of growth in this sector. There is strong underlying demand for housing because in Finland space per person is relatively low and at the same time the number of households is increasing.

About two thirds of housing construction is in flats with a decreasing proportion of these being state-subsidized. Single family house construction is becoming more popular and has doubled since the early 1990s.

Non-residential construction boomed in 1998 with the value of commercial building doubling and office buildings increasing by 70%. Although these increases are spectacular, it must be remembered that they followed a period of continuous decline from 1989 to 1994. Growth is expected to continue because of the general buoyancy of the economy, with high consumer confidence and increasing disposable incomes. The service sector linked to manufacturing is thriving. It is reported that research and development now accounts for about three per cent of GDP – one of the highest in the world and this requires construction of new research facilities.

Public sector construction has been at a low level but, with the current increase in general prosperity, educational building has increased and hospital construction is expected to rise as well as, for example, sports facilities.

On the civil engineering side a steady though modest increase in investment in all infrastructure has been talking place and is expected to continue. Private finance is to some extent being utilized for public sector projects, for example, major road schemes.

Repair and maintenance has been increasing significantly in 1997 and 1998. Much of the housing stock is of an age when it requires increased renovation and repair and finance, in the better economic climate, has been available for this. In spite of state support for repair and maintenance in housing being cut, the level is likely to continue high. Renovation work in non-residential building is also at a high and increasing level. In part this is due to increased rents which provide funds for repair and maintenance. In addition the expected standard of premises is increasing which leads to rehabilitation and renovation work.

Finnish contractors were relatively late starters in the race for foreign contracts. The real breakthrough in Finnish contracting exports took place in the second half of the 1970s. The foreign output of Finnish building contractors amounts to 1 to 2% of the volume of domestic construction, compared with 7% in the late 1980s. The share of foreign construction in Finland diminished as the special trade arrangement between Finland and the USSR was abolished and the clearing system was replaced by trade in freely convertible currencies. Since the early 1990s Finns have become very wary with regard to Russia but they still occupy a significant position as implementers of Western investors' projects. However the economic crisis in Russia in 1998/9 has not affected Finland to a great extent.

Population, 1996 regional distribution

Region	Population (%)
Uudenmaan (Nyland)	26.2
Turun ja Porin (Abo-Björneborg)	13.7
Hameen (Tavastenus)	14.3
Kymen (Kymmene)	6.4
Mikkelin (St Michel)	4.0
Pohjois-Karjalan (Norra Karelen)	3.4
Kuopion	5.0
Keski-Suomen (Mellersta Finland)	5.0
Vaasan (Vasa)	8.7
Oulun (Uleåborg)	8.8
Lapin (Lappland)	3.9
Ahvenanmaa (Aland)	0.5
Total	**100.0**

Sources: Statistical Yearbook 1997, Statistics Finland.

The five growth centres for construction in Finland are Greater Helsinki metropolitan area and the Tampere, Turku, Yyväskylä and Oulu Uleåborg areas. There is considerable internal migration into these areas.

Characteristics and structure of the industry

In 1991 there were some 11,500 contractors in Finland, of which eight firms employed more than 1,000 persons and did over a quarter of the work. This represents a very high degree of concentration compared with most European countries but is lower than in the 1980s. In 1997 there were nine Finnish contractors in *Building* magazine's list of 'Top 300 European Contractors' as shown in the table below. 'Yit Co.' and 'Quattrogemini' are the only Finnish contractor listed in *Engineering News Record's* list of the 'Top 225 International Contractors' for 1997, at rank 95 and 178 respectively.

Major Finnish contractors, 1997

Major contractors	Place in Building's 'Top 300 European Contractors'	Exports (% of turnover)
Yit-Corporation	54	18
Lemminkainen Group	89	22
Skanska Oy	129	25
Ncc Puolimatka	138	7
Polar Corporation	163	–
Viitoset	195	30
Hartela Contractors	228	21
Luja Group	269	–

Source: Building December 1998.

The most prevalent method of contracting in Finland is the main contractor system, where a single contractor bears the overall responsibility for construction. Another common method is design and build where design is also part of the tender.

The separate contracts method is far less common. Under this method the project is divided into several sub-projects. The client, through his consultant, invites separate tenders and enters into contracts directly with various contractors. Most often the contracts are for works by different trades, such as frame construction, prefabricated components, heating, plumbing, air conditioning, electrical services, etc.

In Finland, the architect is the leader of the design team unless the exceptional nature of the project demands other arrangements. The team leader outlines the building project and co-ordinates and supervises the work of other designers. He is also responsible for the development of plans, their implementation and the overall buildability of the design.

The structural design of the building is entrusted to an independent structural engineer. Geotechnical design is treated as a separate field and such services are often needed at the early phases of projects when foundation conditions are considered. Prefabricated component design may either be part of structural design or may be the responsibility of the manufacturer. Heating, plumbing and sewerage, ventilation and electrical design constitute separate fields.

Generally, design offices are small, though there are several which employ 30 to 40 designers. Sometimes the client's organization, especially in the public sector, has design know-how, but most often an independent architect and a design office are engaged. Finnish construction is highly industrialized. The level of prefabrication of structures is high and prefabricated components are widely used.

Selection of design consultants

The client selects qualified designers, usually based on previous experience or recommendations. The scope of the service and fees are agreed after selection. Clients rarely invite competitive tenders from designers.

Architectural competitions, closed or open, are often arranged when designing large or especially demanding projects. Rules approved by the sector organizations are applied to competitions which are usually judged by a jury. In an open competition several proposals are awarded a prize and the one considered best is usually selected for implementation. The number of proposals submitted to an open competition in Finland can be over 200. In a closed competition, that is, one by invitation only, proposals are normally requested from three to five groups of designers and the participating designers all receive equal fees, independent of the quality of their proposal. Normally, the winner is recommended to undertake the project.

Contractual arrangements

The client usually appoints a contractor on the basis of selective competitive tenders, although in certain cases only a single contractor is contacted. In the public sector, tenders must always be invited from at least five and, in large building projects, from at least ten contractors. Contractors are notified of tenders through advertisements in newspapers.

Generally, the tender documents include plans and a work specification. Each contractor prepares his own bill of quantities as a basis for his price quotation. The bill of quantities is not, however, part of the contract.

A table is often drawn up to rank the offers. It is used to study, for instance, the effect of technical solutions on price, delivery times and terms, etc. The most advantageous offer is not necessarily the cheapest one, but quite often the cheapest one is selected. The financial standing of the contractor is also established.

Usually the main contract is a 'lump sum' contract, but the method of setting a target price has become more prevalent. Under the latter arrangement if the main contractor can cut total costs, his remuneration increases but if costs exceed the target level, he is liable for them.

The contract documents include the tender itself, the payment programme, other correspondence between the client and contractor, contract negotiation documents and the contract proper on a standard form.

The state has very few detailed regulations for its own building projects, but among the most important ones are the 'General Conditions of Construction Contract' and the 'Building Information File'. The latter consists of instructions by the National Board of Public Construction that provide clear practical advice, for instance, on labour inputs. The File is intended to facilitate the planning of the works on the basis of data collected from practical experience. The 'General Conditions of Construction Contract' (YSE 1983) have been adopted by both public and private builders. They are not repeated in the contract; a mere reference to them suffices.

Liability and insurance

The designer has both a legal and a moral responsibility. The legal responsibility is based on a design contract and the attached general terms of contract of consultancy. That responsibility extends to, for instance, carrying out the tasks of the design contract in the agreed time and liability for any possible errors. In monetary terms the designer's liability is normally limited to the amount of his design fee. The normal liability period for the designer is 10 years. There is also a product liability law, which shifts some legal liability from the contractor to the designer.

Everyone involved in a project shares in the moral responsibility, the greatest burden being laid on the participants who can most influence the end result. Therefore, the designer's responsibility is more moral in nature and the contractor bears more of the legal responsibility.

Inexperienced clients often have difficulty in evaluating the quality of the design work on the basis of drawings and other documents. Especially for one-off and large projects, clients often hire independent project consultants to ensure the desired end result.

Designers are increasingly taking out professional indemnity insurance. It normally indemnifies losses up to Fmk 1 million with a Fmk 10,000 deductible.

The contractor is liable to the client for any damages resulting from the contractor's failure to satisfy the conditions of contract. In general, the contractor is liable for:

- plans he makes or has made plus information and work performed on the basis of the plans
- measurements he has made
- extra works and modifications
- the fulfilment of contractual obligations by all parties under the main contract.

The contractor is responsible for performing in accordance with the contract for one year, in the case of building and electrical work, and two years, in the case of HVAC work. The defects liability period begins on the date the work passes an acceptance inspection or as the building is taken into use without an inspection.

The contractor is obliged to repair any deficiencies and defects detected during the liability period that he cannot prove to have been caused by factors unrelated to his performance and which are pointed out to him before the end of this period. He is not, however, obliged to perform such repairs that are connected with the regular maintenance of equipment or are required due to faulty operation or neglected or improper maintenance.

To guarantee performance of the works, the contractor must in connection with the acceptance inspection, give a bond, generally equal to 2% of the contract price. The bond is returned to the contractor after the liability period has expired.

The contractor's liability ends at the expiration of the liability period, except for such flaws and deficiencies that the client can prove to be the result of the contractor's gross negligence or intentional breach of contract and which the client could not reasonably have been expected to identify during the acceptance inspection or the defects liability period. The contractor is released from that liability after 10 years have passed from the acceptance inspection.

Development control

A building permit is required for a new building or an extension. Municipalities issue special application forms for that purpose. The application must be accompanied by the following statutory documents:

- proof of the fact that the applicant is in possession of the lot or construction site
- a land registry map
- a ground plan of the lot or construction site in triplicate showing buildings to be erected and demolished (in cities this must show green areas, play-grounds, access roads, parking and loading areas and other site arrangements)
- design documents with explanations in triplicate.

The building committee may also demand that the applicant provides data on specific working and structural drawings, strength calculations, accounts of building materials, quality of foundations, used structures and the soil of the construction site, etc.

In most instances the client obtains the building permit which should be secured before the contract is signed. Exceptions are made in cases of compelling urgency.

Before a decision is made on the permit, the committee must obtain an expert opinion from health, fire or any other authorities as necessary. As building authorities study applications for a building permit, they must ascertain that the building conforms to the ratified plan and site layout plan and see that other rules and regulations of construction are complied with. The 'Finnish National Building Code' specifies the statutory requirements regarding the facts that need to be established before granting a building permit. It includes rules that bind the authorities as well as rules for solutions that conform to the set requirements.

A decision on a building permit is rendered 14 days after presentation of the proposal for public inspection during which period the owners of adjoining properties, industrial buildings and the neighbourhood at large may appeal. A building permit remains in force for three years.

The application process for a building permit lasts three to six months, depending on the municipality, and about 5% of applications are rejected. A large majority of applications are accepted, as the seeking of a building permit constitutes a negotiation process with the authorities; that is, the applicant knows what may be built on his lot according to the city plan and whether a special permit is possible.

Construction cost data

Cost of labour

The figures below are typical of labour costs in on a national average basis at the first quarter 1999. The wage rate is the basis of an employee's income, while the cost of labour indicates the cost to a contractor of employing that employee. The difference between the two covers a variety of mandatory and voluntary contributions – a list of items which could be included is given in section 2.

	Wage rate (per hour) Fmk	Cost of labour (per hour) Fmk	Number of hours worked per year
Site operatives			
Mason/bricklayer	68	115	1,800
Carpenter	66	110	1,800
Plumber	41	69	1,800
Electrician	58	98	1,800
Structural steel erector	69	118	1,800
HVAC installer	58	98	1,800
Skilled labour	51	86	1,800
Semi-skilled worker	50	84	1,800
Equipment operator	57	96	1,800

	Wage rate (per month) Fmk	Cost of labour (per month) Fmk	Number of hours worked per year
Site supervision *			
General foreman	14,700	25,300	–
Trades foreman	12,700	21,300	–
Clerk of works	10,300	17,200	–
Contractors' personnel *			
Site manager	20,300	34,400	–
Resident engineer	15,700	26,300	–
Resident surveyor	15,700	26,300	–
Junior engineer	12,200	20,300	–
Junior surveyor	12,200	20,300	–
Planner	12,200	20,300	–
Consultants' personnel *			
Senior architect	19,200	32,400	–
Senior engineer	20,300	34,400	–
Senior surveyor	18,200	31,400	–
Qualified architect	12,200	20,300	–
Qualified engineer	12,200	20,300	–
Qualified surveyor	12,200	20,300	–

* monthly average salary/labour cost.

Cost of materials

The figures that follow are the costs of main construction materials, delivered to site on a national average basis, as incurred by contractors in the first quarter of 1999. These assume that the materials would be in quantities as required for a medium sized construction project and that the location of the works would be neither constrained nor remote. All the costs in this section exclude value added tax (VAT) which is at 22%.

	Unit	Cost Fmk
Cement and aggregates		
Coarse aggregates for concrete	tonne	15
Fine aggregates for concrete	tonne	8
Ready mixed concrete	m^3	341
Ready mixed concrete	m^3	400
Steel		
Mild steel reinforcement	tonne	2,350
High tensile steel reinforcement	tonne	2,150

	Unit	Cost Fmk
Bricks and blocks		
Common bricks (285 x 85 x 85mm)	1,000	2,790
Good quality facing bricks (285 x 85 x 85mm)	1,000	3,040
Hollow concrete blocks (590 x 290 x 190mm)	1,000	14,200
Solid concrete blocks (390x 300 x 190mm)	1,000	10,100
Precast concrete cladding units	m^2	456
Timber and insulation		
Softwood sections for carpentry (50 x 150mm)	m^3	1,590
Softwood for joinery	m^3	1,490
Exterior quality plywood (15mm)	m^2	80
Plywood for interior joinery (12mm)	m^2	61
Softwood strip flooring	m^2	172
Chipboard sheet flooring	m^2	41
100mm thick quilt insulation	m^2	19
100mm thick rigid slab insulation	m^2	69
Glass and ceramics		
Sealed double glazing units	m^2	1,390
Plaster and paint		
Plasterboard (13mm thick)	m^2	12
Emulsion paint in 10 litre tins	litre	22
Gloss oil paint in 10 litre tins	litre	81
Tiles and paviors		
Clay floor tiles (95 x 195 x 12mm)	m^2	154
Vinyl floor tiles (50 x 220 x 2mm)	m^2	128
Precast concrete roof tiles	1,000	3,460
Drainage		
Lavatory basin complete	each	1,350

Unit rates

The descriptions of work items opposite are generally shortened versions of standard descriptions listed in five languages (English, French, Italian, German and Spanish) in Appendix 3.

Where an item has a two digit reference number (e.g. 05 or 33), this relates to the full description against that number in Appendix 3. Where an item has an alphabetic suffix (e.g. 12A or 34B) this indicates that the standard description has been modified. Where a modification is major the complete modified description is included here and the standard description should be ignored.

Where a modification is minor (e.g. the insertion of a named hardwood) the shortened description has been modified here but, in general, the full description in Appendix 3 prevails.

The unit rates below are for main work items on a typical construction project on a national average basis in the first quarter of 1999. The rates include labour, materials and equipment and 25% to cover allowances for contractor's overheads and profit, preliminary and general items and contractor's profit and attendance on specialist trades. All the rates in this section exclude VAT which is at 22%.

		Unit	Rate Fmk
Excavation			
01	Mechanical excavation of foundation trenches	m^3	15
02	Hardcore filling making up levels	m^2	34
Concrete work			
04	Plain insitu concrete in strip foundations in trenches	m^3	536
05	Reinforced insitu concrete in beds	m^3	492
06	Reinforced insitu concrete in walls	m^3	541
07	Reinforced insitu concrete in suspended floor or roof slabs	m^3	566
08	Reinforced insitu concrete in columns	m^3	542
09	Reinforced insitu concrete in isolated beams	m^3	556
10	Precast concrete slab	m^2	253
Formwork			
11	Softwood formwork to concrete walls	m^3	225
13	Softwood or metal formwork to horizontal soffits of slabs	m^3	274
Reinforcement			
14	Reinforcement in concrete walls (16mm)	tonne	6,570
15	Reinforcement in suspended concrete slabs	tonne	6,700
16	Fabric mat reinforcement in concrete beds	tonne	4,470
Steelwork			
17	Fabricate, supply and erect steel framed structure	tonne	9,120
Brickwork and blockwork			
18	Precast lightweight aggregate hollow concrete block walls (100mm thick)	m^3	226
19	Solid (perforated) concrete bricks (100m thick)	m^3	286
20	Solid perforated sand lime bricks	m^2	410
21	Facing bricks	m^2	461
Roofing			
22	Concrete interlocking roof tiles	m^2	74
23	Plain clay roof tiles	m^2	144
24	Fibre cement roof slates	m^2	100

		Unit	Rate Fmk
25	Sawn soft wood roof boarding	m^2	147
26	Particle board roof covering	m^2	80
27	3 layers glass-fibre based bitumen felt roof covering include chippings	m^3	108
29	Glass-fibre mat roof insulation 150mm thick	m^3	46
30	Rigid sheet loadbearing roof insulation 75mm thick	m^3	98
31	Troughed galvanised steel roof cladding	m^3	90
Woodwork and metalwork			
32	Preservative treated sawn softwood 50 x 100mm	m	67
33	Preservative treated sawn softwood 50 x 150mm	m	874
35	Two panel glazed door in hardwood, size 850 x 2000mm	each	2,420
36	Solid core half hour fire resisting hardwood internal flush doors, size 800 x 2000mm	each	755
37	Aluminium double glazed window, size 1200 x 1200mm	each	1,920
38	Aluminium double glazed door, size 850 x 2100mm	each	3,170
39	Hardwood skirtings	m	12
Plumbing			
48	White vitreous china WC suite	each	1,780
49	White vitreous china lavatory basin	each	463
50	White glazed fireclay shower tray	each	1,710
51	Stainless steel single bowl sink and double drainer	each	1,200
Finishings			
55	2 coats gypsum based plaster on brick walls	m^3	44
56	White glazed tiles on plaster walls	m^3	245
57	Red clay quarry tiles on concrete floor	m^3	221
58	Floor screed. Cement and sand screed to floors	m^2	48
59	Thermoplastic floor tiles	m^2	90
60	Suspended ceiling system	m^2	158
Painting			
62	Emulsion on plaster walls	m^2	64
63	Oil paint on timber	m^2	67

Approximate estimating

The building costs per unit area given overleaf are averages incurred by building clients for typical buildings in Finland as at the first quarter 1999. They are based upon the total floor area of all storeys, measured between external walls and without deduction for internal walls.

Approximate estimating costs generally include mechanical and electrical installations, furniture and loose or special equipment, and external works; they also include fees for professional services. The costs shown are for specifications and standards appropriate to Finland and this should be borne in mind when attempting comparisons with similarly described building types in other countries. A discussion of this issue is included in section 2. Comparative data for countries covered in this publication, including construction cost data, are presented in Part Three.

Approximate estimating costs must be treated with caution; they cannot provide more than a rough guide to the probable cost of building. All the rates in this section exclude VAT which is at 22%.

	Cost Fmk per m^2	Cost Fmk per ft^2
Industrial buildings		
Factories for owner occupation	4,460	414
Factories for owner occupation (heavy use)	3,440	320
Administrative and commercial buildings	4,250	395
Health and education buildings		
Health centres	5,170	480
Primary/junior schools/kindergarten	6,180	574
Secondary/middle schools	5,170	480
Recreation and Arts Buildings		
Sports halls with changing and social facilities	4,660	433
Residential buildings		
Social/economic single family housing	3,950	367
Private single family housing single unit	3,750	348
Purpose designed single family housing	3,850	358
Social/economic apartment low rise	3,650	339
Social/economic apartment high rise	3,340	310
Private sector apartment standard specification	4,250	395

Regional variations

The approximate estimating costs are based on average costs for Finland. Variations would typically be within plus or minus 4% of these costs except in Helsinki and environs where they would be 8% to 10% higher.

Value added tax (VAT)

The standard rate of VAT is currently 22%, chargeable on general building work and materials.

Exchange rates and cost and price trends

The combined effect of exchange rates and inflation on prices within a country and on price comparisons between countries is discussed in section 2.

Exchange rates

The graph below plots the movement of the Finnish markka against sterling, the ECU/Euro, the US dollar and 100 Japanese yen since 1990. The values used for the graph are quarterly and the method of calculating these is described and general guidance on the interpretation of the graph provided in section 2. The average exchange rate in the first quarter of 1999 was Fmk 8.48 to the pound sterling, Fmk 5.09 to the US dollar and Fmk 5.95 to the Euro.

The Finnish markka against sterling, the ECU/Euro, the US dollar and 100 Japanese yen

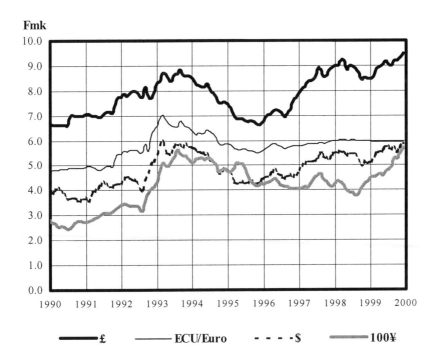

Cost and price trends

The table following presents the indices for consumer prices and building costs, in Finland. The indices have been rebased to 1990=100. Notes on inflation are included in section 2.

Consumer price and residential building cost indices

Year	Consumer prices annual average	change %	Building costs annual average	change %
1990	100.0	6.2	100.0	7.2
1991	104.1	4.1	102.2	2.2
1992	106.8	2.6	100.4	−1.8
1993	109.1	2.2	100.7	0.3
1994	110.3	1.1	102.2	1.5
1995	111.3	0.9	103.5	1.3
1996	112.0	0.6	102.7	−0.8
1997	113.3	1.2	105.2	2.4
1998	114.9	1.4	107.6	2.3

Sources: International Monetary Fund, Yearbook 1998 and
supplements 1999.
Statistics Finland.

The building cost index is a composite index based on a set of monthly input price indices which are themselves based on standard factor costs. It covers new public and private sector construction of the following five types:

- single unit residential buildings

- blocks of flats

- office and commercial buildings

- warehouses and production buildings

- buildings used in agricultural production.

The index includes materials (including their transport to the site) labour, equipment hire, site preparation costs, conveyancing and professional fees, installation and fitting costs, interest on loans and trade margins. The index for the residential construction component includes VAT but other components do not includes VAT.

The striking features of these indices is that the cost of construction has risen considerably less than consumer prices.

Useful addresses

Government and public organizations

Ympäristöministeriö
Ministry of the Environment
PO Box 380
00131 Helsinki
Fax: (358-9) 19919545

Liikenneministeriö
Ministry of Communications and Transport
Eteläesplanadi 16
FIN 00131 Helsinki
Tel: (358-9) 1601
Fax: (358-9) 1602596

Suomen Standardisoimisliitto (SFS)
Finnish Standards Association (SFS)
PO Box 205
00121 Helsinki

Tilastokeskus
Central Statistical Office of Finland
Työpajakatu 13B
00580 Helsinki
Tel: (358-9) 17342219
Fax: (358-9) 17342279

Valtiovarainministeriö
Ministry of Finance
Aleksanterinkatu 3
00170 Helsinki
Tel: (358-9) 16013099
Fax: (358-9) 1604755

Rakennushallitus
National Board of Building
PO Box 237
00531 Helsinki

Trade and professional associations

Suomen Rakennusurakoitsijaliitto
The Association of General
Contractors in Finland
Eteläranta 10
00130 Helsinki

Suomen Arkkitehtiliitto (SAFA)
The Finnish Association of Architects
Yrjonkatu 11 A
00120 Helsinki

Suomen Rakennusinsinöörienliitto
Association of Finnish Civil Engineers
Meritullinkatu 16 A 5
00170 Helsinki

Maanmittausinsinöörien Liitto
The Finnish Association of Surveyors
Kellosilta 10
00520 Helsinki

Rakennustcollisuuden Keskusliitto
Confederation of Finnish
Construction Industries
Unioninkatu 14
00130 Helsinki
Tel: (358-9) 12991
Fax: (358-9) 90 1299214

Other organization

Valtion Teknillinen Tutkimuskeskus
Technical Research Centre of Finland
Box 1802
33101 Tampere
Tel: (358-3) 3163111
Fax: (358-3) 3163497

DAVIS LANGDON ECONOMISTES

Davis Langdon Economistes provides cost consultancy and cost management services tailored to the French construction and property markets.

Our approach is:

*   to be positive and creative in our advice, rather than simply reactive;

*   to concentrate on value for money and value engineering rather than on superficial cost-cutting;

*   to give advice that is matched to the Client's own criteria, rather than to impose standard or traditional solutions;

*   to see cost as one component of a successful design solution, which needs to be balanced with many others, and to work as an integrated member of a design team in achieving that balance;

*   to pay attention to the life-long costs of owning and operating a facility, rather than to the initial capital cost only.

We aim to control cost, limit risk and add value.

DAVIS LANGDON ECONOMISTES

1 Rue Edouard Colonne
Paris, 75001, France

Tel: (33 1) 53 40 94 80
Fax: (33 1) 53 40 94 81

DAVIS LANGDON & SEAH INTERNATIONAL
www.davislangdon.com

France

All data relate to 1998 unless otherwise indicated.

Population
Population	58.6 million
Urban population (1997)	75%
Population under 15	19%
Population 65 and over	16%
Average annual growth rate (1985 to 1998)	0.5%

Geography
Land area	547,030 km^2
Agricultural area	59%
Capital city	Paris

Economy
Monetary unit	Franc (FFr)
Exchange rate (average first quarter 1999) to:	
the pound sterling	FFr 9.35
the US dollar	FFr 5.61
the Euro	FFr 6.56
the yen x 100	FFr 4.75
Average annual inflation (1995 to 1998)	1.3%
Inflation rate	1.0%
Gross Domestic Product (GDP)	FFr 8,520 billion
GDP PPP basis (1997)	$ (PPP) 1,320 billion
GDP per capita	FFr 145,400
GDP per capita PPP basis (1997)	$ (PPP) 22,520
Average annual real change in GDP (1995 to 1998)	2.4%
Private consumption as a proportion of GDP	60%
Public consumption as a proportion of GDP	20%
Investment as a proportion of GDP	17%

Construction
Gross value of construction output	FFr 666 billion
Gross value of construction output per capita	FFr 11,400
Gross value of construction output as a proportion of GDP	7.8%
Average annual real change in gross value of construction output (1994 to 1997)	−0.7%
Annual cement consumption	18.7 million tonnes

PPP Purchasing power parity

The construction industry

Construction output

The value of the gross output of the construction sector in France in 1998 was FFr 666 billion equivalent to 101 billion ECUs (nearly the same as January 1999 Euros). This represents 7.8% of GDP – the lowest proportion in the EU after the UK. In addition, construction undertaken by other sectors, by the black market and by DIY operations is estimated at FFr 155 billion. The table below shows the breakdown of the sector output.

Output of construction sector, 1998 (current prices)

Type of work	FFr billion	ECUs billion	% of total
New work			
Residential building	143	21.7	22
Non-residential building			
Office and commercial	27	4.1	4
Industrial	42	6.4	6
Other	31	4.7	5
Total	100	15.2	15
Civil engineering*	88	13.3	13
Total new work	331	50.2	50
Repair and maintenance			
Residential	175	26.5	26
Non-residential	114	17.2	17
Civil engineering	46	7.0	7
Total repair and maintenance**	335	50.7	50
Total	**666**	**100.9**	**100**

Note: the value of the Euro in January 1999 was very similar to that of the ECU in 1998.
Source: based on de la Morvonnais, P., 'France' in Euroconstruct conference
              proceedings, Prague, June 1999.

France has the third largest construction output in Europe coming just after Italy in value but its output being only rather over half that of Germany.

Total construction output increased in 1998 by nearly 2% after falls in the previous two years.   Growth occurred in all sectors in 1998 except civil engineering. It is expected that overall growth is continuing, even accelerating, during 1999.

Housing starts have been rising especially for one and two family dwellings and building permits are rising faster suggesting good prospects for the future. One and two family dwellings account for about 60% of residential construction units. New residential construction output increased by over 4% in 1998 after falls in the previous two years. Most of the increase is in private sector housing both low and high rise.

The growth in non-residential building is highest in industrial and commercial construction but not spectacular. Public sector construction, especially schools and hospital projects, continue to fall while hospital construction is stagnant after major falls in the past years.

Civil engineering has been at a low level for some years and new work declined marginally in 1998. The only sub-sector with increased output in 1998 is transport infrastructure but all civil engineering sectors may improve in 1999, due to an expected overall increase in government expenditure buoyed up by local authority spending.

Although there is overall growth in renovation, it is growing less quickly than new work and may well continue to lag behind in spite of aging stock especially in civil infrastructure.

The geographical distribution of dwelling starts in 1996 compared with that of population is shown on the next page. While there is considerable agreement between the percentages for each region, dwelling starts in Alsace, Pays de la Loire and Languedoc-Roussillon are a high proportion of the total for all France compared with population. By contrast the populous Ile-de-France has less than its 'fair share' as do most of the northern regions of France.

The work abroad by French contractors present in the top 225 international contractors in Engineering News Record in 1997 was valued at FFr 99 billion or around 15% of total domestic work as shown in the second table on the following page. About half the work abroad is in Europe with substantial amounts in Africa and Asia.

France

Population and dwelling starts 1996 regional distribution

Regions	Population (%)	Dwelling starts (%)
Ile-de-France	18.9	15.1
Champagne-Ardenne	2.3	1.4
Picardie	3.2	1.8
Haute-Normandie	3.1	2.2
Centre	4.2	3.6
Basse-Normandie	2.4	2.1
Bourgogne	2.8	2.3
Nord-pas-de-Calais	6.9	4.3
Lorraine	4.0	2.5
Alsace	2.9	4.3
Franche-Comté'	1.9	2.0
Pays de la Loire	5.4	8.2
Bretagne	4.9	6.5
Poitou-Charentes	2.8	3.3
Aquitaine	4.9	6.5
Mid Pyrénées	4.5	5.8
Limousin	1.2	0.9
Rhône-Alpes	9.6	10.3
Auvergne	2.3	2.4
Languedoc-Roussillon	3.8	6.4
Provence-Alpes-Côte d'Azur	7.6	7.6
Corse	0.4	0.6
Total	**100.0**	**100.0**

Source: INSEE-Institut National de la Statistique et des Etudes Economiques.

French construction work abroad, 1997

Area	FFr billion	% of total
Europe	50.3	50.9
Africa	16.7	16.9
Middle East	4.3	4.4
Asia	16.4	16.6
USA	5.8	5.8
Canada	2.1	2.1
Latin America	3.4	3.3
Total	**99.0**	**100.0**

Source: Engineering News Record 12-8-98.

Characteristics and structure of the industry

Architects in France have a title protected by law and are required to sign applications for building permits. Only individuals wishing to build their own home of less than 170 square metres are exempt from appointing an architect to prepare the application. Architectural practices are generally small. There are over 21,000 firms with about 26,000 employees. The size of building engineering firms is larger, over 6,000 firms with over 28,000 employees, an average per firm of 4.5. There are 1,600 civil engineering firms employing over 9,700 persons, an average per firm of six, with companies ranging from those employing over 1,000 staff to those employing very few.

The French construction industry is dominated by engineers, who are prevalent in client organizations and contractors as well as in design organizations in the public and private sector. Many operate in *Bureaux d'Etudes Techniques* (BETs), which are design organizations which may undertake the whole design, but may also operate as subcontractors to architects and contractors. Unlike that of the architect, the title of engineer is not protected by law.

Quantity Surveyors and Building Surveyors do not exist as such in the French system though there are *métreurs*, generally employed by contractors, and *économistes de la construction*, often simply known as *économistes*, employed within architectural practices, client organizations or within their own firms. There are about 2,900 firms of *économistes de la construction* in France, employing 7,500 people, including 500 *économistes*. The firms work within design teams (31%), directly for clients (24%), or for contractors (29%). They also act as experts in insurance and court cases. Although the context is quite different, the closest professional to the quantity surveyor is the *économiste* especially qualified in *Economie de l'Ingénierie* (EI) by the OPQTECC, but there are only 500 of them. Other *économistes* are specialized in single trades or groups of trades. The *économistes de la construction* profession is now evolving and their number is increasing, especially with the updated definition of design services (June 1994) for the public sector which highlights the necessity to design to budget. Some British quantity surveyors have set up in France but fee levels for cost advice are considerably lower than in the UK.

The traditional system of contracting in France is the separate trades system which favours smaller specialist trade firms. However, in recent years, main contracting has been on the increase, especially for very large projects, with the main contractor in turn subcontracting specialised work.

It is difficult to make a direct comparison of contracting organizations with the UK because the categories are different and because direct employment is low in the UK due to the large amount of sub-contracting.

In France there is a large gap between the major firms and the mass of small contractors. However a number of medium sized contractors have merged to form new larger firms. In 1997 there were 103 French firms out of the 300 firms listed in *Building* magazine's 'Top 300 European Contractors'. Fourteen are in the top 100. The major French contractors in 1997 and their percentage sales abroad are shown below:

Major French contractors, 1997

Major contractors	*Place in* Building's Top 300 European Contractors' 1997	Sales abroad as % of turnover
Bouygues	1	35
SGE	3	34
Group GTM	5	42
Eiffage	8	14
Colas	11	30
Dumez-GTM	17	34
Spie Batignolles	22	30
SOGEA	32	39
Cegelec	33	36
Jean Lefebvre	36	44

Source: Building December 1998.

French contractors have operated overseas for many years and as already mentioned have a broad spread of operations in geographical terms as well as types of work. Ten of *Engineering News Record's* 'Top 225 International Contractors' in 1997 were French.

Clients and finance

Local governments are the most important public sector clients. The state often acts as an initiator of projects which may be financed by private money or very often by a combination of private and public funds in *Sociétés d'Economies Mixtes* (SEMs).

The *Délégation à l'Aménagement du Territoire et à l'Action Régionale* (DATAR), the government agency in charge of implementing regional development policy, may provide grants for investment in new developments according to criteria such as job creation and location. Investors may also receive funds from local authorities or from the European Union.

The majority of business parks and industrial zones in France are owned and operated by municipalities or Chambers of Commerce, which have the status of public authorities and are active in promoting the development of infrastructure.

A common method of financing building is a form of leasing known as *crédit bail*. This includes an option to buy the building for FFr 1 after a 15 to 25 year period. Municipalities can subsidize up to 25% of new building with their own funds. They may also use their better financial conditions for borrowing to give loans at beneficial rates to companies. In some cases land may be given free to entrepreneurs.

Another common source of finance for private work, especially commercial or industrial projects, is banks or insurance companies. In France, these companies are closely connected with business and, indeed, with the building and civil engineering industry. Many of the large contractors are partly owned by banks or insurance companies and the *bureau d'études* not owned by contractors are often closely linked to financial institutions. This can make the process of getting finance for a project easier because *bureaux d'études* may liaise between client and financier. Moreover, French banks have traditionally provided long-term as well as short-term finance. They spread their risks, however, by limiting the size of loans to any one company. As a result many French businesses have accounts at more than one bank to obtain adequate loan funding.

Selection of design consultants

The rules for the selection of the design team in the public sector are clearly laid down. The composition of the team is defined in notices calling for applications. It generally consists of an architect, an engineer and, increasingly, an *économiste*. Other specialists such as experts in acoustics, landscaping or cultural matters, may be requested to be part of the team depending on the nature of the project. The type of project determines which of them is the team leader. The method of selection of the principal designer is by some form of competition (*concours*) which must be announced in the *Bulletin Officiel des Annonces des Marchés Publics* and in the Press, notably in the weekly magazine *Le Moniteur des travaux publics et du bâtiment*.

The type of competition varies according to the size of project. For small projects candidates are selected on the basis of past performance. For medium projects there is a form of simplified public competition and submissions of experience are formally examined by a jury. Large projects require a public call for applications to participate in a design competition. Selected candidates then have to submit a preliminary design with plans, elevations, perspective views, cost estimates, etc. Submission fees are paid to the selected competitors and these are announced in the notice.

In the private sector, competitions are relatively rare. Clients may select architects known to them or whose reputation for the specific type of building is high and then commission them directly. They may invite a contractor to participate in the design at an early stage. Alternatively, a limited competition may be adopted. There are few in-house design teams in the private sector.

Role of design consultants

Since June 1994, new regulations have been implemented regarding the role assigned to the design team in the public sector. The new regulation is known as the '*Décret MOP*' (*Maîtrise d'Ouvrage Publique*). It defines elements of work which are to be included in the contract of the overall design consultant known as a *mission de base*. This includes:

- *Etudes d'esquisses* (ESQ) (Literally: Sketch studies)
- *Etudes d'avant-projet* (AVP) (Literally: Pre-project studies)

Comprising:

The *Avant Projet Sommaire* (APS) (Literally: Pre-project synopsis)

and

The *Avant Projet Définitif* (APD) (Literally: Definitive pre-project)

- *Etudes de projet* (PRO) (Literally: Project studies)
- *Assistance au maître de l'ouvrage pour la passation du ou des contrats de travaux* (ACT) (Literally: Assistance to the client for contracting the works)
- *Etudes d'exécution* (EXE) (Literally: Working studies)

or

Visa des études d'exécutions (VISA) (Literally: Signing and approving of working studies)

- *Direction de l'exécution du contrat de travaux* (DET) (Literally: Management of the fulfilment of the contract)
- *Assistance au maître de l'ouvrage lors des opérations de réception et pendant la période de garantie de parfait achèvement* (AOR) (Literally: Assistance to the client during the operations of handover and during the perfect completion guaranty period)

For the preparation of the working documentation, there are three options in the *mission de base*:

- Production by the design team, as described below: (*Etudes d'exécution* – EXE);
- Prepared by the contractors, in which case, the design team is responsible for approving them. This service is called '*Visa des études d'exécution*';

- Production partly by the design team and partly by the contractors (the design team being still responsible for approval).

The role assigned to the design team in the private sector varies and may include both design and site supervision or only parts of these functions. In the private sector, as in the public sector, there are different practices as to whether the lead consultant, generally the architect, or the client appoints the other professionals.

There are a number of different contracts used in the private sector, notably those issued by the *Ordre des Architectes, the Union Nationale des Syndicats Français d'Architectes* and *the Société Française des Architectes.*

Selection of contractors

As in the case of designer selection, the choice of contractor varies from the public sector, where the rules are clearly defined, to the private sector, where the client can determine his own method.

A *QUALIBAT* system of classification of contractors, based mainly on the criteria of size and qualification, is administered by the Professional Organization for the Qualification and Classification of the Building Industry and Associated Activities (OPQCB *Qualité Bâtiment*). This body comprises representatives of the construction professions, contractors' federations, and technical control bodies. The classification is recognized nationwide in the French building industry. On presentation of substantiated documentary evidence that the contractor has carried out certain types of projects, firms receive a certificate, renewed annually, upon which a 'curriculum vitae' is given. This is the basic document which determines the contractors who may be invited to tender for any project. The *QUALIBAT* database also lists contractors who have received quality insurance certification.

The traditional form of contracting is 'separate trades contracting' (*lots séparés*) or a grouping of contractors. In the public sector, the selection must be by competition, except in a few special circumstances. The call for tender is known as the *appel d'offres.* However, because in France the detailed design work is usually undertaken by the contractor rather than the consultant designer, the design cannot be finalized at the tender stage. However, with the new regulation known as the *Décret MOP* the consultant designers have seen their responsibility increase as they have to fully check and approve the contractor's design.

With the *Code des Marchés Publics* – the regulating document for the public sector – the acceptance of a tender has to be on the basis of the following criteria: construction cost, maintenance cost, technical value, professional guarantees, financial guarantees and programme. Other selection criteria have to be indicated in the tender announcement. The *Code des Marchés Publics* also indicates that when a tender is abnormally low, the tenderer must be asked to give precision to his offer. French contractors often propose options (*variantes*) which may be accepted if they are prepared in accordance with tender rules.

It can happen that when submissions are studied, none of the tenders is found to be satisfactory and the bids may be called *infructueux* (literally, unfruitful). The tender may be cancelled and the project again put out to tender with a different (more detailed) specification, or it may be negotiated with at least half of the tenderers. Another reason why bids may be declared *infructueux* is that there may be a price ceiling for certain types of work, notably public housing. If none of the bids are lower than the ceiling, there will be a new opportunity for negotiation.

Contractors may also be selected on a combination of design and construction factors. This is known as *a concours de conception-réalisation*. But, with the new regulation, this procedure is strictly limited to projects on which the association of the contractor with the design is necessary for technical reasons. This particularly applies when the project requires the utilization of unusual technical equipment.

Negotiation is also permitted for a number of specific reasons listed in the *Code des Marchés Publics*, such as for buildings where a security factor is involved, where there is great urgency, or where buildings are erected as part of a research project.

In the private sector clients may use main contractors or separate trades contractors. The normal method is to have an *appel d'offres* of four to six contractors. They may then negotiate on design, price and time and select the most appropriate contractor for the project. Alternatively, they may negotiate directly with a contractor. A negotiated contract is known as *gré à gré*.

Contractual arrangements

The standard form of contract for public work is the *Cahier des Clauses Administratives Générales Travaux* (CCAG *Travaux*). The CCAG is used as general conditions of contract and is completed by a *Cahier des Clauses Administratives Particulières* (CCAP) which are the supplementary clauses.

In the private sector, the standard form of contract, most commonly used is the *norme* NFP O3-001 published by *l'Association Française de Normalisation* (AFNOR). The *norme* is used as general conditions of contract (CCAG) and it is completed by the *Cahier des Clauses Administratives Particulières* (CCAP).

As a majority of the contracts are lump sum, there is no Standard Method of Measurement as such in France. Although some methods of measurement exist, they are indicative and not compulsory. Contractors, who are responsible for their quantities generally measure the work as built, other elements of the costing being included in the unit rates.

It is the practice in France to include a high proportion of the preliminaries within unit rates.

The vast majority of public and private works contracts are let as *Marché Forfaitaire* (lump sum contracts). Tendering contractors take a varying degree of responsibility for design and commit themselves to a fixed price on the basis of their own quantification. They also often propose alternative technical solutions.

Designers may not always stipulate that a particular method or material should be adopted but will give guidance by referring to a particular manufacturer's model (and asking for an equivalent) in order to indicate the standard to be adopted.

France is generally less claims-orientated than is the case in the UK. However claims do occur within the context of the *Marché Forfaitaire* where work has not been described adequately in the contract documentation.

Some contracts are bid on the basis of *Bordereau de prix* (schedules of unit rates) which are applied to a re-measurement of the works to derive the contractors payment. The unit rates will generally include overheads and profit. Such an approach would normally be used in minor works and repair and maintenance or in situations where the content of the works is not readily definable from the outset.

There are published *Séries de prix* which may be used as the basis for calculating the value of works following an agreed measurement. This kind of approach is limited and is normally used for architecturally protected buildings, or for works with a high traditional craft content.

Contracts usually contain a fluctuation clause, and an agreed formula based upon price indices published by the Ministry of Construction is applied to calculate the additional payments due. The price indices are issued monthly and cover individual trades, thus enabling the formula to be applied to interim payments. Interim payments are normally monthly and subject to a 5% retention. This retention is called the *retenue de garantie*. Its purpose is to guarantee the completion of the work by the contractor in the case of possible defects at handover. This retention can be replaced by a bank guarantee (*Caution bancaire*) issued by the contractor's bank. The guarantee is to be released to the contractor one year after the handover, unless the client officially gives evidence that defects still exist.

Variations to the work described in the contract documents can be authorized. If possible their value is agreed on the contract rates or on a lump sum basis before the work is put in hand. The *norme NFP 03-001* states that the contractor is expected to complete additional work provided the increase does not exceed one quarter of the initial price. This also applies to a decrease up to one fifth of the original sum. Beyond these limits the contractor can request the cancellation of the contract. The same applies to the client if the cause of such variation is not due to him. The final account is prepared by the contractor and checked by the *économiste*.

Liability, insurance and inspection

The French have the most far-reaching, comprehensive and formalized system of liability and insurance in Europe and have a related system of building control.

The most interesting aspect of their system is the special decennial responsibility introduced originally by Napoleon in the Civil Code and then modified over two centuries. The 'Spinetta Law' of 1978 considers all parties involved in the construction process – contractors, architects, engineers, *bureaux de contrôle, économistes* (when they are involved in writing specifications), manufacturers and importers of products and components, even individuals or companies involved in the sale of the finished building – to carry some responsibility. Decennial responsibility commences with the handover of the works to the client, which is a recognized formal event in the process. The liability covers solidity of construction and damages which would leave the works unsuitable for their designated use.

In addition, there is responsibility for perfect achievement for one year from handover; biennial responsibility for perfect functioning from the handover date; third party responsibility under common law starting from the date of damage for 10 years; and contractual and fraudulent responsibility for a maximum period of 30 years.

It is a legal requirement that all parties to the process be insured against the risks associated with their liabilities. The client also has to be insured by a *Dommage-ouvrage*. With this scheme, the client's insurance company covers the repair of damage normally guaranteed by the contractor's or consultant's decennial. It is then the responsibility of the client's insurance to obtain compensation from the contractor's or consultant's insurance. The objective is to finance any necessary works while responsibility is allocated amongst the parties to the process.

There is also the PUC (*Police Unique de Chantier*). With the PUC, the contractor's and consultant's decennial responsibility and the client's '*Dommage-ouvrage*' are grouped in one policy. The client takes this cover on behalf of all consultants and contractors. This policy is generally used on major works. It helps to avoid parties failing to have adequate cover and it simplifies procedures in case of damage. The client may recover the contractor's and consultant's share from the insurance company. The PUC can be extended to cover the biennial responsibility for perfect functioning.

Linked to this insurance process is a system of inspection for construction projects. A *contrôleur technique* (technical inspector), who must be independent of the parties to the process, is engaged by the client to check all stages of the project: general concept, design, detailed specification, working drawings, choice of materials and products, and construction on site. His employment is obligatory when the building is open to the public but, in any case, it facilitates obtaining insurance cover.

Development control and standards

In France, the rules regarding planning are set in the *Code de l'Urbanisme*. There are two major plans for directing development: The *Schéma Directeur d'Aménagement et d'Urbanisme* (SDAU) and the *Plan d'Occupation des Sols* (POS).

The SDAU is defined for a whole *région*, or a *département*, or sometimes for a group of cities closely connected. It provides guidance for the development of the specified area. It indicates areas to be developed, renovated or protected. It also indicates locations for specified activities and outlines the organization of transport systems.

The POS is a local plan which defines strict development rules for the city. The POS indicates different kinds of locations, like the *Zones d'Aménagements Concertés* (ZAC), where the city organizes the development; it sets the *Coefficient d'Occupation des Sols* (COS) which indicates the maximum authorized floor area to be built in relation to the land area; and it gives rules for construction height, building alignment, colours, architecture, etc.

The building authorisation is the *permis de construire* (building permit). There is a clearly laid down procedure for obtaining a *permis de construire* which covers both the planning and control of buildings. Normally the *permis de construire* is issued by the *maire* (mayor) who takes the decision himself after the process of consultation with all interested technical departments and having provided an opportunity for objections to be raised by the public. Consultations with the *maire*, or his *service de l'urbanisme* (planning department) at an early stage are important to ascertain what is and is not possible, especially as there is very little room for discretion when the POS has been approved. If the application is properly completed and complies with all the rules laid down and with the POS, there are very few reasons for refusal of permission.

Mandatory regulations apply to safety, health and energy conservation. In addition there are national regulations applicable to new construction contained in various laws, decrees and orders, amplified where necessary by administrative circulars. The Scientific and Technical Centre for Building, *Centre Scientifique et Technique du Bâtiment* (CSTB), publishes model solutions to many regulations.

The European directives on site safety have been written into French law. The planning supervisor is called the *Coordonateur de sécurité*. His document on site organization and safety issues is called the PGC, *'plan général de coordination en matière de sécurité et protection de la santé'*. Each contractor provides his own method statement including issues called the PPSPS, *'plan particulier de de sécurité et protection de la santé'*. In the event of the site exceeding 10 contractors and 10,000 man days, a safety committee made up of the design team, the local working conditions inspector, *'inspecteur du travail'*, the health insurance agency and a site worker and the manager of each contractor present is assembled to discuss site safety issues. The committee is chaired by the planning supervisor. The committee is called the CISSCT *'collège interentreprises de sécurité, santé et des conditions de travail'*. The planning supervisor manages the assembly of the DIU or DIOU (*dossier d'intervention ultérior sur l'ouvrage*) which assembles them as built information together with his comments. The client holds overall responsibility for site safety. He must be attentive to his contract with his planning supervisor and ensure that the latter has sufficient authority on the project to protect this responsibility.

Standards produced by the French Standards Association, *Association Française de Normalisation* (AFNOR), comprise technical recommendations which builders are advised to follow, though a number are included in the mandatory regulations. In private contracts, standards are binding only if the contracts prescribe their application. In public contracts, application is a statutory requirement. (A majority of the AFNOR standards are available in English).

Research and development

The principal bodies active in research are:

* The Scientific and Technical Centre for Building (CSTB) which is a publicly funded institution and participates in the preparation of building regulations.
* The Inter-Professional Technical Union and its affiliates include the Experimental Research and Study Centre for Building and Public Works. The equipment of this centre is partly financed by the State.
* The *Plan Construction et Architecture* which is an inter-Ministry department. The aim of this organization is to develop research and experimentation in the building sector.

- Various technical professional centres under the aegis of the Ministry of Industrial and Scientific Development. They may receive operational subsidies.

Other research groups include universities.

Construction cost data

Cost of labour

The figures below are typical of labour costs in the Paris area as at the first quarter 1999. The wage rate is the basis of an employee's income, while the cost of labour indicates the cost to a contractor of employing that employee. The difference between the two covers a variety of mandatory and voluntary contributions - a list of items which could be included is given in section 2.

	Wage rate (per month) thousands FFr	Cost of labour (per year) thousands FFr	Number of hours worked per year
Site operatives			
Mason/bricklayer	13.4	261	1,866
Carpenter	14.4	282	1,866
Plumber	16.5	321	1,866
Electrician	16.5	321	1,866
Structural steel erector	14.4	281	1,866
HVAC installer	14.4	281	1,866
Semi-skilled worker	10.3	201	1,866
Unskilled labourer	8.2	161	1,866
Equipment operator	13.4	261	1,866
Watchman/security	8.2	161	1,866
Site supervision			
General foreman	18.5	362	1,866
Trades foreman	18.5	362	1,866
Contractors' staff			
Site manager	20.6	402	1,866
Resident engineer	22.7	442	1,866
Resident surveyor	22.7	442	1,866
Junior engineer	18.5	362	1,866
Junior surveyor	16.5	321	1,866
Planner	22.7	442	1,866

	Wage rate (per month) thousands FFr	Cost of labour (per year) thousands FFr	Number of hours worked per year
Consultant personnel			
Senior architect	25.8	502	1,866
Senior surveyor	25.8	502	1,866
Senior engineer	25.8	502	1,866
Qualified architect	20.6	402	1,866
Qualified engineer	20.6	402	1,866
Qualified surveyor	20.6	402	1,866

Cost of materials

The figures that follow are the costs of main construction materials, delivered to site in the Paris area, as incurred by contractors in the first quarter of 1999. These assume that the materials would be in quantities as required for a medium sized construction project and that the location of the works would be neither constrained nor remote. All the costs in this section exclude value added tax (VAT) which is at 20.6%.

	Unit	Cost FFr
Cement and aggregates		
Ordinary portland cement in 50kg bags	tonne	896
Coarse aggregates for concrete	m^3	243
Fine aggregates for concrete	m^3	212
Ready mixed concrete	m^3	832
Steel		
Mild steel reinforcement	tonne	6,060
High tensile steel reinforcement	tonne	5,750
Structural steel sections	tonne	5,150
Bricks and blocks		
Common bricks (60 x 110 x 220mm)	1,000	1,950
Good quality facing bricks (54 x 105 x 220mm)	1,000	2,380
Hollow concrete blocks (100 x 200 x 550mm)	1,000	3,110
Solid concrete blocks (10 x 200 x 500mm)	1,000	3,870
Precast concrete cladding units	m^2	1,030
Timber and insulation		
Softwood sections for carpentry	m^3	2,060
Softwood for joinery	m^3	2,470

	Unit	Cost FFr
Hardwood for joinery (Oak)	m³	6,110
Exterior quality plywood (16mm)	m²	75
Plywood for interior joinery (19mm)	m²	27
Softwood strip flooring (23mm)	m²	143
Chipboard sheet flooring (22mm)	m²	47
100mm thick quilt insulation	m²	23
100mm thick rigid slab insulation	m²	40
Softwood internal door complete with frames and ironmongery	each	850
Glass and ceramics		
Float glass (6mm)	m²	165
Sealed double glazing units (6-8-6)	m²	428
Good quality ceramic wall tiles (15 x 15 cm)	m²	206
Plaster and paint		
Plaster in 50kg bags	tonne	703
Plasterboard (12.5 mm thick)	m²	21
Emulsion paint in 5 litre tins	litre	31
Gloss oil paint in 5 litre tins	litre	39
Tiles and paviors		
Clay floor tiles (200 x 200 mm)	m²	155
Vinyl floor tiles	m²	108
Precast concrete paving slabs	m²	69
Clay roof tiles	1,000	5,190
Precast concrete roof tiles	1,000	5,850
Drainage		
WC suite complete	each	1,850
Lavatory basin complete (including taps)	each	927
100mm diameter clay drain pipes	m	14
150mm diameter cast iron drain pipes	m	212

Unit rates

The descriptions of work items below are generally shortened versions of standard descriptions listed in five languages (English, French, Italian, German and Spanish) in Appendix 3.

Where an item has a two digit reference number (e.g. 05 or 33), this relates to the full description against that number in Appendix 3. Where an item has an alphabetic suffix (e.g. 12A or 34B) this indicates that the standard description has been modified.   Where a modification is major the complete modified description is included here and the standard description should be ignored.

Where a modification is minor (e.g. the insertion of a named hardwood) the shortened description has been modified here but, in general, the full description in Appendix 3 prevails.

The unit rates below are for main work items on a typical construction project in the Paris area in the first quarter of 1999. The rates include labour, materials and equipment and allowances for contractor's overheads and profit, preliminary and general items and contractor's profit and attendance on specialist trades. All the rates in this section exclude VAT which is at 20.6%.

		Unit	Rate FFr
Excavation			
01	Mechanical excavation of foundation trenches	m^3	93
02	Hardcore filling making up levels	m^2	77
03	Earthwork support	m^2	77
Concrete work			
04	Plain insitu concrete in strip foundations in trenches	m^3	1,090
05	Reinforced insitu concrete in beds	m^3	1,170
06	Reinforced insitu concrete in walls	m^3	1,210
07	Reinforced insitu concrete in suspended floor or roof slabs	m^3	1,160
08	Reinforced insitu concrete in columns	m^3	133
09	Reinforced insitu concrete in isolated beams	m^3	1,310
10	Precast concrete slab	m^2	371
Formwork			
11	Softwood formwork to concrete walls	m^2	168
12	Softwood or metal formwork to concrete columns	m^2	484
13	Softwood or metal formwork to horizontal soffits of slabs	m^2	201
Reinforcement			
14	Reinforcement in concrete walls (16mm)	tonne	11,400
15	Reinforcement in suspended concrete slabs	tonne	11,800
16	Fabric reinforcement in concrete beds	m^2	41
Steelwork			
17	Fabricate, supply and erect steel framed structure	tonne	20,600
Brickwork and blockwork			
18	Precast lightweight aggregate hollow concrete block walls (100mm thick)	m^2	247
19	Solid (perforated) concrete bricks	m^2	247

		Unit	Rate FFr
Roofing			
22	Concrete interlocking roof tiles	m²	464
23	Plain clay roof tiles	m²	494
24	Fibre cement roof slates	m²	232
25	Sawn softwood roof boarding	m²	165
26	Particle board roof covering	m²	175
27	3 layers glass-fibre based bitumen felt roof covering		
	include chippings	m²	211
29	Glass-fibre mat roof insulation	m²	82
30A	Rigid sheet loadbearing roof insulation 100mm thick	m²	134
31	Troughed galvanised steel roof cladding	m²	242
Woodwork and metalwork			
32	Preservative treated sawn softwood 50 x 100mm	m	41
33	Preservative treated sawn softwood 50 x 150mm	m	57
34	Single glazed casement window in tropical hardwood,		
	size 600 x 950mm	each	1,030
35	Two panel glazed door in tropical hardwood,		
	size 800 x 2150mm	each	3,190
36	Solid core half hour fire resisting hardwood internal		
	flush doors, size 800 x 2000mm	each	2,060
37	Aluminium double glazed window, size 1200 x 1200mm	each	4,740
38	Aluminium double glazed door, size 850 x 2100mm	each	3,710
39	Hardwood skirtings (Oak), size 10 x 100mm	m	62
40	Framed structural steelwork in universal joist sections	tonne	18,500
41	Structural steelwork lattice roof trusses	tonne	20,600
Plumbing			
42	UPVC half round eaves gutter	m	175
43	UPVC rainwater pipes	m	149
44	Light gauge copper cold water tubing	m	124
45	High pressure plastic pipes for cold water supply	m	46
46	Low pressure plastic pipes for cold water distribution	m	26
47	UPVC soil and vent pipes	m	129
48	White vitreous china WC suite	each	2,060
49	White vitreous china lavatory basin	each	1,440
50	White glazed fireclay shower tray	each	309
51A	Stainless steel double bowl sink and double drainer	each	1,030
Electrical work			
52	PVC insulated copper sheathed cable	m	20
53	13 amp unswitched socket outlet	each	258
54	Flush mounted 20 amp, 1 way light switch	each	227
Finishings			
55	2 coats gypsum based plaster on brick walls	m²	82
56	White glazed tiles on plaster walls	m²	258

		Unit	Rate FFr
57	Red clay quarry tiles on concrete floor	m^2	299
58	Cement and sand screed to concrete floors 50mm thick	m^2	139
59	Thermoplastic floor tiles on screed	m^2	113
60	Mineral fibre tiles on concealed suspension system	m^2	206
Glazing			
61	Glazing to wood	m^2	258
Painting			
62	Emulsion on plaster walls	m^2	62
63	Oil paint on timber	m^2	93

Approximate estimating

The building costs per unit area given below are averages incurred by building clients for typical buildings in the Paris area as at the first quarter 1999. They are based upon the total floor area of all storeys, measured between external walls and without deduction for internal walls.

Approximate estimating costs generally include mechanical and electrical installations but exclude furniture, loose or special equipment, and external works; they also exclude fees for professional services. The costs shown are for specifications and standards appropriate to France and this should be borne in mind when attempting comparisons with similarly described building types in other countries. A discussion of this issue is included in section 2. Comparative data for countries covered in this publication, including construction cost data, are presented in Part Three.

Approximate estimating costs must be treated with caution; they cannot provide more than a rough guide to the probable cost of building. All the rates in this section exclude value added tax (VAT) which is at 20.6%.

	Cost FFr per m^2	Cost FFr per ft^2
Industrial buildings		
Factories for letting (include lighting, power and heating)	5,150	478
Factories for owner occupation (light industrial use)	5,670	527
Factories for owner occupation (heavy industrial use)	6,700	622
Factory/office (high-tech) for letting (shell and core only)	6,180	574
Factory/office (high-tech) for letting (ground floor shell,		
first floor offices)	6,180	574
Warehouses, low bay (6 to 8m high) for letting (no heating)	3,610	335

	Cost FFr per m²	Cost FFr per ft²
Administrative and commercial buildings		
Civic offices, non air conditioned	6,180	574
Prestige/headquarters office, 5 to 10 storeys, air conditioned	8,240	766
Prestige/headquarters office, high rise, air conditioned	9,270	861
Health and education buildings		
General hospitals	12,400	1,150
Secondary/middle schools	6,180	574
University (arts) buildings	7,210	670
University (science) buildings	8,240	766
Recreation and arts buildings		
Theatres (over 500 seats) including seating and stage equipment	12,400	1,150
Swimming pools (international standard) including changing facilities	10,300	957
National museums including full air conditioning and standby generator	10,300	957
Local museums including air conditioning	9,270	861
City centre/central libraries	10,300	957
Residential buildings		
Private/mass market single family housing 2 storey detached/semidetached (multiple units)	5,150	478
Purpose designed single family housing 2 storey detached (single unit)	6,180	574
Social/economic apartment housing, low rise (no lifts)	4,120	383
Social/economic apartment housing, high rise (with lifts)	4,640	431
Private sector apartment building (standard specification)	5,150	478
Private sector apartment building (luxury)	6,180	574
Homes for the elderly (shared accommodation)	6,180	574
Hotel, 5 star, city centre	12,400	1,150
Hotel, 3 star, city/provincial	10,300	957

Regional variations

The approximate estimating costs are based on average Paris rates. Adjust these costs by the following factors for regional variations:

Alsace	−2%	Centre	−5%
Aquitaine	−3%	Champagne-Ardennes	0%
Auvergne	−6%	Corse	−8%
Bourgogne	−4%	Franche-Comté	−4%
Bretagne	−5%	Languedoc-Roussillon	−5%

Limousin	−6%	Pays de la Loire	−5%
Lorraine	−3%	Picardie	−2%
Midi-Pyrénées	−6%	Poitou-Charente	−4%
Nord-Pas de Calais	−3%	Provence-Alpes Cote	
		d'Azur	0%
Basse Normandie	−2%	Rhône-Alpes	0%
Haute Normandie	−5%		

Value added tax (VAT)

The standard rate of value added tax (VAT) is currently 20.6%, chargeable on general building work and materials.

Exchange rates and cost and price trends

The combined effect of exchange rates and inflation on prices within a country and on price comparisons between countries is discussed in section 2.

Exchange rates

The graph following plots the movement of the French franc against sterling, the ECU/Euro, the US dollar and 100 Japanese yen since 1990. The values used for the graph are quarterly and the method of calculating these is described and general guidance on the interpretation of the graph provided in section 2. The average exchange rate in the first quarter of 1999 was FFr 9.35 to the pound sterling, FFr 5.61 to the US dollar and FFr 6.56 to the Euro.

The French franc against sterling, the ECU/Euro, the US dollar and 100 Japanese yen

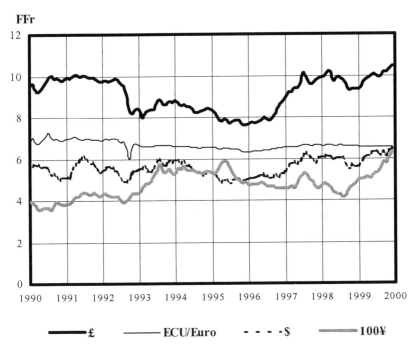

Cost and price trends

The tables overleaf present indices for consumer and residential building prices and also costs of building and civil engineering in France since 1990. The indices have, where necessary, been rebased to 1990=100. The annual change is the percentage change between the average index of consecutive years. Notes on inflation are included in section 2.

The price of residential construction index shows changes in the price paid by the future owner of a new house or buildings with two or more dwellings. It is based on the whole of metropolitan France. It is expensive to collect and may be changed in the future. It is rather confusingly known as the *Indices de Cout de la Construction* (ICC).

The index of the cost of building is based on factor costs for fifty specialized types of building works. The factors included labour, materials, transport, equipment, energy and general expenses. Taxes and design fees are excluded. It is used for deflation of figures in the national account. The index for civil engineering is calculated similarly for thirteen types of work.

Consumer and residential construction price indices

Year	Consumer prices annual average	Consumer prices change %	Prices of residential construction annual average	Prices of residential construction change %
1990	100.0	3.4	100	n.a
1991	103.2	3.2	104.1	4.1
1992	105.7	2.4	105.6	1.4
1993	107.9	2.1	106.8	1.1
1994	109.7	1.7	106.9	0.1
1995	111.6	1.7	106.9	0.0
1996	113.9	2.1	108.8	1.8
1997	115.2	1.1	111.4	2.1
1998	116	0.7	111.6	02

Sources: International Monetary Fund, Yearbook 1997.
Ministiere de l'Equipment et du Logment,
Insistut National de la Statistique et des Etudies Economiques(INSEE),
and Ministere de l'Equipment du Transport et du Tourisme.

Building and civil engineering cost indices

Year	Builidng costs annual average	Builidng costs change %	Civil engineering costs annual average	Civil engineering costs change %
1990	100	n.a	100	n.a
1991	102.1	2.1	100.7	0.7
1992	104.7	2.5	101.9	1.2
1993	108.4	3.5	106.1	4.1
1994	111.7	3.0	108.9	2.6
1995	115.2	3.1	112.0	2.8
1996	116.3	1.0	114.5	2.2
1997	118.6	2.0	117.1	2.3
1998	120.0	1.2	117.4	0.3

Sources: International Monetary Fund, Yearbook 1998 and supplements 1999.
Ministiere de l'Equipment et du Logment,
Insistut National de la Statistique et des Etudies Economiques(INSEE),
and Ministere de l'Equipment du Transport et du Tourisme.

All the construction indices have generally risen roughly in line with the consumer price index.

Useful addresses

Government and public organizations

**Association Française de
Normalisation (AFNOR)**
French Standards Association
Tour Europe
La D éfense
92049 Paris cedex 17
Tel: (33-1) 42915555
Fax: (33-1) 42915656
Minitel: 3616 AFNOR

**Délégation à l'Aménagement du
Territoire et à l'Action Régionale
(DATAR)**
Regional Development Agency
1 Avenue Charles Floquet
75007 Paris
Tel: (33-1) 40651234
Fax: (33-1) 40651234

**Ministère de Culture, de la
Communication et des Grands
Travaux**
Ministry of Culture, Communications
and Major Public Works
3 rue de Valois
75042 Paris
Cedex 01
Tel: (33-1) 40158000

**Commission Centrale des Marchés
(CCM)**
Central Commission for Public
Contracts
185 Rue de Bercy
75012 Paris
Tel: (33-1) 44871717
Fax: (33-1) 53178669

**Ministère de l'Equipement, du
Logement et des Transports**
Ministry of Equipment, Housing and
Transport
La Grande Arche, La Défense
92055 Paris
Tel: (33-1) 40812122
Minitel: 3615 INFOLOGEMENT

**Institut National de Statistiques et
d'Etudes Economiques (INSEE)**
National Institute of Statistics and
Economic Studies
18 Boulevard Adophe-Pinard
75675 Paris
Cedex 14
Tel: (33-1) 41175050
Minitel: 3615 INSEE

Trade and professional associations

Fédération Nationale du Bâtiment (FNB)
National Building Federation
9 Rue la Pérouse
75016 Paris
Tel: (33-1) 40695100
Minitel: 3614 – FNB

Fédération Nationale des Promoteurs Constructeurs (FNPC)
National Federation of Construction Developers
106 Rue de l'Université
75007 Paris
Tel: (33-1) 47054436
Fax: (33-1) 47539273

Chambre des Ingénieurs Conseils de France (CICF)
Federation of French Consulting Engineers
3 Rue Léon Bonnat
75016 Paris
Tel: (33-1) 44304930

Comité Français de la Chambre de Commerce Internationale
French Committee of International Chamber of Commerce
9 Rue Anjou
75000 Paris
Tel: (33-1) 42651266
Fax: (33-1) 42940639

Fédération Nationale des SCOP du Bâtiment et dex Travaux Publics
National Building and Public Works Federation
88 Rue Courcelles
75008 Paris
Tel: (33-1) 55651220
Fax: (33-1) 55651556

Ordre des Architectes – Conseil National
Professional body controlling the register of qualified architects
25 Rue Petit Musc
75004 Paris
Tel: (33-1) 53019555

Organisme Professionnel de Qualification Technique des Economistes et Coordonnateurs de la Construction (OPQTECC)
Professional body controlling the qualification of 'economistes'
41 bis, Boulevard de Latour-Maubourg
75007 Paris
Tel: (33-1) 45569267
Fax: (33-1) 44183526

Other organizations

Centre Scientifique et Technique du Bâtiment (CSTB)
Scientific and Technical Centre for Building
4 Avenue du Recteur Poincare
75016 Paris cedex 16
Tel: (33-1) 40502828

Association Française des Assureurs Construction (AFSA)
Association of French Construction Insurers
26 Boulevard Haussmann
75009 Paris
Tel: (33-1) 42479000

Bâtimat – Blenheim
Organization of the Building Show «BATIMAT»)
22–24 Rue du Président Wilson
92532 Levallois-Perret cedex
Tel: (33-1) 47 56 50 00
Fax: (33-1) 47 56 08 18

Klaus Valet

Consulting Engineer

Bismarckstrasse 28

D-73240 Wendigen

Germany

Tel: +49 7024 3837

Fax: +49 7024 6144

email: KlausValet@aol.com

An independent partner in the investigation and realization of projects including the economic evaluation and planning – providing full responsibility and independence.

Utilization of renewable energies

Project development, project management, investigations, planning and post completion assistance.

Legal expertise, seminars and lectures.

Germany

All data relate to 1998 unless otherwise indicated.

Population

Population	82.1 million
Urban population (1997)	87%
Population under 15	16%
Population 65 and over	16%
Average annual growth rate (1985 to 1998)	2.3%

Geography

Land area	356,910 km^2
Agricultural area	49%
Capital city	Berlin

Economy

Monetary unit	Marks (DM)
Exchange rate (average first quarter 1999)	
the pound sterling	DM 2.79
the US dollar	DM 1.67
the Euro	DM 1.96
the yen x 100	DM 1.42
Average annual inflation (1995 to 1998)	1.4%
Inflation rate	1.0%
Gross Domestic Product (GDP)	DM 3,783 billion
GDP PPP basis (1997)	$ (PPP) 1,740 billion
GDP per capita	DM 46,100
GDP per capita PPP basis (1997)	$ (PPP) 21,190
Average annual real change in GDP (1995 to 1998)	2.1%
Private consumption as a proportion of GDP	57%
Public consumption as a proportion of GDP	19%
Investment as a proportion of GDP	20%

Construction

Gross value of construction output	DM 389 billion
Gross value of construction output per capita	DM 4,740
Gross value of construction output as a proportion of GDP	10.3%
Average annual real change in gross value of construction output (1995 to 1998)	−3.3%
Annual cement consumption	39.0 million tonnes

PPP Purchasing power parity

The construction industry

Construction output

The value of gross construction output in Germany in 1998 was DM 389 billion
or equivalent to $ 197 billion ECUs (nearly the same as January 1999 Euros).
These figures include an estimate for construction by other sectors, the black
economy and DIY but do not include normal repair and maintenance, though
improvement is included. This total represents 10.3% of GDP. The breakdown
for Germany for 1998 is shown below:

Output of construction, (new and renovation) 1998 (current prices)

Type of work	DM billion	ECUs billion	% of total
New work			
Residential building	131.5	66.8	34
Non-residential building			
Office and commercial	38.6	19.5	10
Industrial	10.6	5.4	3
Other	27.4	13.9	7
Total	76.6	38.9	20
Civil engineering	47.7	24.2	12
Total new work	255.8	129.9	66
Renovation			
Residential	84.7	43.0	22
Non-residential	32.1	16.3	8
Civil engineering	15.9	8.1	4
Total renovation	132.7	67.4	34
Total	**388.5**	**197.3**	**100**

Note: the value of the Euro in January 1999 was very similar to that of the ECU in 1998.
Source: based on Russig, V., 'Germany' in Euroconstruct conference paper,
Prague, June 1999.

Germany has the largest construction output of any European country. There is
still a substantial difference in the performance of East and West Germany. East
Germany accounts for rather over a quarter of the construction output of all
Germany.

Output in Germany has fallen in total over the whole period 1995 to 1998,
although it had increased in earlier years. The fall in 1998 was over 4%.

Germany is recovering from a recession, GDP is now increasing and unemployment falling. The main reason for the fall in construction output is that East Germany has not yet adjusted to the market economy so that construction output in East Germany is falling more than increases in output in West Germany. A further factor is that German exports have been sluggish which has affected the level of industrial and commercial construction.

The recession has had a dampening effect on the new housing market. New residential construction fell 6% in 1998. Although the demand for one or two family dwellings is steadily increasing, the demand for flats is falling more. With increased incomes, the effect of government own-home grants and low interest rates, the residential sector is unlikely to fall much further.

There have been falls in output of the types of non-residential buildings of 5 to 8%. This is due to the effects of the recession, including excess capacity in commercial and office buildings. The strict fiscal policies of the Government have limited the amount of new infrastructure work. This affects civil engineering more than the building sector.

Repair and maintenance has also declined but by much less than new work. There is an urgent need for renovation especially in East Germany which is suffering from the long neglect of the housing stock. Government has boosted housing renovation, and especially that related to energy savings, by a loan scheme and this is being extended. Nevertheless the level of renovation is still marking time. In non-residential construction renovation has also been low and falling.

The work abroad by German contractors present in 1997 in the top 225 international contractors in Engineering News Record was valued at DM 6.9 billion or around 4% of total domestic work as follows:

German construction work abroad, 1997

Area	DM billion	% of total
Europe	5.1	30.2
Africa	1.3	8.0
Middle East	0.5	3.2
Asia	4.6	27.4
US	4.4	26.1
Canada	0.3	1.7
Latin America	0.7	3.4
Total	**16.9**	**100.0**

Source: Engineering News Record 12-8-98.

The geographical distribution in Länder (Federal states) of building completions in 1996 compared with that of population is shown below.

Population and construction output 1996 regional distribution

Länder	Population (%)	Construction output (%)
Baden- Württemberg	12.6	13.1
Bayern	14.6	18.7
Berlin	4.3	5.0
Brandenburg	3.1	5.9
Bremen	0.8	0.4
Hamburg	2.1	1.4
Hessen	7.3	5.6
Mecklenburg-Vorpommern	2.2	3.7
Niedersachsen	9.5	6.7
Nordrhein-Westfalen	21.9	14.9
Rheinland-Pfalz	4.9	4.4
Saarland	1.3	0.9
Sachsen	5.6	8.7
Sachsen-Anhalt	3.4	4.0
Schleswig-Holstein	3.3	2.7
Thuringen	3.1	3.9
Total	**100.0**	**100.0**

Sources: Statistiches Jahrbuch 1997, Statistiches Bundesamt.

The *Land* with the highest population – Nordheim-Westfalen – did not fare well in its level of construction. It is likely that Berlin's share of construction will rise considerably in the wake of the movement of the capital there.

Characteristics and structure of the industry

The building contracting system in Germany is substantially that which operated in Western Germany prior to unification.

The majority of work is either let by trade contracts or to general contractors. In the separate trades system, site co-ordination is the responsibility of the architect or independent project manager. Where general contracting is used it is common for several contractors to form a joint company to tender for and construct large contracts. A considerable amount of construction work is undertaken as joint ventures. Many clients favour the system, as it helps to share financial responsibility – if one member of the venture goes into liquidation the other members will usually support the venture through to completion.

In the past most contractors operated in relatively local markets and the differences in regulations from one *Land* to another meant that it was a major decision to expand over the border. Hence there were relatively few large contractors in West Germany. However, more recently the largest West German contractors have expanded greatly and now there are eleven of them among the top 50 European contractors. Moreover, in East Germany contractors were few and they were large in size so that the number of large contractors in the whole of Germany has increased though the total number of contractors has not risen so much. In 1992 there were about 65,700 contractors of which 91 employed more than 500 persons and 520 employed more than 200. Those employing more than 200 did about 23% of the work and those with over 500 employees 8% of the work.

Thirty-eight German contractors are in *Building* magazine's list of top 300 European contractors for 1997. There are 13 German firms in the 1997 list of *Engineering News Record's* 'Top 225 International Contractors'. The principal German contractors and their percentage sales abroad are shown below.

Major German contractors, 1997

Major contractors	Place in Building's Top 300 European Contractors'	Sales abroad as % of turnover
Holzmann	4	38
Hochtief	6	48
Bilfinger + Berger	9	52
Strabag	18	36
Walter Bau	21	25
Dykerhoff & Widmann	24	16
Zublin	26	18
Heilit & Woerner	38	18
Wayss & Freytag	42	24
Teerbau	50	6.1
Heitkamp	65	

Source: Building December 1998.

A feature of the German contracting industry is the high participation of banks and financial institutions in the ownership of construction firms which facilitates German construction firms' access to capital. Frequently there are cross shareholdings between contractors.

The two main professions in the construction and building field are architecture and civil engineering. Education is initially joint but later splits into the two separate disciplines. The role of these professionals is similar to that in the

United Kingdom but there are traditionally no quantity surveyors. Their duties are covered by the architect and engineer.

There are two main streams of education:

- *Technische Hochschule* (universities or technical universities)
- or the *Fachhochschule* (polytechnics).

Students of both the University and the *Fachhochschule* graduate with the degree of *Diplom Ingenieur*.

After two years practical experience in an architect's office an architect registers at the Chamber of Architects in the State in which he lives. The title 'Architect' is protected by law. There are a large number of professionals in West Germany in relation to the population. There are over 60,000 architects, 990 for every million inhabitants, and an estimated 100,000 building engineers, 1,639 for every million inhabitants – figures which are much higher than in most European countries. Architects operate mainly in small practices of two or three architects. Engineers have larger practices. In the East organizations were very large in the past but these are now largely broken up into smaller units.

Clients and finance

Housebuilding is mainly undertaken by small local builders who sell 'off plan' and the buyers make stage payments during the construction process. Development of non-residential buildings is, to a large extent, undertaken by the organisation wishing to occupy the building, using a mixture of cash and loans. It may later sell the building to a property investor. However, banks often employ a developer, while providing the land and finance themselves and, following completion they retain ownership of the development. They thus build up a portfolio of properties. Insurance companies are also important investors in property. Pension funds are of minor importance. Generally, investors prefer to purchase a building which is complete and already let. Large property developers, such as exist in the UK, are relatively few, although speculative development is increasing.

Clients always demand a high standard of construction and it seems to be generally agreed that the end product is of high quality. The laws and regulations for construction are detailed and strictly enforced.

Selection of design consultants

The normal procedure in the public sector is that the client requests several firms to submit proposals. In most cases personal meetings precede the written proposals and sometimes this is the stage at which the designer is chosen, though not formally so. Important criteria for selection are quality of work and special know-how, good references, reliability, a regional presence and personal contacts. There is a growing number of public architectural competitions much disliked by architects. The duties of architects and engineers are established by law. Price is not a key element in selection. Fees have to be calculated on the basis of the *Honorarordnung für Architekten und Ingenieure* (HOAI) (regulations for the setting of architects' and engineers' fees). The normal procedure for private clients is less formal but follows similar lines. Each professional has a separate contract with the client.

Contractual arrangements

Open invitations to tender from an unlimited number of contractors are generally the norm in Germany. For government work the Federal Minister has responsibility for determination of projects for which it is desirable to have selective tendering or direct award. Conditions under which selective tendering is permitted are that:

- an open tender would involve costs disproportionate to the method's advantages
- an open tender has not produced an acceptable result
- other reasons (e.g. urgency or secrecy) make open tendering inexpedient
- only a few contractors are capable of undertaking the work
- the nature of the work would involve very high costs in preparing a tender.

Circumstances for the award of a contract without tender are similar but more extreme – inability to define work, urgency, secrecy or only one contractor competent to undertake the work.

In competitive tender the lowest tender price is not decisive. There may be further negotiation after tendering. Design and build contracts are increasing.

Although it is not mandatory, it is advisable to adopt a similar approach in the private sector because of the nature of the basic practice. All the obligations of the process are subject to the principle of good faith, the *Treu und Glauben*. Specific performance is the main requirement. In the case of poor quality, the contractor has to remedy the problems and, if unable to, may be obliged to reduce the price. In some cases the client will break the contract. Germany has published rules covering contract procedure for all building work, the *Verdingungsordnung für Bauleistungen* (VOB) (Regulations for Building Contracts). It is something like a "code of practice", and does not have the force of law. In the event that a situation is not covered by the VOB, all parts will refer to another type of contract, the *Baugesetzbuch*, which has the force of law and is in accordance with the Civil Code. The VOB, published in three parts, is the main compendium of regulations and a good practice guide for the industry. VOB is the basis of most construction contracts.

- Part A details the procedure for tendering, and is mandatory for public works.
- Part B lists general conditions of contract for the execution of the works. It contains clauses concerning remuneration, execution of the works, termination, payment, liabilities of the parties, and so on.
- Part C comprises 53 DIN standard specifications covering all construction trades arranged in a similar format. These give guidelines for the measurement of quantities and ensure some uniformity of tender documents.

There are additionally standard bills of quantities available where the items of work are properly described, spelling out the specification, the unit and the quantity, and the tenderer has to fill in the unit price and the overall price for each item. In a number of contracts the tenderer is, in addition, required to give details of wage costs and materials costs. The impact of the general and preamble conditions on price are incorporated in the unit rates.

The specification, together with a bill of quantities, necessary drawings and the VOB (Parts B and C) and any other special conditions, are collected together to form the tender documents; these are supplemented by an exchange of letters placing and accepting the contract. The conditions of contract may be as VOB Part B, but these are frequently supplemented with amended or additional clauses to suit the particular circumstances of each scheme. Another type of contract, the *Verdingungsordnung für Lieferungen* (VOL) is added to deal with the supply of materials and components.

The most common type of contract is a 'remeasurement contract' based on a specification and approximate bills of quantities (subject to remeasurement on completion). Most contracts are fixed price, though alterations to rates are usually permitted when there are quantity adjustments on unvaried work in excess of 10%. Interim payments are usually paid to a contractor upon receipt of a detailed interim account from him for the work executed at that time. In the case of delays by the contractor, he is liable for penalties, rather than liquidated damages.

Variations are enforceable providing they do not change the nature of the work; the contractor prepares details of the cost of the variation for agreement with the architect, based as far as possible on rates in the contract documents, but new rates are possible for non-related work. It is intended that the cost of variations be agreed before the work is undertaken on site but this often fails to occur.

Most contracts make provision for retention, usually 5% to 10% of the contract sum, which is partially released upon agreement of the final account. The remainder is released after the defects liability period – usually five years. Each contractor, however, is liable for the maintenance of his part of the work for anything between two and five years after completion. It is the duty of the contractor to prepare the final account which, in the case of Government work, is subject to audit by the Federal Auditor. The audit must be carried out within five years and the account may be re-opened within this time.

Liability, insurance and inspection

In the German system the formal acceptance of work is crucial. The responsibility for building failure rests with the contractor. However, the relevant building authority appoints an independent assessor (*Prüfungs-ingenieur*) – often an academic or independent consultant – to certify that the building is structurally sound and that the overall requirements have been met. He then accepts responsibility and is obliged to have professional indemnity insurance. Owners' rights to claim damages are limited to five years and then only for damage caused directly by the fault of the builder. In the case of design faults, a five year limit applies, but owners also have the right to claim for consequential loss.

The builder according to his contract is, with some exceptions, liable only for two or five years from handover. Thus there is no statutory guarantee period for architects, engineers or contractors. Some latent defects insurance is available, notably for new housing. Architects and engineers cannot currently indemnify themselves against calculation errors, e.g. cost advice or quantities.

Development control and standards

Planning law comes from both the Federal Government and the *Länder* (Federal States) and procedures are based on a system of statutory development plans. The 'Land Use Plan' is fixed and binding on all parties. The 'Building Plans' for the local areas are similar to building regulations. Building regulation approval is covered at the same time as development control. In Germany, every kind of building work needs a permit. Applications must be submitted to the municipality by a qualified architect or engineer registered in the region. The municipality then passes them on as necessary to the next superior level. The application must be accompanied by a report on the structural soundness of the proposed construction and any other relevant reports. Authorities have to approve all applications which comply with the land use plans; failure to do so may result in legal action by the applicant in the administrative courts. No time limit is specified for giving a decision but, after three months, application may be made to the administrative courts on the grounds of undue delay so that, in practice, this is taken as representing adequate time. The construction process is also controlled. One copy of the permit and of the approved drawings must be available on site to be inspected at any time. Contravention may result in the stopping of the works or even demolition. Typically a fifth of applications comply fully with the regulations, while almost all the rest deviate to some extent and the authority will prescribe changes to be made. A very small percentage deviate so greatly that permission is refused. In practice a wide range of informal consultation takes place in addition to the formal procedures. A building permit is valid for two years.

Germany has a comprehensive range of standards covering technical details and information on installation for a wide range of products used in construction. These are published and kept up to date by DIN (*Deutsches Institut für Normung* – German Standards Institute), which also has a materials testing division. DIN also maintain a databank of all standards, technical rules, and regulations – DITR (*Deutsches Informationszentrum für technische Regeln*) – which is available in a variety of forms – by microfiche, on-line computer connection, or as a telephone service, available to subscribers. Individual *Länder* each have their own set of Building Regulations which are based on a set of national 'model by laws'. The variations between individual *Länder* are generally minor. Indeed 90% of the regulations are now harmonised.

There is also an *Agrément* Institute, which is solely concerned with the safety of materials.

The standards are not legally binding but they are generally regarded as stating good practice. However, deviations from standards are permitted subject to approval of the building control authority, through the issue of a certificate or consent given in individual cases. Most contracts insist on current standards.

Both the legal conditions and the construction standards apply to both building and civil engineering construction, though there are additional regulations for certain types of construction, e.g. bridges. In the case of public works the building control is exercised by the relevant competent authority.

Research and development

Research and development work in building and civil engineering in more than 30 governmental and private institutions is co-ordinated by the recently established Association for Building Research (*Arbeitsgemeinschaft für Bauforschung*) at the Federal Ministry for Regional Planning, Building and Urban Development. Most funds are spent on research and development in building materials, components and building processes.

Construction cost data

Cost of labour

The figures below are typical of labour costs in the Frankfurt – Hessen area as at the first quarter 1999. The wage rate is the basis of an employee's income, while the cost of labour indicates the cost to a contractor of employing that employee. The difference between the two covers a variety of mandatory and voluntary contributions – a list of items which could be included is given in section 2.

	Wage rate (per hour) DM	Cost of labour (per hour) DM	Number of Hours worked per year
Site operatives			
Mason/bricklayer	22	56	1,755
Carpenter	22	56	1,755
Plumber	24	61	1,755
Electrician	24	61	1,755
Structural steel erector	22	56	1,755
HVAC installer	24	61	1,755
Semi-skilled worker	21	53	1,755
Unskilled labourer	20	51	1,755
Equipment operator	25	62	1,755
Watchman/security	18	45	1,755
Site supervision			
General foreman	28	70	1,755
Trades foreman	26	64	1,755
Clerk of works	20	51	1,755
Contractors' personnel	*(per year)*	*(per year)*	
Site manager	152,000	303,000	1,716
Resident engineer	121,000	242,000	1,716
Resident surveyor	111,000	222,000	1,716
Junior engineer	60,600	121,000	1,716
Junior surveyor	55,600	111,000	1,716
Planner	90,900	182,000	1,716
Consultants' personnel			
Senior architect	121,000	242,000	1,716
Senior engineer	126,000	253,000	1,716
Senior surveyor	111,000	222,000	1,716
Qualified architect	80,800	162,000	1,716
Qualified engineer	85,900	172,000	1,716
Qualified surveyor	70,700	141,000	1,716

Cost of materials

The figures that follow are the costs of main construction materials, delivered to site in the Frankfurt – Hessen area, as incurred by contractors in the first quarter of 1999. These assume that the materials would be in quantities as required for a medium sized construction project and that the location of the works would be neither constrained nor remote. All the costs in this section exclude value added tax (VAT) which is at 15%.

	Unit	Cost DM
Cement and aggregates		
Ordinary portland cement in 50kg bags	tonne	182
Coarse aggregates for concrete	m³	152
Fine aggregates for concrete	m³	152
Ready mixed concrete (mix B25)	m³	206
Ready mixed concrete (mix B35)	m³	209
Steel		
Mild steel reinforcement	tonne	747
High tensile steel reinforcement	tonne	778
Structural steel sections	tonne	1,010
Bricks and blocks		
Common bricks (240 x 365 x 238mm)	1,000	3,430
Hollow concrete blocks (240 x 175 x 213mm)	1,000	808
Solid concrete blocks (240 x 175 x 113mm)	1,000	909
Timber and insulation		
Softwood sections for carpentry	m³	293
Softwood for joinery	m³	278
Exterior quality plywood (21mm)	m²	41
Plywood for interior joinery (19mm)	m²	12
Softwood strip flooring (13mm)	m²	5
Chipboard sheet flooring (19mm)	m²	11
100mm thick quilt insulation	m²	21
100mm thick rigid slab insulation	m³	101
Softwood internal door complete with frames and ironmongery	each	242
Glass and ceramics		
Float glass (4mm)	m²	120
Float glass (5mm)	m²	141
Float glass (6mm)	m²	164
Float glass (8mm)	m²	233
Sealed double glazing units (885 x 500mm)	m²	298
Good quality ceramic wall tiles (350 x 350 x 1.6mm)	m²	63

	Unit	Cost DM
Plaster and paint		
Plaster in 50kg bags	tonne	273
Plasterboard (12.5mm thick)	m^2	5
Emulsion paint in 5 litre tins	litre	5
Gloss oil paint in 5 litre tins	litre	8
Tiles and paviors		
Clay floor tiles	m^2	61
Vinyl floor tiles	m^2	15
Precast concrete paving slabs (500 x 500 x 50mm)	m^2	23
Clay roof tiles	1,000	1,920
Precast concrete roof tiles	1,000	1,620
Drainage		
WC suite complete	each	808
Lavatory basin complete	each	140
125mm diameter clay drain pipes (1.25m lengths)	m	27
100mm diameter cast iron drain pipes (3m lengths)	m	41

Unit rates

The descriptions of work items below are generally shortened versions of standard descriptions listed in five languages (English, French, Italian, German and Spanish) in Appendix 3.

Where an item has a two digit reference number (e.g. 05 or 33), this relates to the full description against that number in Appendix 3. Where an item has an alphabetic suffix (e.g. 12A or 34B) this indicates that the standard description has been modified. Where a modification is major the complete modified description is included here and the standard description should be ignored.

Where a modification is minor (e.g. the insertion of a named hardwood) the shortened description has been modified here but, in general, the full description in Appendix 3 prevails.

The unit rates following are for main work items on a typical construction project in the Munich area in the first quarter of 1999. The rates include labour, materials and equipment and allowances for contractor's overheads and profit (2%) preliminary and general items (7%) and contractor's profit and attendance on specialist trades (5%). All the rates in this section exclude VAT which is at 15%.

		Unit	Rate DM
Excavation			
01	Mechanical excavation of foundation trenches	m³	27
02	Hardcore filling making up levels	m²	30
Concrete work			
04	Plain insitu concrete in strip foundations in trenches	m³	288
05	Reinforced insitu concrete in beds	m³	288
06	Reinforced insitu concrete in walls	m³	288
07	Reinforced insitu concrete in suspended floor or roof slabs	m³	288
08	Reinforced insitu concrete in columns	m³	288
09	Reinforced insitu concrete in isolated beams	m³	288
Formwork			
11	Softwood formwork to concrete walls	m²	81
12	Softwood or metal formwork to concrete columns	m²	111
13	Softwood or metal formwork to horizontal soffits of slabs	m²	81
Reinforcement			
14	Reinforcement in concrete walls	tonne	2,020
15	Reinforcement in suspended concrete slabs	tonne	2,020
16	Fabric reinforcement in concrete beds	m²	6
Steelwork			
17	Fabricate, supply and erect steel framed structure	tonne	4,830
Brickwork and blockwork			
18	Precast lightweight aggregate hollow concrete block walls	m²	81
19	Solid (perforated) concrete bricks	m²	91
20	Solid (perforated) sand lime bricks	m²	97
21	Facing bricks	m²	101
Roofing			
22	Concrete interlocking roof tiles	m²	51
23	Plain clay roof tiles	m²	51
24	Fibre cement roof slates	m²	51
25	Sawn softwood roof boarding	m²	40
26	Particle board roof covering	m²	35
27	3 layers glass-fibre based bitumen felt roof covering include chippings	m²	45
28	Bitumen based mastic asphalt roof covering	m²	101
30	Rigid sheet loadbearing roof insulation 75mm thick	m²	40
31A	Troughed aluminium roof cladding	m²	91

		Unit	Rate DM
Woodwork and metalwork			
32	Preservative treated sawn softwood 50 x 100mm	m	5
33	Preservative treated sawn softwood 50 x 150mm	m	5
34	Single glazed casement window in hardwood, size 650 x 900mm	each	354
35	Two panel glazed door in hardwood, size 850 x 2000mm	each	1,520
36	Solid core half hour fire resisting hardwood internal flush doors, size 800 x 2000mm	each	3,030
37	Aluminium double glazed window, size 1200 x 1200mm	each	859
38	Aluminium double glazed door, size 850 x 2100mm	each	732
39	Hardwood skirtings	m	8
40	Framed structural steelwork in universal joist sections	tonne	4,830
41	Structural steelwork lattice roof trusses	tonne	4,830
Plumbing			
42	UPVC half round eaves gutter	m	35
43	UPVC rainwater pipes	m	45
44	Light gauge copper cold water tubing	m	33
45	High pressure plastic pipes for cold water supply	m	10
46	Low pressure plastic pipes for cold water distribution	m	8
47	UPVC soil and vent pipes	m	45
48	White vitreous china WC suite	each	455
49	White vitreous china lavatory basin	each	303
50	White glazed fireclay shower tray	each	152
51	Stainless steel single bowl sink and double drainer	each	505
Electrical work			
52	PVC insulated copper sheathed cable	m	6
53	13 amp unswitched socket outlet	each	30
54	Flush mounted 20 amp, 1 way light switch	each	25
Finishings			
55	2 coats gypsum based plaster on brick walls	m^2	25
56	White glazed tiles on plaster walls	m^2	94
57	Red clay quarry tiles on concrete floor	m^2	131
58	Cement and sand screed to concrete floors 50mm thick	m^2	25
59	Thermoplastic floor tiles on screed	m^2	40
60	Mineral fibre tiles on concealed suspension system	m^2	121
Glazing			
61A	Glazing to wood (3mm glass)	m^2	66
Painting			
62	Emulsion on plaster walls	m^2	27
63	Oil paint on timber	m^2	57

Approximate estimating

The building costs per unit area given overleaf are averages incurred by building clients for typical buildings in the Stuttgart (Baden-Württemberg) area as at the first quarter 1999. They are based upon the total floor area of all storeys, measured between external walls and without deduction for internal walls.

Approximate estimating costs generally include mechanical and electrical installations but exclude furniture, loose or special equipment, and external works; they also exclude fees for professional services. The costs shown are for specifications and standards appropriate to Germany and this should be borne in mind when attempting comparisons with similarly described building types in other countries. A discussion of this issue is included in section 2. Comparative data for countries covered in this publication, including construction cost data, are presented in Part Three.

Approximate estimating costs must be treated with caution; they cannot provide more than a rough guide to the probable cost of building. All the rates in this section exclude VAT which is at 15%.

	Cost DM per m²	Cost DM per ft²
Industrial buildings		
Factories for letting (include lighting, power and heating)	1,180	110
Factories for owner occupation (light industrial use)	1,180	110
Factories for owner occupation (heavy industrial use)	1,520	141
Factory/office (high-tech) for letting (shell and core only)	1,360	126
Factory/office (high-tech) for letting (ground floor shell, first floor office)	1,580	147
Factory/office (high-tech) for owner occupation (controlled environment, fully furnished)	1,670	155
Warehouses, low bay (6 to 8m high) for letting (no heating)	1,140	106
Warehouses, low bay for owner occupation (including heating)	1,230	114
Warehouses, high bay for owner occupation (including heating)	1,230	114
Cold stores/refrigerated stores	1,310	122
Administrative and commercial buildings		
Civic offices, non air conditioned	2,410	224
Civic offices, fully air conditioned	2,770	257
Offices for letting, 5 to 10 storeys, non air conditioned	2,410	224
Offices for letting, 5 to 10 storeys, air conditioned	2,770	257
Offices for letting, high rise, air conditioned	2,770	257
Offices for owner occupation, 5 to 10 storeys, non air conditioned	2,410	224
Offices for owner occupation, 5 to 10 storeys, air conditioned	2,770	257
Offices for owner occupation, high rise, air conditioned	3,070	285
Prestige/headquarters office, 5 to 10 storeys, air conditioned	3,950	367
Prestige/headquarters office, high rise, air conditioned	4,830	449

	Cost DM per m²	Cost DM per ft²
Health and education buildings		
General hospitals	2,640	245
Teaching hospitals	2,510	233
Private hospitals	2,930	272
Health centres	2,190	203
Nursery schools	1,930	179
Primary/junior schools	1,930	179
Secondary/middle schools	2,260	210
University (arts) buildings	2,640	245
University (science) buildings	2,640	245
Management training centres	2,190	203
Recreation and arts buildings		
Theatres (over 500 seats) including seating and stage equipment	2,640	245
Theatres (less than 500 seats) including seating and stage equipment	2,860	266
Concert halls including seating and stage equipment	2,640	244
Sports halls including changing and social facilities	2,320	216
Swimming pools (international standard) including changing facilities	3,280	305
Swimming pools (schools standard) including changing facilities	3,280	305
National museums including full air conditioning and standby generator	2.640	245
Local museums including air conditioning	1,930	179
City centre/central libraries	2,350	218
Branch/local libraries	2,280	212
Residential buildings		
Social/economic single family housing (multiple units)	1,180	110
Private/mass market single family housing 2 storey detached/semidetached (multiple units)	1,230	114
Purpose designed single family housing 2 storey detached (single unit)	1,490	138
Social/economic apartment housing, low rise (no lifts)	1,310	122
Social/economic apartment housing, high rise (with lifts)	1,340	124
Private sector apartment building (standard specification)	1,290	120
Private sector apartment building (luxury)	1,820	169
Student/nurses halls of residence	1,410	131
Homes for the elderly (shared accommodation)	1,620	151
Homes for the elderly (self contained with shared communal facilities)	1,620	151
Hotel, 5 star, city centre	2,990	278
Hotel, 3 star, city/provincial	2,480	230
Motel	1,310	122

Regional variations

The approximate estimating costs are based on average Stuttgart (Baden-Württemberg) rates. Adjust these costs by the following factors for regional variations:

Bayern	3%	Niedersachsen	-13%
Berlin	23%	Nordrhein-Westfalen	-4%
Bremen	-9%	Rheinland-Pfalz	-3%
Frankfurt-Hessen	-3%	Saarland	-4%
Hamburg	-3%	Schleswig-Holstein	-10%

Value added tax (VAT)

The standard rate of value added tax (VAT) is currently 15%, chargeable on general building work and materials.

Exchange rates and cost and price trends

The combined effect of exchange rates and inflation on prices within a country and on price comparisons between countries is discussed in section 2.

Exchange rates

The graph opposite plots the movement of the German deutschmark against sterling, the ECU/Euro, the US dollar and 100 Japanese yen since 1990. The values used for the graph are quarterly and the method of calculating these is described and general guidance on the interpretation of the graph provided in section 2. The average exchange rate in the first quarter of 1999 was DM 2.79 to the pound sterling, DM 1.67 to the US dollar and DM 1.96 to the Euro.

Cost and price trends

The tables on the next two pages present the indices for consumer prices and building costs for dwellings, offices, commercial properties and industrial building inflation in Germany since 1990. The indices have been rebased to 1990=100. The annual change is the percentage change between the average index of consecutive years. Notes on inflation are included in section 2.

The German deutschmark against sterling, the ECU/Euro, the US dollar and 100 Japanese yen

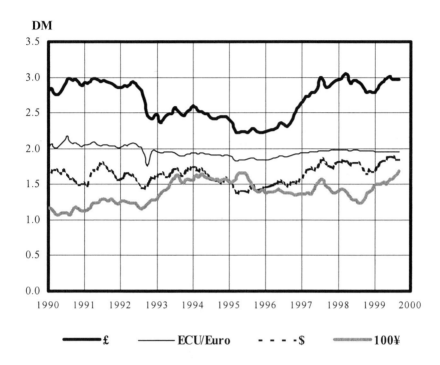

Consumer price, residential and office building cost indices

Year	Consumer prices annual average	change %	Residential building costs annual average	change %	Office building costs annual average	change %
1990	100.0	2.7	100.0	6.5	100.0	5.9
1991	103.6	3.6	106.7	6.7	106.4	6.4
1992	108.9	5.1	112.8	5.7	111.9	5.2
1993	113.7	4.4	116.9	3.6	116.2	3.8
1994	116.8	2.7	119.3	2.1	118.5	2.0
1995	119.0	1.9	122.1	2.3	121.3	2.3
1996	120.8	1.5	122.0	-0.1	121.5	0.2
1997	122.9	1.7	121.5	-0.4	121.2	-0.3
1998	124.1	1.0				

Source: International Monetary Fund, Yearbook 1998 and supplements 1999.
OECD, Quarterly National Accounts.

Commercial and industrial building cost indices

Year	Commercial building costs annual average	change %	Industrial building costs annual average	change %
1990	100.0	6.2	100.0	6.3
1991	106.3	6.3	106.0	6.0
1992	111.8	5.2	111.2	4.9
1993	115.4	3.2	114.5	3.0
1994	117.5	1.8	116.5	1.8
1995	120.2	2.3	119.2	2.3
1996	120.7	0.4	119.9	0.6
1997	120.5	-0.2	119.7	-0.2

Source: OECD, Quarterly National Accounts..

Over the period as a whole construction costs have roughly kept pace with consumer prices. However they exceeded the rate of inflation from 1990 to about 1992 and then fell behind in the recession.

Useful addresses

Government and public organizations

Bundesbaudirektion (BBD)
Federal Construction Directorate
Fasanenstrasse 87
10623 Berlin
Tel: (49-30) 315890

Deutsches Institut für Normung (DIN)
German Standards Institute
Burggrafenstrasse 6
10787 Berlin

Institut für Bautechnik
Institute for Building Technology
Kolonnenstrasse 30
10829 Berlin
Tel: (49-30) 787300
Fax: (49-30) 78730320

Arbeitsgemeinschaft für das Bau, Wohnungs und Siedlungswesen zuständigen Minister/Senatoren der Länder (ARGEBAU)
Consortium of the State Ministers Senators Responsible for Building, Housing and Development
Deichmann Aue 31-37
53179 Bonn
Tel: (49-228) 2283370
Fax: (49-228) 3373060

Statistisches Bundesamt Deutschland
Federal Statistical Office of Germany
Gustav-Stresemann-Ring 11
65180 Wiesbaden
Tel: (49-611) 752405
Fax: (49-611) 753330

Trade and professional associations

Bund Deutscher Architekten (BDA)
Federation of German Architects
Ippendorfer Allee 14b
53127 Bonn
Tel: (49-228) 631381

**Bund Deutscher Baumeister,
Architekten und Ingenieure e.V.
(BDB)**
Federation of German Contractors,
Architects and Engineers
Kennedyallee 11
53175 Bonn
Tel: (49-228) 376784

**Hauptverband der Deutschen
Bauindustrie e.V. (HDB)**
Central Federation for the German
Construction Industry
Abraham Lincoln Strasse 30
65189 Wiesbaden
Tel: (49-61) 217220

**Verband Beratender Ingenieure e.V.
(VBI)**
Association of Consulting Engineers
Halbe Höhe 59
45147 Essen

Verein Deutscher Ingenieure (VDI)
Association of German Engineers
Ahrtrasse 45
53175 Bonn

Greece

All data relate to 1997 unless otherwise indicated.

Population

Population	10.6 million
Urban population	60%
Population under 15	17%
Population 65 and over	16%
Average annual growth rate (1984 to 1997)	0.5%

Geography

Land area	131,940 km^2
Agricultural area	68%
Capital city	Athens

Economy

Monetary unit	Drachma (Dr)
Exchange rate (average first quarter 1999) to:	
the pound sterling	Dr 468
the US dollar	Dr 281
the Euro	Dr 329
the yen x 100	Dr 238
Average annual inflation (1994 to 1997)	7.6%
Inflation rate	5.5%
Gross Domestic Product (GDP)	Dr 32,750 billion
GDP PPP basis	$ (PPP) 137.4 billion
GDP per capita	Dr 3,084,000
GDP per capita PPP basis	$ (PPP) 12,940
Average annual real change in GDP (1994 to 1997)	2.7%
Private consumption as a proportion of GDP	76%
Public consumption as a proportion of GDP	15%
Investment as a proportion of GDP	20%

Construction

Gross value of construction output	Dr 3,930 billion
Gross value of construction output per capita	Dr 371,000
Gross value of construction output as a proportion of GDP	12%
Annual cement consumption	7.8 million tonnes

PPP Purchasing power parity

The construction industry

Construction output

The value of gross construction output in Greece in 1997 was about Dr 3,930 billion equivalent to about 12.7 billion ECUs in 1997. This represents about 12% of GDP. The breakdown by type of work in 1997 was estimated as follows:

Output of construction, 1997

Type of work	Dr billion	ECUs billion	% of total
Residential building	932	3.0	24
Non-residential building			
Office and commercial	425	1.4	11
Industrial	393	1.3	10
Other	598	1.9	15
Total	1,415	4.6	36
Civil engineering	1,584	5.1	40
Total work	**3,931**	**12.7**	**100**

Source: based on Euroconstruct conference paper, Berlin, December 1998.

Construction output declined in real terms from the late 1980s until 1994. From that year the sector started growing again and, due to new infrastructure projects partially funded by the European Union, it is expected to grow substantially for some years to come, especially in the view of the celebration of the Olympic Games in 2004. Nevertheless, all the official figures must be treated with caution as it has been estimated that the unofficial economy provides the equivalent of around 35% of GDP. The speed of convergence towards European integration might be a heavy constraint to Government investment for the future.

Athens accounts for a large amount of the building work in Greece. The table opposite shows the percentage of permits issued for private new buildings in January 1997. In addition to the expected dominant position of Athens the large proportion of permits in Macedonia is noteworthy.

Value of private new building and extension permits by geographical area – percentages, January 1997

Regions	% of total
Attica	26.8
of which Greater Athens	(21.6)
Rest of Central Greece and Evia	6.5
Peloponnese	7.8
Ionian Is.	1.7
Ipiros	2.8
Thessaly	3.3
Macedonia	42.1
Thrase	2.2
Aegean Is	3.5
Crete	3.3
Total	**100.0**

Source: National Statistical Service of Greece, Monthly Statistical Bulletin 1998.

The public investment programme is mainly composed of new infrastructure building, such as the new Athens airport, motorway construction and underground projects for Athens and Salonika. Moreover, new work is expected for the 2004 Olympic Games, which will be hosted in Athens. Large contractors will undertake most of this work. Public work contributes to around 85% of the quoted companies' turnover; therefore the procurement of this work is vital for these companies. Most of these projects are Build Operate Transfer (BOT).

Between 1998 and 1999 Greece had to spend 65% of the second EU Community Support Framework (CSF) or 4.5 trillion drachmas. The first part of the third CSF involves expenditure of 1.5 to 2 trillion drachmas and the second part in around 6 trillion drachmas. The following table shows breakdowns for the Community planned expenditure from 1994 to 1999.

Community support framework planned expenditure, 1994 to 1999 (Dr billion)

	Community grants	National participation	Private expenditure	Total cost
Basic infrastructure				
Transport	887	448	858	2193
Communications	71	57	0	128
Environment and water	176	59	0	234
Health and care	188	73	0	262
Energy	244	246	107	597
Athens and Thessaloniki underground	241	241	40	521
Total	1,806	1,124	1,005	3935
Human resources				
Education and training	530	177	0	707
Training-employment	349	116	69	535
Public administration	48	39	0	86
Total	927	332	69	1328
Productive environment				
Agriculture and rural development	616	248	301	1165
Industry and services	284	159	803	1246
Research and technology	93	31	44	169
Tourism and culture	195	95	223	513
Total	1,189	533	1,371	3093
Technical assistance	20	5	0	25
Total	**3,942**	**1,994**	**2,445**	**8,381**

Totals do not always sum exactly due to rounding.
Source: Ministry of National Economy.

In addition to the funds available under the CSF, the Commission has earmarked ECU 15,150m as a special Cohesion Fund designed to help finance international transport and environmental projects in the four poorest member states of the EU – Spain, Portugal, Ireland and Greece. Greece stands to gain 16 to 20% of this fund. Cohesion Fund grants can cover up to 80% of the cost of a project and 100% of the cost of studies. The majority of the investment from the CSF will be channelled through the Ministry of National Economy's Investment Incentive Programme. The ministry has received permission from the European Commission to grant higher than ordinary subsidies to investments by local authorities and cooperatives, to firms establishing in industrial parks and to hospitals, physical rehabilitation centres and professional training centres.

Characteristics and structure of the industry

The engineer is extremely important in Greece. The main representative body of engineers is the Hellenic Association of Consulting Firms (HACF), also known as 'Hellasco'. There is a requirement for Greek engineers to register with the Technical Chamber of Greece. The Chamber has approximately 50,000 professional members, of whom about 7,500 are officially registered as individual consultants, the others being either employed by public bodies or elsewhere in the construction (or other) industries. The average consultancy is medium-sized, having, say, two to three partners and four or five engineers. There are also a number of small firms and a very few large ones. Most consulting engineers are based in Athens although the larger firms have branch offices or arrangements with engineers in the regions. In the late 1980s a new type of organization was established: the public sector consultancy which is publicly owned but operates much as a private firm. These public sector consultancies have significant advantages over private firms. They may enter into particular agreements, called programmed or policy agreements, with any public authority whereas private consultancies are not permitted to do this. There are about 15 public sector companies, some of them with several hundred employees.

The title of architect is protected by law (No. 4663 of 1930) for all members of the Technical Chamber of Greece, which is responsible for granting licenses to practice. The employment of architects is not obligatory. Architects normally work in close association with consulting engineers, who have overall responsibility including that for the aesthetic aspects of buildings. Private architects are graduates of universities such as the National Technical University of Athens. Local government architects graduate from technical universities but can also become certified within the public sector by an examination of the Ministry of Public Works.

The Greek construction industry is highly fragmented and consists of about 2,600 firms, of which most of them have less than 50 employees, though in many companies the numbers of employees fluctuate dramatically according to the number of contracts being undertaken. There are large numbers of very small, unregistered jobbing builders and trades persons who hire themselves out to registered companies and undertake small local contracts. Around 400 companies are *Societés Anonymes*. In 1997 no Greek companies were listed in *Building* magazine's 'Top 300 European Contractors'. 'Consolidated Contractors Intl. Co SAL' of Athens was number 23 in the *Engineering News Record's* list of 'Top 225 International Contractors' in 1997. All of its work was abroad, where it undertook building, industrial/petroleum, sewerage and transportation works. The large number of Greek contractors has led to intense competition. Quite recently, there has been an increasing trend towards joint ventures as a mean of compensating for scarcity of technology and skilled local resources.

To obtain public sector work contractors must hold the appropriate licence for the class of work involved. To be licensed, a contractor must have among his staff sufficient numbers of suitably qualified and licensed engineering and architectural personnel, and experience in one of four specific classes of work.

The number of construction companies listed on the Athens Stock Exchange has increased recently because several of them are seeking to modernize their equipment and increase staff to be able to bid for the projects expected to arise from the Community Support Framework. The principal Greek contractors ranked according to their 1997 turnover are shown below.

Major Greek contractors, 1997

Major contractors
Hellenic Technodomiki
Aegek
Michaniki
Gek
Aktor
Themeliodomi
Technodomi
Sarandopoulos
Athina
Avax
Alte
Tev

Source: Directory of Hellenic Construction Enterprises, Association of Greek Contracting Companies.

Clients and finance

About 75% of housing in Greece is privately owner-occupied, the remainder being let by the public or private sector. Housing is financed by private funds.

About half of non-residential private building consists of hotels and other tourist facilities financed by a variety of organizations including some of the Greek construction companies.

After a period of tight controls on public investment in new infrastructure projects, the Greek government announced in 1987 a 12-year public works programme, then valued at Dr 1,000 billion, including the controversial plan to divert the Acheloos river, the upgrading of Greece's national road system, the construction of two subway lines in Athens, a new airport for the capital and a suspension bridge across the neck of the Gulf of Corinth. Many of the projects have been transformed into build-own-operate (BOO), or build-own-operate-transfer (BOOT) schemes. Others, like the highways expansion and railways' upgrade, have been incorporated in the Community support programmes. Construction for the Athens metro, the airport and the Gulf of Corinth suspension bridge is well under way.

Selection of design consultants

All buildings must be designed by a registered engineer or, depending on the value of the project, an appropriate engineering consultancy qualified in architecture, planning, civil engineering or building installations engineering. The selection procedures are very closely regulated. In the public sector a distinction is made between design and consultancy. Design fees are related to the cost of the project whereas consultancy is the giving of advice unrelated to the cost of the project. A designer must be registered with the Government Design Register and a consultant in the Consultants' Register. It is not possible to be registered in both.

Sometimes projects are started on the basis of the preliminary design which is completed later. This causes problems as the work cannot progress efficiently. Bill of quantities are often inaccurate. As the quantities are actually higher than those estimated, the cost rises quickly and financial and legal issues contribute to slow down the realization of the project.

All government agencies or state-controlled enterprises such as foundations, banks, public utilities, etc. are obliged by law publicly to announce most of their requirements for design work and to allot them to professionals registered in the Government Design Register. The Government then organizes a competition for contracts. Open tendering is the most common in Greece. In announcements for some projects, design and construction are combined. However, for many public sector projects, especially in the municipalities, the public engineering companies often obtain work without competition. In the case of some large projects the Government sometimes contacts a company direct. In these situations, foreign companies are often used because of the need for specific experience. The projects are often combined with financial arrangements, for example, joint venture between a foreign and a Greek company.

In the private sector, registration with the Technical Chamber of Greece is required. There is no compulsory legal requirement for projects in the private sector to be advertised. However, some clients procure projects more or less on the basis of the same guidelines as hold good for public authorities. Very few, and usually only highly specialized projects, are advertised in daily newspapers or technical periodicals.

The best way for engineering consultants to receive information about forthcoming projects is through contacts. As a general rule only invited consulting engineers are allowed to bid for a project. In a typical private sector project the designer is employed to provide both architectural and engineering services. This includes briefing, definition of the scope of the project, primary design, sketch design, working drawings, final planning and supervision of construction.

Some of these arrangements are changing to bring them in line with European Union directives.

Contractual arrangements

In Greece there are few registered companies and a large number of unregistered sub-contractors. Most of the significant projects are undertaken by a few contractors, who carry a licence by the Government to do the work, usually on a lump sum basis. Greek legislation promotes construction projects by giving concessions and incentives to the contractor. Recently, most important infrastructure projects such as power stations, water and gas installations have benefited from this particular arrangement. The idea is to attract foreign capital to provide funds for infrastructure projects. This method is also used because the former public sector has been largely privatized. New private bodies have been created to create competition and improve both the provision of services and products.

Greek construction contracts are not based on fixed standard forms of terms and conditions, but are closely related to the tender documents. Where the contracts relate to the execution of public works there is a very close relationship between the tender conditions and the specific legal provisions contained within the public works law. Nevertheless the specific nature of different projects and the method of award make the existence of differences between the legislation and the terms and conditions of any contract very likely. Sometimes, very significant contracts must be approved by the Greek Parliament and ratified by law. The nature of the contract and the responsibilities of the parties do not change, but in this way the contractor is protected by a formal recognition and acceptance of the terms and conditions of the contract.

Following completion by the consulting engineers of design and specifications and its checking by the public sector client, the project goes to tender. Tenders for government contracts in Greece are usually invited by public advertisement. The rules governing public works contracts are set out in 'Law 1286/6.10.72', 'Presidential Decree 475/76' and 'amendment 724/79'. Tender documents usually comprise:

- drawings
- general specifications
- technical specifications or technical description
- work item description and unit price list
- schedule of quantities or cost estimate
- form of tender.

In the case of building contracts, prices included in the unit price list are normally determined by the tendering contractor, whilst for civil engineering projects the contractor states only his required percentage adjustment to prices already established by the client. The submitted tenders are subject to the 'uniformity of pricing test', *Omalotita Timon*, which restricts the acceptable adjustments to prices for each category of work to within 0.80 to 1.10 times the average percentage adjustment for the overall project. For public works contracts, fluctuations in prices are governed by procedures included in 'Law 889/79'. Quarterly indices for these purposes are published by the Ministry of Public Works, the base quarter being the calendar quarter in which the contract is signed.

Tenders received are considered by a committee comprising the architects, design engineers and the client's representatives, who will assess the technical and cost implications of the bid.

The bidder is allowed to offer an unrestricted percentage discount, which sometimes reaches a discount of 75%. The law does not foresee any discount restriction, and hence the projects are awarded to the bidder who offers the biggest discount. Recently, a new method has been applied by law, which uses a formula which is expected to produce the best offer by averaging all offers. This new system should avoid the degrading of the quality of both work and materials used for it. Most of the work lacks detailed specification and this, in turn, leads to higher financial expenses and delays.

The 'Construction Management Contract' method of contracting is usually used in public works of moderate size, or for hotels, shopping centres, office blocks etc. The client chooses a construction company for the project, which will usually assign the design and specification process to a consulting engineer. After completion and approval of this by the client, the construction company takes on the role and responsibilities of the contractor. This method usually means more rapid completion of the project, but not necessarily at the lowest cost. It is estimated that 20% of projects are carried out in this way.

Private building contracts are not subject to the same degree of regulation as public works. For private work many contracts are negotiated. Contractors need not be licensed and the documents supplied to tenderers may vary considerably from those regarded as typical for public works. In some cases the documentation is very limited. Negotiations will normally take place with the tenderers, so the contract is not automatically awarded to the tenderer who submits the lowest initial bid. For larger private sector contracts, the client takes into consideration previous experience with the contractor, the company's financial profile, other projects currently being undertaken by the contractor and the price tendered.

All construction work must be supervised by a licensed engineer, whose duties also include the recording of the condition and extent of any part of the works that would subsequently be concealed from view. The supervising engineer also has the authority to make interim and final payments. Quantities of 'as built' work are ascertained, and used by the contractor to prepare final cost sheets which are subject to the approval of the supervising engineer. In the event of variations creating new items of work, prices are generally calculated in accordance with a set of rules issued by the Ministry of Public Works for the costing of building works.

There is no provision for quality tests to be executed by independent laboratories and the state owned laboratories are very few. Normally tests are undertaken by the contractors' laboratories and are often doubted by the supervisors. Sometimes, tests are undertaken regularly and according to the right procedures but they cause delay and are often available only after the part of the work under scrutiny has been finished. This usually results in the owner of the project imposing penalties or reducing the payment to the contractor on grounds of a lack of systematic checking of the quality

A performance bond is usually required and the client normally retains 10% of the contract sum until the works are complete. After completion of the works a warranty guarantee bond is held until the expiry of the maintenance period (usually 12 months) during which time the contractor makes good any defects.

Liability

Liabilities are defined by the Civil Code, by 'law 696' of 1974 and 'decree no. 723' of 1979. Liability for architects is limited to 10 years after handover. Insurance is not obligatory for architects. The contractor carries liability for any damage, defects in construction, type and quality of materials and for all employees' activities.

Development control and regulations

Current planning legislation is covered by three main laws of 1923, 1979 and 1983, each of which extends and amends the previous laws. Since 1983 the planning programme comprises:

- General Town Plans and Implementation Town Plans
- Master Plans for Athens and Salonika
- planning and rehabilitation of main coastal resort areas
- defining of boundaries and building regulations for about 10,000 rural settlements
- planning and extension of 1,800 rural settlements
- reports for 51 prefectures.

Planning may be undertaken by central government or local authorities and may be assigned by them to private planning offices. In practice, facilities in the local authorities are insufficient to cope with the requirements which include consultation. Any new building or any renovated building extended by more than 2% of the original structure must be given a 'works permit' by the appropriate local authority. The works permit is issued if all drawings and specifications conform to existing regulations. The local planning authority is responsible for checking that consequent work conforms to the permitted work.

The new planning programme is slowly being implemented but neither the industry nor the people are used to the system, especially the consultation procedures. There are substantial penalties for illegal development but nevertheless some illegal housing construction takes place.

The 'General Building Regulations' were issued in 1985 and partly updated in 1988 by the Ministry of Public Works and the Ministry of Industry, Energy and Technology. The regulations cover health and safety, the soundness of buildings, insulation, etc. In addition there is a 'Regulation for Buildings (59/3289)' covering all elements of construction, safety requirements and building services requirements, including fire protection.

Greek standards are almost entirely in accordance with the European and American standards, (e.g. DIN, VDE, and ISO). Some public companies (e.g. 'Astrofos Engineering') have their own standards which are a mixture of the above.

Exchange rates and cost and price trends

The combined effect of exchange rates and inflation on prices within a country and on price comparisons between countries is discussed in section 2.

Exchange rates

The graph on the next page plots the movement of the Greek drachma against sterling, the ECU/Euro, the US dollar and 100 Japanese yen since 1990. The values used for the graph are quarterly and the method of calculating these is described and general guidance on the interpretation of the graph provided in section 2. The average exchange rate in the first quarter of 1999 was Dr 468 to the pound sterling, Dr 281 to the US dollar and Dr 329 to the Euro

The Greek drachma against sterling, the ECU/Euro, the US dollar and 100 Japanese yen

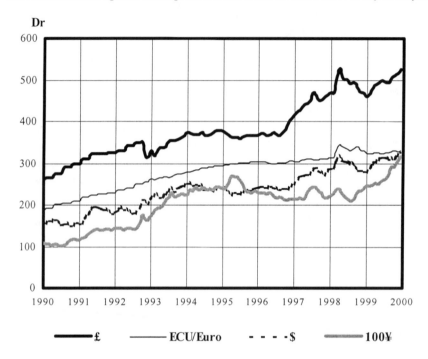

Cost and price trends

The table following presents the indices for consumer prices, residential building costs and material costs in Greece since 1990, rebased to 1990=100. The annual change is the percentage change between the average index of consecutive years. Notes on inflation are included in section 2.

Consumer price, labour and building materials cost indices

Year	Consumer prices annual average	change %	Residential building costs annual average	change %	Building materials costs annual average	change %
1990	100.0	20.3	100.0	22.0	–	–
1991	119.5	19.5	119	19.0	–	–
1992	138.4	15.8	133	11.8	–	–
1993	158.4	14.5	150	12.8	150.2	
1994	175.7	10.9	163	8.7	161.0	7.2
1995	191.4	8.9	175	7.4	172.9	7.4
1996	207.1	8.2	186	6.3	185.7	7.4
1997	218.6	5.6	–	–	198.4	6.8
1998	228.9	4.7	–	–	–	–

Sources: International Monetary Fund, Yearbook 1998 and supplements 1999.
European Mortgage Federation, Hypostat 1986 to 1996.
Ministry of National Economy – National Statistical Service.

Useful addresses

Government and public organizations

Ministry of Environment, Planning and Public Works
17 Amaliados Str
Athens
Tel: (30-1) 6449113

National Statistical Service of Greece (ESYE)
14–16 Lykourgou Str
10166 Athens
Tel: (30-1) 3289389
Fax: (30-1) 3244748/3241102

General Secretariat of Public Works, Department of Building and Construction
1 P. Tsaldari
Kallithea
Athens

Trade and professional associations

Association of Greek Architects
3 Ipitov Street
Athens

Technical Chamber of Greece
4 Karageorgi Servias Street
10248 Athens
Tel: (30-1) 3254591
Fax: (30-1) 3223145

Hellenic Association of Consulting Firms
106 Themistocleavs Str
10557 Athens
Tel: (30-1) 3616008
Fax: (30-1) 3615043

Association des Architectes Diplomés
Association of Qualified Architects
15 Vrisakiou & Kladou
10555 Athens
Tel/Fax: (30-1) 3215147

Panhellenic Union of Iron and Industrial Materials Merchants
15 Nikou Street
10560 Athens
Tel: (30-1) 3215733
Fax: (30-1) 3211294

Athens Industrial Chamber of Commerce
7 Akademias Street
10671 Athens
Fax: (30-1) 3607897

Association Panhellenique des Ingénieurs Diplomés Entrepreneurs de Travaux Publics
Panhellenic Association of chartered engineers and public works contractors
23 rue Asklipiou
144 Athens
Tel: (30-1) 3614978
Fax: (30-1) 3641402

Other organization

**Hellenic Organization for
Standardization (ELOT)**
313 Acharnon
11145 Athens
Tel: (30-1) 2015025
Fax: (30-1) 2025917

Paterson Kempster & Shortall

Chartered Quantity Surveyors
24 Lower Hatch Street
Dublin 2
Ireland
Tel: 00353 1-6763671
Fax: 00353 1-6763672
E-mail: pks@pksdub.ie
Web Site: http:// www.pksdub.ie

Irish Associate of:

Davis Langdon & Seah International

Ireland

All data relate to 1998 unless otherwise indicated.

Population

Population	3.6 million
Urban population (1997)	58%
Population under 15	17%
Population 65 and over	16%
Average annual growth rate (1985 to 1998)	0.2%

Geography

Land area	70,280 km^2
Agricultural area	81%
Capital city	Dublin

Economy

Monetary unit	Punt (IR£)
Exchange rate (average first quarter 1999) to:	
the pound sterling	IR£ 1.12
the US dollar	IR£ 0.67
the Euro	IR£ 0.79
the yen x 100	IR£ 0.57
Average annual inflation (1995 to 1998)	1.8%
Inflation rate	2.4%
Gross Domestic Product (GDP)	IR£ 54.5 billion
GDP PPP basis (1997)	$ (PPP) 59.9 billion
GDP per capita	IR£ 15,040
GDP per capita PPP basis (1997)	$ (PPP) 16,590
Average annual real change in GDP (1995 to 1998)	9.1%
Private consumption as a proportion of GDP	52%
Public consumption as a proportion of GDP	14%
Investment as a proportion of GDP	20%

Construction

Gross value of construction output (1997)	IR£ 9.4 billion
Gross value of construction output per capita	IR£ 2,610
Gross value of construction output as a proportion of GDP	17.2%
Average annual real change in gross value of construction output (1995 to 1998)	13.4%
Annual cement consumption	2.8 million tonnes

PPP Purchasing power parity

The construction industry

Construction output

The value of gross output of the construction sector in Ireland in 1998 was about IR £9.4 billion equivalent to 11.9 billion ECUs (nearly the same as January 1999 Euros). No information is available on the amount of work done by other sectors, the black economy or DIY. The breakdown by type of work is shown in the table below:

Output of construction sector, 1998 (current prices)

Type of work	IR£ million	ECU'S billion	% of total
New work			
Residential building	3.3	4.2	35
Non-residential building			
Office and commercial	1.1	1.4	12
Industrial	0.6	0.8	7
Other	0.6	0.8	7
Total	2.4	3.0	25
Civil engineering	1.2	1.5	13
Total new work	6.8	8.6	73
Repair and maintenance			
Residential	1.7	2.1	18
Non-residential	0.6	0.7	6
Civil engineering	0.3	0.4	3
Total repair and maintenance	2.6	3.3	27
Total	**9.4**	**11.9**	**100**

Note: the value of the Euro in January 1999 was very similar to that of the ECU in 1998.
Source: based on Hughes,.A., 'Ireland' in Euroconstruct conference paper, Prague, June 1998.

Ireland has the highest level of construction output as a percentage of GDP in the EU. It has enjoyed very high levels of growth of construction output for many years. From 1995 to 1998 growth was 13.4% – again the highest of any country in the EU. The highest growth in 1998 was in civil engineering but this sector has for long had a very low percentage of total construction output. Every sector in 1998 enjoyed double digit growth in new work. Government is concerned that the construction industry is overheating; prices have risen substantially and there are skill shortages.

The residential sector grew by about 50% from 1995 to 1998. Some supply constraints, in terms of skill shortages and lack of land, may appear in some areas driving up prices. House prices have risen dramatically. Stable growth in GDP and low interest rates will be the engine of further growth. The residential sector has the largest share of construction output and the proportion of owner-occupation is very high compared to other European countries. High economic growth and employment levels, immigration, low but increasing property prices, a finance system depending on building societies and government subsidies and tax relief are the major reasons for the strong residential output in Ireland. In 1998 78% of housing constructed was in once or two family dwellings as opposed to flats.

The private non-residential sector grew by more than 75% from 1995 to 1998, due to strong growth in the industrial and office sectors, but growth may now be slowing. Agriculture is in a state of recession and construction levels are falling. Public non-residential output grew strongly, due to new investment in hospitals and education.

New civil engineering output grew by about 15% in 1998 rather less than in 1997. Telecommunications programmes are the major contributors to the growth of the sector. There is a backlog of work for roads and water supply and the increase in civil engineering work may well continue.

Repair and maintenance has increased but at a low rate compared to new work.

The geographical distribution of construction output in 1996 compared with that of population is shown below. In general construction output has followed population.

Population and construction output 1996 regional distribution

Regions	Population (%)	Construction output (%)
Border	11.0	10.0
Dublin	29.0	31.0
Mid-East	10.0	12.0
Midland	6.0	5.0
Mid-West	9.0	9.0
South-East	10.0	10.0
South-West	15.0	13.0
West	10.0	10.0
Total	**100.0**	**100.0**

Sources: Construction Industry Review'97 Outlook '98.

Characteristics and structure in the industry

The main engineering organization in Ireland is the Institution of Engineers of Ireland (IEI) with a membership of about 6,000, of whom half are chartered engineers or fellows. Engineers occupy key positions of authority in central, and local government and in industry. Up to 50% are employed in the public sector.

The organization for architects is the Royal Institute of the Architects of Ireland (RIAI). The title of architect is protected by law.

Almost all quantity surveyors are members of the Society of Chartered Surveyors in Ireland which offers them full membership of the Royal Institution of Chartered Surveyors in the UK. Many professionals have trained or gained experience in the UK or USA.

There are about 4,000 building and civil engineering contracting firms in Ireland. The majority of them are of medium and small size with working proprietors who also act as site agents and general foremen. In the whole country there are probably not more than ten contractors with the resources and experience to manage and construct major projects in excess of say IR£10.0 million. These contractors have shed directly employed tradesmen in recent years but retained key management personnel which has led to a growth of subcontracting firms.

Generally, roadbuilding maintenance is carried out by local authorities employing direct labour, and relying on plant hire for the provision of equipment, in order to avoid tying up capital during slack periods. All newbuild roads are tendered.

In 1998 no Irish companies were listed in *Building* magazine's 'Top 300 European Contractors'. 'John Sisk & Son was number 80 in the *Engineering News Record's* list of 'Top 225 International Contractors' in 1997. The table below shows the principal Irish contractors and their main office location.

Major Irish contractors 1997

Major contractors	Head office
John Sisk & Son Ltd.	Dublin
Ascon Ltd.	Dublin
Pierse Contracting Ltd.	Dublin
P.J. Hegarty & Sons	Cork
G&T Crampton Ltd.	Dublin
John Paul & Co. Ltd.	Dublin
P.J. Walls Holdings & Co. Ltd.	Dublin
SIAC Construction LTd.	Dublin
Michael Mc Namara & Co.	Dublin
P. Elliott & Co. Ltd.	Cavan

Sources: Construction Industry Review'97 Outlook '98.

Clients and finance

Around 60% of the output of the Irish construction industry is privately funded. Government expenditure is increasing with funds from the European Union.

Around 90% of housing is financed privately. Central Government also plays a part in private housing, with new housing grants and water and sewerage grants. The proportion built by local authorities has been falling rapidly. In 1983 it was 25% but by 1988 it had fallen to 7%. By 1994, it had however risen again to 13%. The stock of local authority housing requires a large expenditure by local authorities in repair and maintenance, accounting for around a quarter of all housing maintenance.

Selection of consultants

For all major state funded projects the appointment of consultants is made after public advertisement. For projects exceeding the EU limit such advertisements are included in the EU Journal.

Following detailed technical submissions a short list of consultants is interviewed. Subsequently negotiations with regard to fees are entered into with the preferred consultant and, if a satisfactory agreement cannot be reached, such negotiations are continued with the second and third preferred consultant.

For the private sector the client or project manager normally appoints consultants on the basis of experience and personal contact. Consultants include an architect, structural, mechanical and electrical services engineers and a quantity surveyor. Only a few private employers and some public agencies use in-house staff. Consulting engineers undertake all design and cost control for civil engineering projects.

Developer competitions have emerged as an alternative method of procuring some state buildings, including decentralized government offices and new civic offices. Bids are invited from developers/contractors who are responsible for the design, construction and, in some cases, financing of the project.

On some recent projects a design competition has been held and the construction carried out in a traditional manner or by 'novation' with a developer/contractor adopting the winning design as part of a 'design and construct' package.

Contractual arrangements

In the past, contractual procedures in Ireland were similar to those traditional in the UK. Now as in the UK there are many different procedures. Project managers are often used. Construction procedures are also increasingly influenced by foreign systems as foreign firms increase their importance in the market.

Tendering procedures are governed by a Code of Practice published by the Liaison Committee and are strictly policed by the Construction Industry Federation (CIF) which represents most contractors and sub-contractors. Most major contracts are based on bills of quantities issued to a selected list of contractors – four to six in the private sector but increasingly eight to ten for public authority tenders. Bills are submitted with tenders. Tenders are usually based on fluctuations but fixed price contracts are often negotiated after receipt of tenders.

The nationally accepted basis for the presentation of cost information is the 'National Building Elements' published by the Irish testing and research organization *An Foras Forbartha*, now known as *'Forbairt'*, and broadly based on CI/Sfb. Estimates for public sector contracts must be based on the 'National Building Elements' and they are widely used in the private sector.

Most estimating is carried out by surveyors who are members of the Society of Chartered Surveyors (SCS). The SCS publishes cost analyses of existing buildings for its members and a Building Cost Index which is used throughout the industry. Most major quantity surveying firms also keep their own building cost data bank, based on cost information from current and previous projects. Moreover, surveyors prepare tender documents and advise the architect.

The CIF and SCS have jointly produced an 'Agreed Rules of Measurement' for use in Ireland: SMM7 and the 'UK Common Arrangement' system are not used in Ireland.

Sometimes multinational construction managers supervise the main contractor in a project. The Liaison Committee, which consists of representatives of architects, surveyors and contractors, drafts conditions of contract, with or without quantities, which are published by the RIAI. There is not a standard form of contract for management contracting or design and build. Usually, particular conditions are drafted in the framework of a traditional standard form. Sometimes UK based contracts are adopted, although fewer variations are permitted compared to an UK contract.

The Construction Industry Federation (CIF) form is used if a private client builds a house. The form is issued by the solicitors' and contractors' institutes and confirms that all works have been completely designed and specified. Payment is made on the basis of interim payments, at different stages of work. Additional work may be required and the contractor may ask extra payment for the changes before they take place. The CIF form contains an arbitration clause whereby issues minor defects are settled by an independent expert.

The GDLA82 form of contract is used for all construction work undertaken for government and for work paid from public funds. The Department of Finance must approve it, after consultation with the contractors', architects' and surveyors, institutes. There are two versions of the contract. In the first one detailed priced bills of quantities are part of the contract. In the second one quantities are not included in the contract.

The architect and specialist engineers design and supervise the works. The architect on the advice of a quantity surveyor certifies all payments. Any issue about variations, cost increases, payments, subcontracting, delays, defects, etc. can require detailed negotiation. The main contractor is responsible for the sub-contractors but if the sub-contractor defaults on debt before completion of work the client pays for additional costs for getting a new sub-contractor.

Contractors have the right to demand interest on liquidated amounts but not on late payments or delayed certification.

The main form of RIAI 1988 is used for most private construction contracts apart from single houses or house improvements contracts. The architects' institute along with the contractors' and surveyors' institutes issue the contract. There are two versions of the contract. In the first, detailed priced bills of quantities are part of the contract. In the second, quantities are not included in the contract. It is very similar to the GDLA 82 form, although some differences exist. Under this contract, if the main contractor defaults, the client may employ another contractor and may deduct the costs of so doing from the amounts due to the previous contractor. However this power does not extend to other contracts, unlike the GLDA 82.

According to the RIAI 1988 form, if materials and plant are brought onto site they still belong to the contractor, unlike from the GLDA 82 which states that if materials and plant are brought onto site they are immediately considered to be the property of the client.

Another difference is that the RIAI 1988 has no restriction about when cost increases may be applied, while in the GLDA 82 only cost increases prior to the completion of the contract or any date stated by the employer are considered.

The RIAI SF 1988 is used for small one-off projects, such as house and small shop alterations. It allows for flexibility.

The CIF/SCSA 1989 is intended for sub-contracts. Two types are used: the first one is related to the RIAI 1988 while the second one refers to the GLDA 82. The contractors' institute, along with the sub-contractors' institute, issue the contract. According to this contract, sub-contractors signing the form are entitled to provisions. The main contractor has to hold monies on trust for the sub-contractors and pass them on to them, after they have been appointed. Payments are due after 14 days of the receipt by the contractor of the certificate of the architect or engineer.

The RIAI 1988 contract is adopted where the client wants to have a direct contract with a nominated sub-contractor. The architects' along with the surveyors' and contractors' associations issue the contract. The form is supposed to safeguard the client in the event that a sub-contractor does not perform his work correctly or to safeguard the sub-contractor against the main contractor not paying him. This type of contract is not widely used, because of the additional administrative costs involved with its use.

Development control and standards

The Local Government (Planning and Development) Acts 1963 to 1990, together with supporting legislation, provide the legislative framework for planning control. The local authorities are responsible for planning permission for property development and construction within their area, and for the enforcement of planning control. Local Authorities must produce a development plan for their area, reviewed at least every five years. This outline plan gives guidelines for the control of future development in the area. Planning decisions must be made within two months of the original application.

The design and construction of buildings – including significant alterations to existing buildings – is governed by the Building Regulations of 1992 (enacted under the Building Control Act 1990). These regulations replaced the old system of bye-law control. The regulations cover such issues as site suitability, stability and safety of structures and the fire safety of buildings and structures. They provide for a system of self-certification of compliance. This allows certificates of compliance to be submitted to the Building Control Authority (usually the local authority) prior to the completion of work. The Building Control Authority is responsible for the issue of certificates relating to overall building control approval and fire safety approval.

The National Standards Authority of Ireland (NSAI) draws up and promulgates standard specifications for building products, and it is normal for manufacturers to comply. Architects and other building professionals generally have regard to Irish Standards and Codes of Practice, or, in their absence, British Standards or Codes of Practice. The Construction Industry Federation operates a National Housebuilding Guarantee Scheme, which guarantees houses built by contractors registered under the scheme against major structural defects for six years.

Construction cost data

Cost of labour

The figures opposite are typical of labour costs in the Dublin area as at the first quarter 1999. The wage rate is the basis of an employee's income, while the cost of labour indicates the cost to a contractor of employing that employee. The difference between the two covers a variety of mandatory and voluntary contributions - a list of items which could be included is given in section 2.

	Wage rate (per hour) IR£	Cost of labour (per hour) IR£	Number of hours worked per year
Site operatives			
Mason/bricklayer	6.90	13.70	1,872
Carpenter	6.90	13.70	1,872
Plumber	6.90	13.70	1,872
Electrician	8.00	15.50	1,872
Structural steel erector	6.90	13.70	1,872
HVAC installer	7.00	13.80	1,872
Skilled labour	6.30	12.20	1,872
Semi-skilled worker	5.90	11.70	1,872
Unskilled labourer	5.50	11.00	1,872
Equipment operator	6.40	12.60	1,872
Site Supervision			
General foreman	9.30	18.00	1,872

Cost of materials

The figures that follow are the costs of main construction materials, delivered to site in the Dublin area, as incurred by contractors in the first quarter of 1999. These assume that the materials would be in quantities required for a medium sized construction project and that the location of the works would be neither constrained nor remote. All the costs in this section exclude value added tax (VAT) which is at 12.5% for general building work.

	Unit	Cost IR£
Cement and aggregates		
Ordinary portland cement in 50kg bags	tonne	98.90
Coarse aggregates for concrete	tonne	12.40
Fine aggregates for concrete	tonne	12.40

	Unit	Cost IR£
Ready mixed concrete (mix 11.5kN)	m³	49.40
Ready mixed concrete (mix 30kN)	m³	55.60
Steel		
Mild steel reinforcement	tonne	330.00
High tensile steel reinforcement	tonne	330.00
Structural steel sections	tonne	422.00
Bricks and blocks		
Common bricks (215 x 102.5 x 65mm)	1,000	196.00
Good quality facing bricks (215 x 102.5 x 65mm)	1,000	330.00
Medium quality facing bricks	1000	237.00
Hollow concrete blocks (440 x 225 x 215mm)	1,000	700.00
Solid concrete blocks (440 x 215 x 102.5mm)	1,000	443.00
Precast concrete cladding units with exposed aggregate finish	m²	160.00
Timber and insulation		
Softwood sections for carpentry	m³	237.00
Softwood for joinery (Red Deal)	m³	330.00
Hardwood for joinery (Iroko)	m³	686.00
Exterior quality plywood (18mm)	m²	9.43
Plywood for interior joinery (18mm)	m²	9.43
Softwood strip flooring (22mm) tongue and grooved	m²	6.86
Chipboard sheet flooring (22mm)	m²	3.51
100mm thick quilt insulation	m²	1.39
100mm thick rigid slab insulation	m²	7.67
Softwood internal door complete with frames and ironmongery	each	113.00
Glass and ceramics		
Float glass (6mm)	m²	18.50
Sealed double glazing units	m²	63.90
Good quality ceramic wall tiles (150 x 150mm)	m²	18.80
Plaster and paint		
Plaster in 50kg bags	tonne	118.00
Plasterboard (12.5mm thick)	m²	2.27
Emulsion paint in 5 litre tins	litre	2.67
Gloss oil paint in 5 litre tins	litre	4.36
Tiles and paviors		
Precast concrete paving slabs (600 x 600 x 50mm)	m²	8.00
Clay roof tiles	1,000	474.00
Precast concrete roof tiles	1,000	639.00
Drainage		
WC suite complete	each	92.70
Lavatory basin complete	each	31.90
100mm diameter clay drain pipes	m	6.18

Unit rates

The descriptions of work items opposite are generally shortened versions of standard descriptions listed in five languages (English, French, Italian, German and Spanish) in Appendix 3.

Where an item has a two digit reference number (e.g. 05 or 33), this relates to the full description against that number in Appendix 3. Where an item has an alphabetic suffix (e.g. 12A or 34B) this indicates that the standard description has been modified. Where a modification is major the complete modified description is included here and the standard description should be ignored.

Where a modification is minor (e.g. the insertion of a named hardwood) the shortened description has been modified here but, in general, the full description in Appendix 3 prevails.

The unit rates below are for main work items on a typical construction project in the Dublin area in the first quarter of 1999. The rates include labour, materials and equipment and an allowance for contractor's overheads and profit (5%). No allowance has been made to cover contractor's profit and attendance on specialist trades (typically 5%) or preliminary and general items (typically 9%). All the rates in this section exclude VAT which is at 12.5% for general building work and 21% for professional fees and fittings.

		Unit	Rate IR£
Excavation			
01	Mechanical excavation of foundation trenches	m³	6.70
02	Hardcore filling making up levels (Clause 804)	m³	17.70
Concrete work			
04	Plain insitu concrete in strip foundations in trenches	m³	82.40
05	Reinforced insitu concrete in beds	m³	91.00
06	Reinforced insitu concrete in walls	m³	98.40
07	Reinforced insitu concrete in suspended floor or roof slabs	m³	96.30
08	Reinforced insitu concrete in columns	m³	98.40
09	Reinforced insitu concrete in isolated beams	m³	96.30
10	Precast concrete slab	m²	35.90
Formwork			
11	Softwood formwork to concrete walls	m²	20.30
12	Softwood or metal formwork to concrete columns	m²	20.30
13	Softwood or metal formwork to horizontal soffits of slabs	m²	22.50
Reinforcement			
14	Reinforcement in concrete walls (16mm)	tonne	482.00
15	Reinforcement in suspended concrete slabs	tonne	482.00
16	Fabric reinforcement in concrete beds	m²	2.40

		Unit	Rate IR£
Steelwork			
17	Fabricate, supply and erect steel framed structure	tonne	1,550.00
Brickwork and blockwork			
18	Precast lightweight aggregate hollow concrete block walls (100mm thick)	m^2	18.70
19	Solid (perforated) concrete blocks (100mm thick)	m^2	18.70
21	Facing bricks	m^2	58.90
Roofing			
22	Concrete interlocking roof tiles	m^2	27.80
23	Plain clay roof tiles	m^2	27.80
24	Fibre cement roof slates	m^2	24.60
25	Sawn softwood roof boarding	m^2	21.40
26	Particle board roof covering	m^2	19.30
27	3 layers glass-fibre based bitumen felt roof covering include chippings	m^2	21.40
28	Bitumen based mastic asphalt roof covering	m^2	37.50
29	Glass-fibre mat roof insulation	m^2	9.60
30	Rigid sheet loadbearing roof insulation 75mm thick	m^2	11.20
31	Troughed galvanised steel roof cladding	m^2	20.90
Woodwork and metalwork			
32	Preservative treated sawn softwood 50 x 100mm	m	4.80
33	Preservative treated sawn softwood 50 x 150mm	m	6.40
34	Single glazed casement window in Iroko hardwood, size 650 x 900mm	each	182.00
35	Two panel glazed door in Iroko hardwood, size 850 x 2000mm	each	380.00
36	Solid core half hour fire resisting internal flush door	each	428.00
37	Aluminium double glazed window, Iroko hardwood size 1200 x 1200mm	each	385.00
38	Aluminium double glazed door, Iroko hardwood size 850 x 2100mm	each	856.00
39	Hardwood skirtings in Iroko hardwood	m	11.80
40	Framed structural steelwork in universal joist sections	tonne	1,610.00
41	Structural steelwork lattice roof trusses	tonne	1,710.00
Plumbing			
42	UPVC half round eaves gutter	m	12.70
43	UPVC rainwater pipes	m	11.20
44	Light gauge copper cold water tubing	m	10.40
45	High pressure UPVC cold water supply	m	9.10
46	Low pressure UPVC cold water distribution	m	9.10
47	UPVC soil and vent pipes	m	12.70
48	White vitreous china WC suite	each	161.00
49	White vitreous china lavatory basin	each	169.00
51	Stainless steel single bowl sink and double drainer	each	303.00

		Unit	Rate IR£
Electrical work			
52	PVC insulated and PVC sheathed copper cable core and earth	m	3.60
53	13 amp unswitched socket outlet	each	19.20
54	Flush mounted 20 amp, 1 way light switch	each	25.70
Finishings			
55	2 coats gypsum based plaster on brick walls	m^2	11.80
56	White glazed tiles on plaster walls	m^2	39.60
57	Red clay quarry tiles on concrete floor	m^2	37.50
58	Cement and sand screed to concrete floors 50mm thick	m^2	9.10
59	Thermoplastic floor tiles on screed	m^2	21.40
60	Mineral fibre tiles on concealed suspension system	m^2	24.10
Glazing			
61	Glazing to wood	m^2	48.20
Painting			
62	Emulsion on plaster walls	m^2	3.80
63	Oil paint on timber	m^2	4.60

Approximate estimating

The building costs per unit area given on the opposite page are averages incurred by building clients for typical buildings in the Dublin area as at the first quarter 1999. They are based upon the total floor area of all storeys, measured between external walls and without deduction for internal walls.

Approximate estimating costs generally include mechanical and electrical installations but exclude furniture, loose or special equipment, and external works; they also exclude fees for professional services. The costs shown are for specifications and standards appropriate to Ireland and this should be borne in mind when attempting comparisons with similarly described building types in other countries. A discussion of this issue is included in section 2. Comparative data for countries covered in this publication, including construction cost data, are presented in Part Three.

Approximate estimating costs must be treated with caution; they cannot provide more than a rough guide to the probable cost of building. All the rates in this section exclude VAT which is at 12.5% for general building work and 21% for professional fees and fittings.

	Cost m² IR£	Cost ft² IR£
Industrial buildings		
Factories for letting (include lighting, power and heating)	385	36
Factories for owner occupation (light industrial use)	439	41
Factories for owner occupation (heavy industrial use)	492	46
Factory/office (high-tech) for letting (shell and core only)	492	46
Factory/office (high-tech) for letting (ground floor shell, first floor offices)	578	54
Factory/office (high-tech) for owner occupation (controlled environment, fully furnished)	696	65
High tech laboratory (air conditioned)	872	81
Warehouses, low bay (6 to 8m high) for letting (no heating)	294	27
Warehouses, low bay for owner occupation (including heating)	385	36
Warehouses, high bay for owner occupation (including heating)	460	43
Cold stores/refrigerated stores	524	49
Administrative and commercial buildings		
Civic offices, non air conditioned	990	92
Civic offices, fully air conditioned	1,150	107
Offices for letting, 5 to 10 storeys, non air conditioned	749	70
Offices for letting, 5 to 10 storeys, air conditioned	856	80
Offices for owner occupation, 5 to 10 storeys, non air conditioned	990	92
Offices for owner occupation, 5 to 10 storeys, air conditioned	1,150	107
Prestige/headquarters office, 5 to 10 storeys, air conditioned	1,500	139
Health and education buildings		
General hospitals	1,200	111
Private hospitals	1,770	164
Health centres	910	85
Nursery schools	669	62
Primary/junior schools	669	62
Secondary/middle schools	669	62
University (arts) buildings	1,150	107
University (science) buildings	1,660	154
Management training centres	1,340	124
Recreation and arts buildings		
Theatres (over 500 seats) including seating and stage equipment	1,390	129
Theatres (less than 500 seats) including seating and stage equipment	1,260	117
Sports halls including changing and social facilities	696	65
Residential buildings		
Social/economic single family housing (multiple units)	749	70
Private/mass market single family housing 2 storey detached/semidetached (multiple units)	562	52
Purpose designed single family housing 2 storey detached (single unit)	803	75
Social/economic apartment housing, low rise (no lifts)	856	80

	Cost m² IR£	Cost ft² IR£
Social/economic apartment housing, high rise (with lifts)	910	85
Private sector apartment building (standard specification)	803	75
Private sector apartment building (luxury)	910	85
Student/nurses halls of residence	963	90
Homes for the elderly (shared accommodation)	910	85
Homes for the elderly (self contained with shared communal facilities)	963	90
Hotel, 5 star, city centre	1,610	150
Hotel, 3 star, city/provincial	1,280	119
Motel	1,020	95

Regional variations

The approximate estimating costs are based on average Dublin rates. Adjust these costs by the following factors for regional variations:

Cork	−2% to −3%	Midlands	−2% to − 3%
Galway	0%	Dundalk	−2% to −3%
Limerick	−1%		

Value added tax (VAT)

The standard rate of value added tax (VAT) is currently 21%, chargeable on fittings and professional fees. A special rate of 12.5% applies to general building work.

Exchange rates and cost and price trends

The combined effect of exchange rates and inflation on prices within a country and on price comparisons between countries is discussed in section 2.

Exchange rates

The graph on the following page plots the movement of the Irish punt against sterling, the ECU/Euro, the US dollar and 100 Japanese yen since 1990. The values used for the graph are quarterly and the method of calculating these is described and general guidance on the interpretation of the graph provided in section 2. The average exchange rate in the first quarter of 1999 was IR£ 1.12 to the pound sterling, IR£ 0.67 to the US dollar and IR£ 0.79 to the Euro.

The Irish punt against sterling, the ECU/Euro, the US dollar and 100 Japanese yen

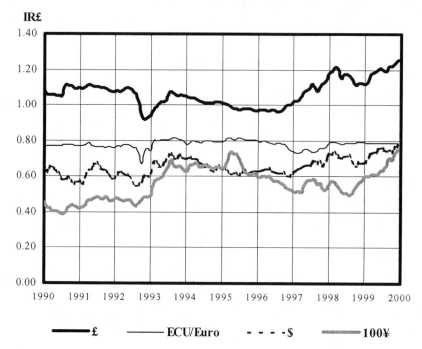

Cost and price trends

The table on the next page presents the indices for consumer prices, house building costs and building material and labour costs in Ireland since 1990. All the indices have been rebased to 1990=100. The annual change is the percentage change between the last quarter of consecutive years. Notes on inflation are included in section 2.

The house building cost index relates to a public sector standard house in Dublin. It is based on costs of materials and labour which together do not exceed 65% of the total price of the house.

The labour and materials cost index estimates the cost of construction by combining a price of materials index with hourly wage rates from a survey of earnings and hours worked. It is also known as the 'capital goods price index.'

Construction costs rose faster than consumer prices in 1997 and 1998 partly because of the boom in construction. Indeed some sources suggest that the rise in all construction costs has been greater than shown by labour and material indices above.

Consumer price, housebuilding and material and labour costs indices

Year	Consumer prices annual average	change %	Housebuilding costs annual average	change %	Material and labour costs annual average	change %
1990	100.0	3.3	100	4.8	100	n.a
1991	103.2	3.2	104.3	4.3	102.7	2.7
1992	106.4	3.1	106.2	1.8	104.8	2.0
1993	107.9	1.4	109.0	2.6	107.5	2.6
1994	110.4	2.3	113.1	3.8	110.0	2.3
1995	113.2	2.5	116.3	2.8	113.8	3.5
1996	113.9	0.8	117.7	1.2	115.2	1.2
1997	113.3	−0.5	122.1	3.7	119.2	3.5
1998	116.0	2.4	126.7	3.8	122.5	2.7

Sources: International Monetary Fund, Yearbook 1998 and supplements 1999
Central Statistical Office.

Useful addresses

Government and public organizations

Central Statistics Office
Skehard Road
Cork
Tel: (353-21) 359000
Fax: (353-21) 359090

Department of the Environment
Custom House
Dublin 1
Tel: (353-1) 6793377
Fax: (353-1) 8742710

National Standards Authority of Ireland
Glasnevin
Dublin 9
Tel: (353-1) 8073800
Fax: (353-1) 8073838

Trade and professional associations

Royal Institute of the Architects of Ireland (RIAI)
8 Merrion Square
Dublin 2
Tel: (353-1) 6761703
Fax: (353-1) 6610948

Association of Consulting Engineers Ireland (ACE)
51 Northumberland Road
Dublin 4
Tel: (353-1) 6600374
Fax: (353-1) 6682595

The Construction Industry Federation (CIF)
Construction House, Canal Road
Dublin 6
Tel: (353-1) 4977487
Fax: (353-1) 4966953

Union of Construction Allied Trades and Technicians
56 Parnell Square West
Dublin 1
Tel: (353-1) 8731599

The Society of Chartered Surveyors in the Republic of Ireland (SCS)
5 Wilton Place
Dublin 2
Tel: (353-1) 6765500
Fax: (353-1) 6761412*

The Institution of Engineers of Ireland (IEI)
22 Clyde Road
Ballsbridge
Dublin 4
Tel: (353-1) 6684341
Fax: (353-1) 6685508

Other organization

The Building Information Centre
Dominic Court
41 Lower Dominic St
Dublin 1
Tel: (353-1) 8732329
Fax: (353-1) 8729126

Italy

All data relate to 1998 unless otherwise indicated.

Population

Population	57.5 million
Urban population (1997)	67%
Population under 15	15%
Population 65 and over	17%
Average annual growth rate (1985 to 1998)	0.1%

Geography

Land area	301,230 km^2
Agricultural area	56%
Capital city	Rome

Economy

Monetary unit	Lira (L)
Exchange rate (average first quarter 1999) to:	
the pound sterling	L 2,760
the US dollar	L 1,660
the Euro	L 1,940
the yen x 100	L 1,400
Average annual inflation (1995 to 1998)	2.4%
Inflation rate	1.7%
Gross Domestic Product (GDP)	L 2,067,000 billion
GDP PPP basis (1997)	$ (PPP) 1,240 billion
GDP per capita	L 35,940,000
GDP per capita PPP basis (1997)	$ (PPP) 21,560
Average annual real change in GDP (1995 to 1998)	1.7%
Private consumption as a proportion of GDP	63%
Public consumption as a proportion of GDP	17%
Investment as a proportion of GDP	17%

Construction

Gross value of construction output	L 200,200 billion
Gross value of construction output per capita	L 3,480,000
Gross value of construction output as a proportion of GDP	9.7%
Average annual real change in gross value of construction output (1995 to 1998)	2.0%
Annual cement consumption	35.7 million tonnes

PPP Purchasing power parity

The construction industry

Construction output

The value of gross output of the construction sector in Italy in 1998 was L 200,200 billion equivalent to 102.9 billion ECUs (nearly the same as January 1999 Euros). This represents 9.7% of GDP. In addition it is estimated that a further L 34,400 billion of construction work is built by other sectors, the black economy and DIY. This is equal to an extra 17% of construction output. The breakdown by type of work for 1998 is shown below.

Output of construction sector 1998 (current prices)

Type of work	L billion	ECUs billion	% of total
New work			
Residential building	33,500	17.2	17
Non-residential building			
Office and commercial	7,000	3.6	3
Industrial	12,300	6.3	6
Other	11,300	5.8	6
Total	30,500	15.7	15
Civil engineering	19,100	9.8	10
Total new work	83,300	42.8	42
Repair and maintenance			
Residential	58,900	30.3	29
Non-residential	33,700	17.3	17
Civil engineering	24,500	12.6	12
Total repair and maintenance	116,900	60.1	58
Total	**200,200**	**102.9**	**100**

Note: the value of the Euro in Januaryy 1999 was very similar to that of the ECU in 1998.
Source: Based on Bellicini, L., 'Italy' in Euroconstruct conference proceedings
        Prague, June 1999.

Italy has the second largest construction output in Europe, coming well below Germany but rather above France. The output of the industry has been rising slowly over the last few years reaching a 3% increase in 1998.

The new residential sector has had a period of decline but the number of units in building permits rose in 1998 which should eventually be reflected in an increase in output of residential buildings. This is helped by a fall in interest rates. About three quarters of residential building consist of flats. It is possible that new legislation will encourage social housing.

Private non-residential investment increased strongly in 1995 and 1996, due, among other things, to special tax incentives which caused companies to bring forward their construction programmes. The result is a fall in 1998. Office and commercial building continued to grow but industrial building fell. Public non-residential investment performed well, due to additional public spending for health and education.

Spending in civil engineering work enjoyed substantial growth in 1998 which took place in all sectors.: telecommunications, energy, water and miscellaneous. Further investment will come from the works for the 2000 Jubilee in Rome. Italy is a large recipient of European Investment Bank (EIB) loans which are devoted to infrastructure improvement.

Italy has a very high proportion (60%) of repair and maintenance in the output of the construction sector. Moreover, a considerable proportion of the 'informal' output is likely to be in repair and maintenance. A law of 1998 gave tax incentives for repair of residential buildings and also for certain categories of non-residential buildings. Civil engineering renovation has had a high rate of increase both in 1997 and 1998.

Construction investment is spread fairly evenly over the country, but the proportion accounted for by the public sector tends to be high in relation to population in the south of the country (the *Mezzogiorno*) despite the recent downtrend, whereas private construction expenditure is concentrated in the industrial north. However, public work is relatively fairly distributed.

The geographical distribution of building completions in 1996 compared with that of population is shown on the next page.

Population and building completions 1996 regional distribution

Regions	Population (%)	Building completions (%)
Piemonte	7.5	8.6
Valle d'Aosta	0.2	1.7
Lombardia	15.6	13.7
Trentino-Alto Adige	1.6	5.6
Veneto	7.7	8.4
Friuli-Venezia Giulia	2.1	3.1
Liguria	2.9	3.5
Emilia-Romagna	6.9	7.1
Toscana	6.1	4.6
Umbria	1.4	1.4
Marche	2.5	2.2
Lazio	9.1	10.5
Abruzzo	2.2	1.9
Molise	0.6	1.0
Campania	10.1	4.8
Puglia	7.1	5.2
Basilicata	1.1	2.1
Calabria	3.6	4.1
Sicilia	8.9	6.0
Sardegna	2.9	4.3
North-Centre	63.6	70.4
South	36.4	29.6
Total	**100.0**	**100.0**

Sources: Annuario Statistico Italiano 1997, ISTAT.

The work abroad by Italian contractors present in the top 225 international contractors in Engineering News Record in 1997 was valued at L 11,098 billion or more than 6% of total domestic work as follows:

Italian construction work abroad, 1997

Area	L billion	% of total
Europe	2,521.6	22.7
Africa	1,617.5	14.6
Middle East	1,993.5	18.0
Asia	2,135.8	19.2
USA	695.5	6.3
Canada	0.4	0.0
Latin America	2,133.7	19.2
Total	**11,098**	**100.0**

Source: Engineering News Record 12-8-98.

Characteristics and structure of the industry

The State plays a major role in the construction industry in Italy as a client for public works and as a regulator of the process including planning, registration of contractors, standards etc. However, administration of the intentions of government and of laws is not always adequate.

The dominant professions in the construction industry are engineers and architects and either one or the other, if not both, must be involved in all major construction projects. They are subject to precise legal constraints that are related to the extent of their responsibility. Practising architects must be members of the National Society of Architects (*Ordine Nazionale Architetti*) and engineers must be in the Engineers' National Register (*Ordine Nazionale Ingegneri*). The third important consultant is the *geometra* who has a more technical role, but his role is nevertheless significant. Originally he was a land surveyor but now he is responsible for a large proportion of small scale construction work. Architects operate in studios (which are usually fairly small), in public sector offices, in contracting firms and in commercial consultancy companies. A law of 1939 prohibits the formation of companies of professionals in favour of associations of professionals. Thus major multi-disciplinary offices have developed primarily as public sector or contractor in-house design departments or those undertaking work abroad. The *geometra* works mainly in private practice as a sole practitioner but may also work in architectural practices or with contractors.

Cost estimates for projects are produced as part of the normal plan of work, in broad terms at the *progetto di massima* (outline design) and detailed at the *esecutivo* (working drawing) stage. Estimates are prepared either by the responsible professional or frequently by a *geometra*.

The *direttore dei lavori* (director of works) may be the designer or another professional. He is appointed by the client to whom he is responsible. He acts as a project manager and is responsible for making sure that the work is undertaken according to the design. He acts as the client's representative in relation to the contractor in the supervision of the works and on the client's behalf strictly with regard to technical issues.

The construction industry employs around 1.6 million people but the structure of the industry is somewhat different from that of similar European countries. There are many small to medium sized contracting firms (both in terms of employees and capital employed) and a few large firms. In the beginning of the 1990's there were around 500 thousand contracting firms registered with the Chamber of Commerce. Around 500 firms employ over 100 people and probably around 20 to 30 employ over 1,000. Companies employing 20 or more – although only around 1% of firms – carry out more than two-thirds of total work. Most firms are private but there are a few large public sector companies and co-operative companies. Private companies do about 80% of the work – the rest being divided more or less equally between public sector companies and co-operatives. Usually companies are family controlled, partly because of long-established tradition and partly because the stock market is not well developed so that funding usually comes from bank lending.

Most Italian contractors prefer to undertake substantial amounts of construction work themselves and directly employ the craftsmen with the skills to enable them to do so. They subcontract only the mechanical and electrical installation work and some specialist trades.

Italian contractors have operated overseas for many years and have a broad spread of operations in geographical terms as well as in types of work. In 1997 they accounted for over 11% of work done abroad by European contractors, and for 23 out of the 300 firms of contractors listed in *Building* magazine's 'Top 300 European Contractors', of whom four are in the top 100. They are very important as exporters. Fifteen of *Engineering News Record's* 'Top 225 International Contractors' in 1997 were Italian firms – more than any other European country – though they did less work than the British, German or French. Many Italian firms are very specialized. The principal Italian contractors are shown in the following table.

Major Italian contractors 1997

Major contractors	Place in Building's 'Top 300 European Contractors'	% of sales abroad
Impregilo	40	57
Alstaldi	72	47
Ferrocemento	84	41
Caltagirone	99	2.4
Condotte d'Acqua	150	17
Icla	173	5.4
Italstrade	182	27
CMC	190	39
Trevi	213	82
Vianini Lavori	214	8.8

Source: Building December 1998.

Moreover, four consortia exist, two in the private sector, namely *Consorzio Argo* and *Consorzio Grandi Opere*, one in the public sector, *Il Gruppo IRI* (the biggest industrial public holding) and one as a co-operative. These groups are intended to create organizations large and powerful enough to operate on a European scale. Nevertheless, companies belonging to *Il Gruppo IRI* are currently being privatized and the consortium itself is supposed to dissolve in few years according to the EU directives.

Contractors must be registered with the *Albo Nazionale dei Costruttori* (the National Contractors Register) to be permitted to carry out public work. The classification covers twenty types of work and ten value ranges.

The organization representing both public and private contractors is the *Associazione Nazionale Costruttori* Edili (ANCE) (National Association of Building Contractors).

Clients and finance

A feature of the Italian system is the use of *concessioni* (concessions) for the execution and maintenance of public works projects such as motorways and airports. Separate companies are established, which may have a state, regional or municipal shareholding, which bid for the right to run the particular *concessioni*.

A well developed speculative property market with major developers as found in the UK does not exist in Italy. Generally, private sector projects are for individual client's own use or for sale. Over half the office stock is owner-occupied. The remainder is owned by insurance companies and banks, and to a lesser extent by private organizations or individuals.

The retail sector, dominated by small, often family run firms, has prevented the construction of major out-of-town shopping centres but in the last five years the situation is changing with the building of many shopping centres, often under co-operative ownership. Manufacturing industry also has a number of small family firms who own their own factories and often use package deals for procurement of new premises.

Selection of design consultants

In the public sector the appointment of professional advisers is regulated by government decrees. Selection of the architect and/or engineer is made directly or by design competition which may be for an outline design or for detailed proposals including a cost estimate. Influence and appropriate contacts often play an important part in awarding design contracts. The winner of a competition is not necessarily commissioned, though his design becomes the property of the public authority. The majority of public works design is undertaken by outside consultants. In-house technical departments concentrate more on checking and approval of designs.

In the private sector appointments are regulated by the provisions of the *Codice Civile* (Civil Code) and there is a statutory fee scale. Contracts must be signed by a registered professional; non-compliance voids all contracts and on discovery will result in legal proceedings. Signing on behalf of someone who is not registered does occur but brings with it full legal responsibility.

Contractual arrangements

The normal form of contracting involves the employment of a main contractor, though clients may seek separate tenders for portions of the work, normally specialist work.

In the public sector contracts are usually let by open tender or selective tender. In the private sector some contracts may be negotiated. It is estimated that around 80% of projects go to open tender, 15% to selective tender and 5% are negotiated. Usually open and selective tender methods are used by the public sector while negotiation is used for private work.

For public works, it is common for tenderers to quote a discount or increase on a base price that is usually an estimate based on pre-priced bills. The bills may be calculated by a professional, who can be an engineer, an architect or a *geometra*. The winning tender may be the lowest, the average or something else, calculated on a pre-arranged formula. The number of tenderers for open tenders varies substantially.

For private sector work only, contractors may, prior to submitting tenders, have to pay a deposit of 2.5% to 5.0% to the client which is refunded, less a nominal tax, to unsuccessful tenderers.

A Decree of 16.7.62 established general contract conditions for public works and the *Associazione Nazionale Costruttori Edili* (ANCE), the building contractors federation, publishes a contract for the private sector. There is no widely used form of contract and parties tend to define their own clauses. However, a document provides detailed clauses covering most occurrences known as the Special Conditions of Contract for Private Building Works.

Contract documents normally comprise the following:

- drawings
- general contract
- specification clauses
- special specification clauses, applicable to the particular contract
- form of tender.

Three copies of the contract documents are made. The client and the contractor each hold one and the third is deposited with the Public Registrar. Contracts may be let to a general contractor who will be responsible to the client for the whole works, including the work of his sub-contractors. Alternatively contracts may be let on a trade-by-trade basis; in this case the client's consultants are responsible for co-ordination.

Most contracts are firm price. In some instances fluctuations are permitted, with a formula for valuing any adjustment set out in the special specification clauses, usually related to a published national index. Fluctuation clauses on contracts of less than one year's duration are forbidden by law.

Variations are permitted and, given the fact that the costs of these are contained within certain limits, they are valued at rates pro-rata to the contract. The final account is prepared by the contractor along with the employer. Interim payments are made to the contractor and are subject to a retention of 5% or 1%, until satisfactory completion of the contract. A performance bond may be demanded before the works start (5% or 1% of the contract amount). Within some days (specified in the contract) from the engineering inspection, the balance of the final account is due, the retention is given back and the performance bond is released. If these sums are not paid, the contractor will receive interest at a variable rate on the amount withheld. Any disputes arising out of the contract are referred to a panel of three arbitrators; one being appointed by the client, one by the contractor and one by the Courts or a similar body, as provided for in the contract documents. The arbitrators' majority decision is binding on both parties. Any dispute arising out of the tendering process is dealt with according to the Civil Code.

Liability and insurance

Liability is determined by the provisions of the civil and criminal codes which cover obligations assumed under contract and to third parties both by professionals and contractors. All construction work is subject to a ten year liability period for both the contractor and the professionals involved which starts after the issue of the *certificato di abitabilità* (the building occupancy permit).

The *direttore dei lavori* (director of works) is personally responsible with the building approval holder, the client and the contractor for compliance with the planning and building regulations and the terms of the building approval. He is also responsible for site supervision, cost control and the approval of variations. The contractor is responsible for undertaking the works according to the regulation in force, to the contract and to the instructions of the director of the works. When the works are finished, the client chooses an inspection engineer, whose responsibility is to investigate the works and check that they have been implemented correctly.

Professional indemnity insurance is not obligatory and is rarely used. The construction industry is not prone to litigation, largely because of the cost and the length of judicial procedures. Problems are usually resolved by agreement between the parties concerned, unless the contractor or the consultant is clearly responsible for gross negligence.

Construction insurance may be taken out for the period of construction, the maintenance period after completion (normally 12 months, with a limit of 24) and the ten year period of liability. All insurance premiums are subject to a tax. A main contractor insures the construction work on the basis of his valuation of the risks involved. This normally includes subcontractors who do not usually carry separate cover. Small contractors working at a domestic scale do not usually hold insurance other than third party cover. The use of all risks policies (*assicurazione* – CAR) is increasing strongly. Usually major contractors hold general cover for all their activities which is redefined for specific projects, and this practice is on the increase. An important fact for public works contracts is that, while insurance is not obligatory, damage deriving from natural calamity needs to be accepted as such by the authorities within a period of five days, otherwise the liability remains with the contractor.

Contractors usually maintain insurance cover for the whole maintenance period and this is extended to the work of subcontractors. Insurance for the 10 year liability period is normally determined by agreement with the client and may either be held by the contractor or by the client who is interested to cover the risk of a contractor going into liquidation.

Development control and standards

The planning laws are a matter for the regions and there are considerable differences from one region to another. Administration is further devolved to the *Provincia* and the *Comune* and the latter is the most important for land use planning and building control. There are a whole series of plans ranging from broad strategies to detailed plans for specific types of development. Application for planning approval is complex and the information required is comprehensive, covering standards of construction as well as use. The application must be signed by the owner or title holder and designer. When permission is given, work must be started within one year and completed within three years after which, if work is still not completed, a new application has to be made. The building must conform exactly to that which is approved.

It is difficult to assess exactly how all the regulations are enforced and this will vary by locality and over time. It is known that a substantial amount of building work, including new housing, is undertaken without proper permission, particularly in the South.

There are requirements for testing structural materials in accordance with legislation of 1972, which must be carried out in laboratories recognized by the *Ministero dei Lavori Pubblici,* and fire tests which must be carried out in laboratories recognized by the *Ministero dell'Interno* on materials for use in public places. In addition, there are national standards issued by the *Ente Nazionale per l'Unificazione nell'Industria* (UNI) which cover such aspects as safety associated with electrical and gas installations.

Italian specification practice is descriptive in character rather than prescriptive and rarely is a reference made to performance criteria or specific standards. Generally, contractors enjoy substantial freedom of interpretation with regard to the selection of materials, and consultants are prepared to concede a considerable level of autonomy to a trusted contractor. The consultant/ contractor relationship tends to be less adversarial than in the UK – there is a level of mutual respect and pride in the quality of the end result.

New techniques and materials are not subject to compulsory approval procedure. However, the *Instituto Centrale per l'Industrializione e la Technologia Edilizia* (ICITE, Central Institute for Building Industrialization and Technology) and the *Consiglio Nazionale delle Ricerche* (National Council of Research) can issue a technical approval after laboratory tests.

The absence of a comprehensive, widely applied body of standards for building products has placed Italian manufacturers, many of whom are small family run concerns, at a disadvantage in the European market in view of the 'Construction Products Directive'. Steps are underway to remedy this with new arrangements for the accreditation of testing laboratories and certification bodies.

Construction cost data

Cost of labour

The figures overleaf are typical of labour costs in Rome as at the first quarter 1999. The wage rate is the basis of an employee's income, while the cost of labour indicates the cost to a contractor of employing that employee. The difference between the two covers a variety of mandatory and voluntary contributions – a list of items which could be included is given in section 2.

	Wage rate (per hour) L	Cost of labour (per hour) L	Number of hours worked per year
Site operatives			
Mason/bricklayer	20,100	38,200	1,600
Carpenter	20,100	38,200	1,600
Plumber	19,300	36,700	1,600
Electrician	19,300	36,700	1,600
Structural steel erector	19,300	36,700	1,600
HVAC installer	19,300	36,700	1,600
Semi-skilled worker	19,000	36,100	1,600
Unskilled labourer	17,600	33,400	1,600
Equipment operator	20,100	38,200	1,600
Watchman/security	17,300	32,800	1,600
Site supervision		(per yea)r	
General foreman	–	160,000,000	–
Trades foreman	–	112,000,000	–
Clerk of works	–	105,000,000	–
Contractors' staff			
Site manager	–	260,000,000	–
Resident engineer	–	154,000,000	–
Resident surveyor	–	154,000,000	–
Junior engineer	–	112,000,000	–
Junior surveyor	–	105,000,000	–
Planner	–	105,000,000	–
Consultants' personnel			
Senior architect	–	154,000,000	–
Senior engineer	–	154,000,000	–
Senior surveyor	–	154,000,000	–
Qualified architect	–	105,000,000	–
Qualified engineer	–	105,000,000	–
Qualified surveyor	–	91,000,000	–

Cost of materials

The figures that follow are the costs of main construction materials, delivered to site in the Milan area, as incurred by contractors in the first quarter of 1999. These assume that the materials would be in quantities as required for a medium sized construction project and that the location of the works would be neither constrained nor remote. All the costs in this section exclude value added tax (VAT) which is at 20%.

	Unit	Cost L
Cement and aggregate		
Ordinary portland cement in 50kg bags	tonne	140,000
Coarse aggregates for concrete	m³	28,000
Fine aggregates for concrete	m³	33,000
Ready mixed concrete (25N/mm³)	m³	95,000
Ready mixed concrete (35N/mm³)	m³	110,000
Steel		
Mild steel reinforcement	tonne	515,000
High tensile steel reinforcement	tonne	525,000
Structural steel sections	tonne	1,200,000
Bricks and blocks		
Common bricks (250 x 120 x 60mm)	1,000	250,000
Good quality facing bricks (250 x 120 x 60mm)	1,000	600,000
Hollow concrete blocks (250 x 300 x 190mm)	each	1,500
Solid concrete blocks (625 x 250 x 200mm)	each	6,000
Timber and insulation		
Softwood sections for carpentry	m³	350,000
Softwood for joinery	m³	850,000
Hardwood for joinery	m³	2,000,000
100mm thick quilt insulation (22kg/m³ density)	m²	13,000
100mm thick polystyrene insulation self extinguishing (25kg/m³)	m²	18,000
Softwood internal door complete with frames	each	250,000
Glass and ceramics		
Float glass (6mm)	m²	65,000
Good quality ceramic wall tiles (200 x 200mm)	m²	25,000
Plaster and paint		
Plaster in 50kg bags	tonne	180,000
Plasterboard (13mm thick)	m²	5,500
Tiles and paviors		
Clay floor tiles (200x 200 x 10mm)	m²	14,000
Vinyl floor tiles (250 x 250 x 1.8mm)	m²	11,000

	Unit	Cost L
Clay roof tiles	1,000	500,000
Precast concrete roof tiles	m^2	12,000

Unit rates

The descriptions work items below are generally shortened versions of standard descriptions listed in five languages (English, French, Italian, German and Spanish) Appendix 3.

Where an item has a two digit reference number (e.g. 05 or 33), this relates to the full description against that number in Appendix 3. Where an item has an alphabetic suffix (e.g. 12A or 34B) this indicates that the standard description has been modified. Where a modification is major the complete modified description is included here and the standard description should be ignored.

Where a modification is minor (e.g. the insertion of a named hardwood) the shortened description has been modified here but, in general, the full description in Appendix 3 prevails.

The unit rates below are for main work items on a typical construction project in the Milan area in the first quarter of 1999. The rates include all necessary labour, materials and equipment and, where appropriate, allowances for contractor's overheads and profit, preliminary and general items and contractor's profit and attendance on specialist trades. All the costs in this section exclude VAT which is at 20%.

		Unit	Rate L
Excavation			
01	Mechanical excavation of foundation trenches	m^3	15,000
02	Hardcore filling making up levels	m^2	8,000
Concrete work			
04	Plain insitu concrete in strip foundations in trenches	m^3	150,000
05	Reinforced insitu concrete in beds	m^3	165,000
06	Reinforced insitu concrete in walls	m^3	175,000
07	Reinforced insitu concrete in suspended floor or roof slabs	m^3	180,000
08	Reinforced insitu concrete in columns	m^3	185,000
09	Reinforced insitu concrete in isolated beams	m^3	185,000
10	Precast concrete slab	m^2	55,000

		Unit	Rate L
Formwork			
11A	Softwood formwork to concrete walls	m^2	40,000
12A	Softwood formwork to concrete columns	m^2	45,000
13A	Softwood formwork to horizontal soffits of slabs	m^2	45,000
Reinforcement			
14	Reinforcement in concrete walls	tonne	1,500,000
15	Reinforcement in suspended concrete slabs	tonne	1,500,000
16	Fabric reinforcement in concrete beds	m^2	5,000
Steelwork			
17	Fabricate, supply and erect steel framed structure	tonne	3,250,000
Brickwork and blockwork			
18	Precast lightweight aggregate hollow concrete block walls	m^2	45,000
Roofing			
22	Concrete interlocking roof tiles 430 x 380mm	m^2	40,000
23	Plain clay roof tiles 260 x 160mm	m^2	35,000
25	Sawn softwood roof boarding	m^2	36,000
28	Bitumen based mastic asphalt roof covering	m^2	28,000
29	Glass-fibre mat roof insulation 160mm thick	m^2	32,000
30	Rigid sheet loadbearing roof insulation 75mm thick	m^2	50,000
31	Troughed galvanised steel roof cladding	m^2	37,000
Woodwork and metalwork			
32	Preservative treated sawn softwood 50 x 100mm	m	8,000
33	Preservative treated sawn softwood 50 x 150mm	m	10,000
35A	Two panel glazed door in Chestnut hardwood, size 850 x 2000mm	each	1,250,000
36	Solid core half hour fire resisting hardwood internal flush doors, size 800 x 2000mm	each	600,000
37	Aluminium double glazed window, size 1200 x 1200mm	each	850,000
38	Aluminium double glazed door, size 850 x 2100mm	each	1,000,000
39A	Hardwood skirtings (Chestnut hardwood)	m	45,000
40	Framed structural steelwork in universal joist sections	tonne	3,250,000
41	Structural steelwork lattice roof trusses	tonne	3,500,000
Plumbing			
43	UPVC rainwater pipes	m	25,000
44	Light gauge copper cold water tubing	m	12,000
47	UPVC soil and vent pipes	m	30,000
48	White vitreous china WC suite	each	550,000
49	White vitreous china lavatory basin	each	400,000
50	Glazed fireclay shower tray	each	300,000

		Unit	Rate L
Electrical work			
52	PVC insulated and copper sheathed cable	m	15,000
53	13 amp unswitched socket outlet	each	115,000
54	Flush mounted 20 amp, 1 way light switch	each	135,000
Finishings			
55A	2 premixed gypsum based plaster on brick walls	m^2	17,000
56A	White glazed tiles on plaster walls (200 x 200 x 4mm)	m^2	60,000
57	Red clay quarry tiles on concrete floor	m^2	60,000
58	Cement and sand screed to concrete floors	m^2	16,000
60A	Mineral fibre tiles on concealed suspension system	m^2	55,000
Glazing			
61	Glazing to wood	m^2	25,000
Painting			
62	Emulsion on plaster walls	m^2	75,000

Approximate estimating

The building costs per unit area given opposite are averages incurred by building clients for typical buildings in the Lombardia area as at the first quarter 1999. They are based upon the total floor area of all storeys, measured between external walls and without deduction for internal walls.

Approximate estimating costs generally include mechanical and electrical installations, furniture and loose or special equipment and external works; they also include fees for professional services. The costs shown are for specifications and standards appropriate to Italy and this should be borne in mind when attempting comparisons with similarly described building types in other countries. A discussion of this issue is included in section 2. Comparative data for countries covered in this publication, including construction cost data, are presented in Part Three.

Approximate estimating costs must be treated with caution; they cannot provide more than a rough guide to the probable cost of building. All the rates in this section exclude VAT which is generally at 20%.

	Cost thousands L per m²	Cost thousands L per ft²
Industrial buildings		
Factories for letting	900	84
Factories for owner occupation (light industrial use)	1,100	102
Factories for owner occupation (heavy industrial use)	1,200	111
Factory/office (high-tech) for letting (shell and core only)	950	88
Factory/office (high-tech) for letting (ground floor shell, first floor offices)	1,050	98
Factory/office (high-tech) for owner occupation (controlled environment, fully furnished)	1,320	125
High tech laboratory (air conditioned)	2,000	186
Warehouses, low bay (6 to 8m high) for letting (no heating)	620	58
Warehouses, low bay for owner occupation (including heating)	720	67
Warehouses, high bay for owner occupation (including heating)	850	79
Administrative and commercial buildings		
Civic offices, non air conditioned	2,100	195
Civic offices, fully air conditioned	2,600	242
Offices for letting, 5 to 10 storeys, non air conditioned	2,000	186
Offices for letting, 5 to 10 storeys, air conditioned	2,300	214
Offices for letting, high rise, air conditioned	2,500	232
Offices for owner occupation, 5 to 10 storeys, non air conditioned	2,800	260
Offices for owner occupation, 5 to 10 storeys, air conditioned	3,000	279
Prestige/headquarters office, 5 to 10 storeys, air conditioned	3,300	307
Health and education buildings		
General hospitals (100 beds)	2,500	232
Teaching hospitals (100 beds)	2,600	242
Private hospitals (100 beds)	3,000	279
Health centres	2,200	204
Nursery schools	2,400	223
Primary/junior schools	2,100	195
Secondary/middle schools	2,000	186
University (arts) buildings	2,100	195
University (science) buildings	2,300	214
Management training centres	2,200	204
Recreation and arts buildings		
Theatres (over 500 seats) including seating and stage equipment	3,900	362
Theatres (less than 500 seats) including seating and stage equipment	3,000	279
Concert halls including seating and stage equipment	6,000	557

	Cost thousands L per m²	Cost thousands L per ft²
Sports halls including changing and social facilities	1,900	177
Swimming pools (international standard) including changing and social facilities	3,200	297
Swimming pools (schools standard) including changing facilities	2,400	223
National museums including full air conditioning and standby generator	5,700	530
Local museums including air conditioning	2,600	242
City centre/central libraries	2,600	242
Branch/local libraries	2,200	204
Residential buildings		
Social/economic single family housing (multiple units)	1,100	102
Private/mass market single family housing 2 storey detached/ semidetached (multiple units)	1,500	139
Purpose designed single family housing 2 storey detached (single unit)	1,650	153
Social/economic apartment housing, low rise (no lifts)	1,300	121
Social/economic apartment housing, high rise (with lifts)	1,450	135
Private sector apartment building (standard specification)	1,700	158
Private sector apartment building (luxury)	2,200	204
Student/nurses halls of residence	1,900	177
Homes for the elderly (shared accommodation)	1,800	167
Homes for the elderly (self contained with shared communal facilities)	1,700	158
Hotel, 5 star, city centre	4,100	381
Hotel, 3 star, city/provincial	3,000	279
Motel	2,800	260

Regional variations

The approximate estimating costs are based on average Milan rates. Adjust these costs by the following factors for regional variations:

Rome	−4%	Naples	−7%
Florence	−3%	Palermo	−5%
Bologna	−6%		

Value added tax (VAT)

The standard rate of VAT is currently 20%, chargeable on general building work and materials. Housing and public building may qualify for lower rates, for example first homes incur a charge of 4%.

Exchange rates and cost and price trends

The combined effect of exchange rates and inflation on prices within a country and on price comparisons between countries is discussed in section 2.

Exchange rates

The graph below plots the movement of the Italian lira against sterling, the ECU/Euro, the US dollar and 100 Japanese yen since 1990. The values used for the graph are quarterly and the method of calculating these is described and general guidance on the interpretation of the graph provided in section 2. The average exchange rate in the first quarter of 1999 was L 2,760 to the pound sterling, L 1,660 to the US dollar and L 1,940 to the Euro.

The Italian lira against sterling, the ECU/Euro, the US dollar and 100 Japanese yen

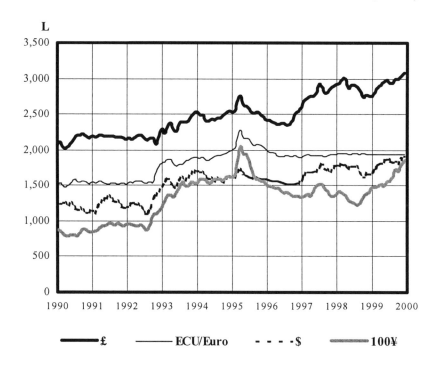

Cost and price trends

The table overleaf presents the indices for consumer prices and residential building costs in Italy since 1990. The indices have been rebased to 1990. The annual change is the percentage change between the average index of consecutive years. Notes on inflation are included in section 2.

Until 1995 residential building costs more or less kept pace with inflation but in 1995 and 1996 they fell behind.

Consumer price and residential building cost indices

Year	Consumer prices annual average	change %	Residential building costs annual average	change %
1990	100.0	6.5	100.0	10.1
1991	106.3	6.3	108.0	8.0
1992	111.7	5.1	114.0	5.6
1993	116.7	4.5	117.0	2.6
1994	121.4	4.0	121.0	3.4
1995	127.8	5.3	123.0	1.7
1996	132.8	3.9	125.0	1.6
1997	135.5	2.0	128.0	2.4
1998	137.9	1.7	–	–

Sources: International Monetary Fund, Yearbook 1998 and supplements 1999.
OECD, Quarterly National Accounts.

Useful addresses

Government and public organizations

Ministero dei Lavori Pubblici
Ministry of Public Works
Piazzale di Porta Pia
00198 Roma
Tel: (39-06) 44121

Istituto Nazionale di Statistica (ISTAT)
National Statistical Office
Via Cesare Balbo 16
00184 Roma
Tel: (39-06) 4673

Ente Nationale Italiano di Unificazione
Italian National Standards Body
Via Battistotti Sassi 11B
20133 Milano
Tel: (39-02) 700241
Fax: (39-02) 70105

Trade and professional associations

Associazione Nazionale Costruttori Edili (ANCE)
National Association of Building
Contractors
Via Guattani 16
00161 Roma
Tel: (39-06) 84881
Fax:(39-06) 44232766

Associazione delle Organizzazioni di Ingegneria Consulenza Tecnico-Economica (OICE)
Association of Engineering and
Technical-Economic
Consultancy Companies
Via Adda 55
00198 Roma
Tel: (39-06) 8558797

Associazione Ingegneri e Architetti
Association of Engineers and
Architects
Via Pantelleria 1
00015 Monterotondo (Roma)
Tel: (39-06) 9065441

Sindicato Nazionale degli Ingegneri Liberi Professionisti, (SNILPI)
Union of Independent Professional
Engineers
Via Salaria 292
00199 Roma
Tel: (39-06) 8549796

Camera di Commercio Industria Artigianato e Agricoltura di Milano
Milan Industry, Artisanry and
Agriculture
Chamber of Commerce
Via Meravigli 9/b
20123 Milano
Tel: (39-02) 876981
Fax: (39-02) 86461885

Other organizations

Consiglio Nazionale delle Ricerche
National Council of Research
Piazzale Aldo Moro 7
00185 Roma
Tel: (39-06) 49931
Fax: (39-06) 4461954

CRESME Ricerche
Building Research Institute
Via Sebenico 2
00198 Roma
Tel: (39-06) 8540158
Fax: (39-06) 8415795

CRI Business Information

Name: CONSTRUCTION RESEARCH INSTITUTE
(KENSETSU BUKKA CHOSAKAI)

Representative: Jiro Taguchi, President

Established: 1 September 1947

Authorised as Foundation: 23 June 1955 (authorised as a Juridical Foundationa by the Ministry of Construction)

Head Office: 11-8, Odenma-Cho, Nihonj, Chuo-ku, Tokyo 103-0011, Japan
Tel: (03) 3663 3891
Fax: (03) 3663 2066

Objectives: The aim of CRI is to contribute to the progress and development of the construction industry, by surveying current construction costs, material prices and labour wage rates related to civil engineering and building construction, so that these costs, prices and wage rates can be used as model information for estimation.

(1) To conduct surveys and studies on current construction costs, material prices and labour wage rates to civil engineering and construction works in major cities

(2) To conduct surveys and studies in order to understand the present state of economic affairs in the construction industry

(3) To distribute reports and to publish books based on the surveys and studies carried out

(4) To be commissioned to conduct surveys and studies

(5) Other activities necessary to accomplish these objectives.

Japan

All data relate to 1998 unless otherwise indicated.

Population

Population	125.9 million
Urban population (1997)	78%
Population under 15	15%
Population 65 and over	16%
Average annual growth rate (1985 to 1998)	0.3%

Geography

Land area	377,835 km^2
Agricultural area	14%
Capital city	Tokyo

Economy

Monetary unit	Yen (¥)
Exchange rate (average first quarter 1999) to:	
the pound sterling	¥ 196
the US dollar	¥ 118
the Euro	¥ 138
Average annual inflation (1995 to 1998)	0.8%
Inflation rate	0.7%
Gross Domestic Product (GDP)	¥ 527,800 billion
GDP PPP basis (1997)	$ (PPP) 3,080 billion
GDP per capita	¥ 4,192,000
GDP per capita basis (1997)	$(PPP) 24,502
Average annual real change in GDP (1995 to 1998),	0.8%
Private consumption as a proportion of GDP	61%
Public consumption as a proportion of GDP	10%
Investment as a proportion of GDP	29%

Construction

Gross value of construction output	¥ 75,570 billion
Gross value of construction output per capita	¥ 600,200
Gross value of construction output as a proportion of GDP	14.3%
Average annual real change in gross value of construction output (1995 to 1998)	−4.1%
Annual cement production (1996)	82.4 million tonnes

PPP Purchasing power parity

The construction industry

Construction output

The value of construction works completed in 1998 was around ¥ 76 trillion of GDP equivalent to 512 billion ECUs (447 billion Euros in January 1999 exchange rates). This represents 14.3% of GDP. Unlike other countries this output excludes most repair and maintenance. The Japanese economy is experiencing a deep economic and financial crisis and the construction industry is suffering in terms of new orders and output. The breakdown by type of work in 1998 is shown below:

Output of new construction, 1998 (current prices)

Sector/type of work	¥ billion	ECUs billion	% of total
Private			
Building			
Residential building	24,182	163.9	32
Mining and manufacturing building	2,267	15.4	4
Commercial and services building	6,046	41.0	8
Total	32,495	220.3	43
Civil engineering	6,044	41.0	8
Total private	38,539	261.3	51
Public			
Building			
Residential building	1,511	10.2	2
Non-residential building	4,534	30.7	6
Total	6,045	41.0	8
Civil engineering	30,983	210.0	41
Total public	37,030	251.0	49
All construction	**75,569**	**512.3**	**100**

Source: Monthly Statistics of Japan, May 1999.

Construction output in 1997 decreased substantially while it increased slightly in 1998. The evolution of the trend in construction in 1997 and 1998 is shown opposite:

Construction output % real change, 1996–97 and 1997–98

	1996–97	1997–98
Type of work	%	%
New work		
Residential building	−16.2	−2.8
Public non-residential building	−12.9	−11.8
Private non-residential building	−1.9	−1.9
Civil-engineering	−6.8	2.8
Total	**−9.7**	**0.7**

Source: Construction investment 1998, Construction Research Institute.

Low interest rates and tax incentives to consumers had contributed to increased investment in private residential building in 1996 but, despite the fact that the same incentives applied to 1997 as well, demand slumped. Apart from the slowdown in the whole economy, reasons for the decrease in investment are the end of earthquake-related demand, increasing difficulty in finding good land and previous oversupply. In 1998 the economy was in crisis, with an actual decrease in GDP, and residential construction fell further.

Private non-residential building had recovered in 1996, due to a partial recovery in industrial and office building, but decreased marginally in 1997 and again in 1998. Public non-residential investment stagnated in 1996 but slumped in 1997 and 1998, due to a reduction in public works projects.

Civil engineering building decreased in 1997, due to less investment in electricity but recovered somewhat in 1998.

The construction industry is experiencing a crisis in new building and the proportion of reconstruction and rehabilitation is currently low but is expected to grow. The proportion of routine repair and maintenance is also low, perhaps around 10 to 15% of total output. This is partly because of the type of housing constructed and its very recent construction, as well as the traditional Japanese 'scrap and build' approach to construction with newbuild clearly preferred over refurbishment. The share of new building investment is very high compared to other countries. A reason for an increase of repair and maintenance investment is the need to improve the resistance of buildings to earthquakes.

Over 50% of construction activity is concentrated in the Kanto and Kinki regions which encompass Tokyo and Osaka, Japan's two biggest conurbations.

Japan has an adequate number of dwellings related to the number of households, but the average dwelling size is only 92 square metres (about half that of the USA). The government recognizes that the country's social capital stock lags behind that of other industrialized nations and therefore that the nation's economic strength is not reflected in its housing and other social capital. In 1990 a 'Public Investment Basic Plan' was formulated to bring these facilities up to the level of other industrialized countries by the year 2000. In particular the average house floor area is planned to increase to 100 square metres.

Although the basic materials and methods of construction used in Japan are similar to those of other developed countries, there are considerable differences in detail. Generally, structures consist of heavy reinforced concrete, steel framed reinforced concrete, or steel frame with spray-applied fireproofing for large buildings, whereas for housing, timber frame is predominant.

The work abroad by Japanese contractors present in the top 225 international contractors in Engineering News Record in 1997 was valued at Yen 1,785 billion or 3% of total domestic work as follows:

Japan's construction work abroad, 1997

Area	¥ Billion	% of total
Europe	51.8	2.9
Africa	94.8	5.3
Middle East	81.2	4.5
Asia	1,293.4	72.5
USA	230.8	12.9
Canada	7.8	0.4
Latin America	25.3	1.5
Total	**1,785.1**	**100.0**

Source: Engineering News Record 12 August 1998.

Characteristics and structure of the industry

Construction companies in Japan usually carry out both design and construction for private projects. Design departments of contractors may have as many as 1,000 professionals.

For public sector projects design is normally undertaken by public in-house design departments and the construction contract is then let. In particular most civil engineering work is publicly sponsored and the public sector offices either have their own design sections or hire specialized consultant firms to design their projects. It is quite usual for firms of consultant engineers to employ over 500 people. The largest architectural consultancy firm has over 1,000 employees and many have several hundred. There is co-operation as well as competition between consultants and contractors; contractors may be invited to participate in consultants' design work or vice versa.

In March 1997 there were over 565,000 construction contractors licensed either by the Ministry of Construction or by the Governors of Prefectures. Most contractors are small but there are some large companies. There are three types of licences that are common though the list is not exhaustive: general civil engineering, general building and scaffolding, earthworks and concrete. Around 60% of all contractors hold one type of licence and the rest hold more than one. Around 60% of all licence holders have 10 million yen or less in capital. Around 6,000 firms, equivalent to around 1% of all contractors, have 100 million yen or more of capital.

In March 1997, only 80 foreign contractors were present in Japan, mainly from the USA and South Korea.

Employment decreased widely in the 1980s but it started recovering in the 1990s The following table shows the breakdown of construction employment in 1996.

Employees by job category, 1996

Type of occupation	Number of employees (thousands)	% of total
Skilled and on-site workers	4,500	67.0
Clerical workers and sales staff	1,240	18.0
Professional, technical and managing workers	790	12.0
Others	170	3.0
Total	**6,700**	**100.0**

Source: Management and Coordination Agency, 1996 Annual
Report on the Labour Force.

There are 19 Japanese firms in the 1997 list of *Engineering News Record's* 'Top 225 International Contractors'; a substantial decrease from 28 firms in 1996. The principal Japanese contractors and their ranking are shown on the next page.

For the last few decades local contractors have had an expanding domestic market because Japan lags behind western developed countries in the provision of infrastructure, most notably roads, sewerage, housing and the city environments. During the 1980s the economy expanded more or less continuously. Government policy is used to regulate the economy and this helps to give the construction industry a relatively even work flow. The negative growth of construction output over the last few years has seriously affected the Japanese industry.

Major Japanese contractors, 1997

Company	Place in ENR's Top 225 International Contractors
JGC Co.	16
Toyo Engineering Co.	17
Kajima Co.	20
Obayashi Co.	26
Nishimatsu Construction Co.	30
Aoki Co.	33
Taisei Co.	35
Takenaka Co.	37
Penta-Ocean Construction Co.	38
Taikisha Ltd.	49

Source: Engineering News Record 17 August 1998.

The 'Contract Construction Business Law' requires contractors to obtain a licence to start a construction business. Nearly all site work in Japan is undertaken by trade contractors who maintain a special relationship with a general contractor, known as a *zenecon*. Under this relationship the general contractor will endeavour to provide continuous employment for his subcontractors, in return for which each subcontractor will allow the general contractor to stipulate a contract price, and to monitor both his financial and project performance. The very large companies do not have a permanent workforce, but a family of subcontractors who are loosely connected to them. The average number of employees per firm was 12 in 1998 lower than any other Asian country except Hong Kong and China.

Major Japanese construction firms are developing in a number of directions: internationally; diversifying into other businesses in some way linked to construction; strengthening the total engineering competence by research and development; and providing construction related finance. The Japanese have both the expertise and experience to compete with Western European and American contractors. During the 1980s Japanese international contractors have increasingly directed their efforts to the industrialized regions of the world. Some of these companies are very specialized, for example in petrochemicals, but most have a broad range of operations. However it may be seen from the table opposite that overseas turnover is under 10% of the overall turnover of large contractors. Japanese contractors have worked abroad to service Japanese investors, though usually they use mostly local construction companies and suppliers. Partly to ensure an even workload they have then become developers abroad and have gradually gained non-Japanese clients, often for 'design and build' projects.

One of the features of contracting organizations in Japan is that they undertake a considerable amount of research and development work. The range of research is very wide, from soil testing to air supported domes. Earthquake engineering is important and the Japanese are generally regarded as world leaders in both research on the use of robots in construction and the development of intelligent buildings. Direct expenditure on research and development by large construction firms is about 1% of turnover, but the firms also fund a considerable amount of outside research.

For private sector projects the law requires that the contractor shall check all design and products to be used in a project and report to the client any possible failures. For public sector projects the public authorities are in charge of design and bear responsibility. The high degree of responsibility placed on the contractor for the success of projects is one reason why in-house research and development departments are needed.

The Ministry of Construction (MOC) oversees all aspects of construction. Research institutes and other organizations are under its control although each research institute has a large degree of autonomy. Construction is also monitored by other ministries such as the Ministry of Agriculture and Transport. The government is concerned that a number of sub-standard unqualified construction companies have entered the market; the smaller companies tend to have less stable management and tend to be smaller due to the shortage of young Japanese workers. In response, the Ministry of Construction formulated the 'Structural Adjustment Promotion Program' to run from fiscal year 1989 for three years and aimed at improving and upgrading industrial training and management.

Only recently have foreign contracting firms been allowed a licence to operate in Japan. As a result of pressure from the USA in 1988 the first Japan-US Construction Agreement permitted registration of foreign firms. By May 1992 contractor permits had been issued to 27 foreign contractors: 13 from the USA, 10 from South Korea and one each from France, Australia, Switzerland and the Netherlands.

There are three types of architects in Japan: first class architects, second class architects and wooden building architects. First class architects must have passed an examination set by the Minister of Construction and be licensed. The other two categories are dealt with on a similar basis by prefectural governors. A contractor with at least one architect can register as an architects office and offer design services. About 50% of first class architects and most second class architects work for contractors.

The Japanese Institute of Architects is limited in its membership to independent architects and has not extended its membership to those employed by contractors. The *Kenchikushi* is a unique qualification of combined architect and building engineer, held by a number of construction supervisors as well as designers.

Clients and finance

In 1996 the clients for new housing construction were as indicated in the table below:

Ownership of new dwellings, 1996

Type	% of dwelling units	Average floor areas
– private rented accommodation	38	50 m²
– company housing for rent	2	79 m²
– private speculative housing for sale or rent	22	91 m²
– owner occupied housing	38	140 m²
All housing	**100**	**90 m²**

Selection of design consultants

For projects not let by 'design and build', architects, engineers and cost consultants are usually appointed by the client either directly or after some form of competition. Other consultants are chosen by one of the main consultants. The most important basis for selection is track record, with price a secondary factor. Personal contacts and recommendations are sometimes relevant in the private sector but rarely in the public sector. In some cases, however, when the contractor is responsible for the construction only, the contractor is appointed first and he will ask the client to appoint an architect – often one of his selection. The architect would still be paid by the client.

The Ministry of Construction publication 'Public Announcement No. 1206' includes guidelines for the appointment of consultants. The professional associations publish recommended – but not mandatory – fee scales. If the price for the design is fixed fee it is sometimes altered during the course of the project. Designers normally retain copyright.

Contractual arrangements

In the public sector, construction companies of the appropriate category and experience are invited to bid. In selecting those invited, central and local governments rank construction firms according to past orders obtained, sales, financial status and technological capabilities. The contract is then awarded to the lowest bidder. In the private sector the client may appoint a specific contractor or invite selected contractors to bid – the latter is the more common system. Many projects – some 30 to 40% of work – are also undertaken on a 'design and build' basis where the architect is employed by the contractor. There is little 'construction management'.

The Japanese contractual system is based on trust and mutual understanding. It is very important for both parties to maintain a good and long-term relationship. The Japanese rarely bring a lawyer into negotiations – that implies mistrust – and litigation is undertaken only as a last resort. Clients tend to work regularly with a contracting firm, and will often have in-house staff with knowledge of building design and construction who will have prepared outline drawings of the proposed works. The contractor generally prepares the working drawings, except for building services, which are prepared by the specialist contractor.

The two contract forms in most common use are the 'Standard Form of Agreement and General Conditions of Government Contract for Works of Building and Civil Engineering', prepared and recommended by the Construction Industry Council of Japan, and the 'General Conditions of Construction Contracts' (GCCC) approved by a number of architects' and contractors' associations. Contract documents, which are relatively short, normally consist of the written contract, general conditions, the design drawings and the specification. There is no bill of quantities but the contractor submits an itemized list of prices (including quantities). Liquidated damages are payable if a project is delayed, and there is a guarantee period of two years for brick or concrete buildings and one year for timber structures. The employer is given express rights to vary the work and negotiations take place on dates and costs. Claims are rare.

Liability and insurance

Although the designer has primary responsibility for faults, in practice the contractor will normally correct the defect in order to retain the confidence of the client. Architects do not therefore normally carry insurance.

The Registration Organization for Warrantied Houses, administered by the Construction Ministry, provides a warranty scheme. This gives a ten year guarantee on the durability of structural components, including foundations, floors, walls and roofs plus a five year warranty on the weather resistance of roofs. The scheme is available to single unit housebuilders using traditional Japanese housebuilding techniques. Prefabricated house builders, who compete with the single unit home builders, also provide a ten year protection on structural components. Some condominium builders have recently started a similar ten year guarantee. In response, the Housing and Urban Development

Corporation, the government-managed house supplier, has, since 1983, developed a long-term warranty programme for some condominiums with warranties of ten years for structural elements, including the roof, and five to ten years for other elements.

Before this long-term warranty of houses and buildings can apply to all builders, a number of problems must be solved regarding such issues as design responsibilities, insurance systems, business profitability and so on. The principle behind long-term warranty is not to guarantee free repair services for ten years, but to build structures in which defects will not occur for at least ten years. Since this puts greater importance on quality, a long-term warranty is a necessity for all construction companies and an inevitable outcome in today's quality-conscious market. Those companies unable to offer such a warranty will eventually lose out in a competitive market.

Construction cost data

Cost of labour

The figures following are typical of labour costs in the Tokyo area as at the first quarter 1999. The wage rate is the basis of an employee's income a further description of which is given in section 2.

	Wage rate (per day) ¥
Site operatives	
Mason/bricklayer	29,600
Carpenter	22,000
Plumber	20,300
Electrician	20,300
Structural steel erector	20,200
HVAC installer	18,600
Semi-skilled worker	15,700
Unskilled labourer	13,400
Equipment operator	19,400
Watchman/security	12,200

	Wage rate (per day) ¥
Contractors' personnel	
Site manager	35,700
Resident engineer	28,900
Resident surveyor	28,900
Junior engineer	14,400
Junior surveyor	14,400
Planner	14,400
Consultants' personnel	
Senior architect	54,300
Senior engineer	54,300
Senior surveyor	54,300
Qualified architect	37,700
Qualified engineer	37,700
Qualified surveyor	37,700

Cost of materials

The figures that follow are the costs of main construction materials, delivered to site in the Tokyo area, as incurred by contractors in the first quarter of 1999. These assume that the materials would be in quantities as required for a medium sized construction project and that the location of the works would be neither constrained nor remote. All the costs in this section exclude value added tax (VAT) which is at 5%.

	Unit	Cost ¥
Cement and aggregate		
Ordinary portland cement in 25kg bags	tonne	16,800
Coarse aggregates for concrete	m^3	4,250
Fine aggregates for concrete	m^3	4,800
Ready mixed concrete (21N/cm^2)	m^3	12,300
Ready mixed concrete (18 N/cm^2)	m^3	12,000
Steel		
Mild steel reinforcemnt R25 16-25mm	tonne	42,000
High tensile steel reinforcement D295A 19mm	tonne	31,500
Structural steel sections H200 x 100	tonne	36,000
Bricks and blocks		
Common bricks (210 x 100 x 60mm)	1,000	80,000
Good quality facing bricks (210 x 100 x 60mm)	1,000	180,000

	Unit	Cost ¥
Hollow concrete blocks (190 x 190 x 390mm)	no.	266
Solid concrete blocks (450 x 300 x 150mm)	no.	760
Precast concrete cladding units with exposed aggregate finish	m^2	11,300
Timber and insulation		
Softwood sections for carpentry	m^3	53,000
Softwood for joinery	m^3	108,000
Hardwood for joinery	m^3	175,000
Exterior quality plywood (12mm)	m^2	1,000
Plywood for interior joinery (5.5mm)	m^2	725
Softwood strip flooring (14mm)	m^2	10,200
Chipboard sheet flooring (15mm)	m^2	790
100mm thick quilt insulation	m^2	300
100mm thick rigid slab insulation	m^2	1,990
Softwood internal door complete with frames and ironmongery	each	75,900
Glass and ceramics		
Float glass (5mm)	m^2	1,360
Sealed double glazing units (FL3+A6+FL3)	m^2	4,390
Good quality ceramic wall tiles (227 x 60mm)	m^2	3,300
Plaster and paint		
Plaster in 20kg bags	tonne	45,000
Plasterboard (9.5mm thick)	m^2	154
Emulsion paint in 5 litre tins	litre	250
Gloss oil paint in 5 litre tins	litre	540
Tiles and paviors		
Clay floor tiles (200 x 100mm)	m^2	3,450
Vinyl floor tiles (2 x 305 x 305mm)	m^2	9,040
Precast concrete paving slabs (300 x 300 x 60mm)	m^2	3,780
Clay roof tiles	1,000	102,000
Precast concrete roof tiles	1000	110,000
Drainage		
WC suite complete	each	37,300
Lavatory basin complete	each	46,200
100mm diameter clay drain pipes	m	2,360
150mm diameter cast iron drain pipes	m	8,670

Unit rates

The descriptions of work items on the next page are generally shortened versions of standard descriptions listed in five languages (English, French, Italian, German and Spanish) in Appendix 3.

Where an item has a two digit reference number (e.g. 05 or 33), this relates to the full description against that number in Appendix 3. Where an item has an alphabetic suffix (e.g. 12A or 34B) this indicates that the standard description has been modified. Where a modification is major the complete modified description is included here and the standard description should be ignored.

Where a modification is minor (e.g. the insertion of a named hardwood) the shortened description has been modified here but, in general, the full description in Appendix 3 prevails.

The unit rates below are for main work items on a typical construction project in the Tokyo area in the first quarter of 1999. The rates include all necessary labour, materials and equipment, but exclude allowances for contractor's overheads and profit (14%) and preliminary and general items (10%). No allowance to cover contractor's profit and attendance on specialist trades has been made. All the rates in this section exclude VAT which is at 5%.

		Unit	Rate ¥
Excavation			
01	Mechanical excavation of foundation trenches	m³	900
02	Hardcore filling making up levels	m³	6,550
Concrete work			
04	Plain insitu concrete in strip foundations in trenches (20 N/mm²)	m³	14,400
05	Reinforced insitu concrete in beds (21 N/mm²)	m³	14,400
06	Reinforced insitu concrete in walls (21 N/mm²)	m³	14,400
07	Reinforced insitu concrete in suspended floor or roof slabs (21 N/mm²)	m³	14,400
08	Reinforced insitu concrete in columns (21 N/mm²)	m³	14,400
09	Reinforced insitu concrete in isolated beams (21 N/mm²)	m³	14,400
10	Precast concrete slab	m³	45,100
Formwork			
11	Softwood formwork to concrete walls	m²	3,400
12	Softwood formwork to concrete columns	m²	4,850
13	Softwood formwork to horizontal soffits of slabs	m²	3,150
Reinforcement			
14	Reinforcement in concrete walls (16mm)	tonne	85,500
15	Reinforcement in suspended concrete slabs	tonne	72,500
16	Fabric mat reinforcement in concrete beds	m²	650
Steelwork			
17	Fabricate, supply and erect steel framed structure	tonne	213,000

		Unit	Rate ¥
Brickwork and blockwork			
18	Precast lightweight aggregate hollow concrete block walls (100mm thick)	m²	5,400
19	Solid (perforated) clay bricks (100mm thick)	m²	13,700
21	Facing bricks, in gauged mortar, flush pointed	m²	16,500
Roofing			
22	Concrete interlocking roof tiles	m²	4,350
23	Plain clay roof tiles	m²	7,400
24	Fibre cement roof slates	m²	4,000
25	Sawn softwood roof boarding	m²	2,000
26	Particle board roof covering	m²	2,950
27	3 layers glass-fibre based bitumen felt roof covering include chippings	m²	8,300
28	Bitumen based mastic asphalt roof covering	m²	4,150
29	Glass-fibre mat roof insulation	m²	1,700
31	Troughed galvanised steel roof cladding	m²	3,350
Woodwork and metalwork			
32	Preservative treated sawn softwood 50 x 100mm	m²	4,300
35	Two panel glazed door in hardwood, size 850 x 2000mm	each	233,000
36	Solid core half hour fire resisting hardwood internal flush doors, size 800 x 2000mm	each	67,600
37	Aluminium double glazed window, size 1200 x 1200mm	each	98,000
38	Aluminium double glazed door, size 850 x 2100mm	each	261,000
39	Hardwood skirtings	m	2,200
40	Framed structural steelwork in universal joist sections	tonne	151,000
41	Structural steelwork lattice roof trusses	tonne	176,000
Plumbing			
42	UPVC half round eaves gutter (diameter 105mm)	m	1,500
43	UPVC rainwater pipes	m	3,400
44	Light gauge copper cold water tubing	m	1,550
45	High pressure plastic pipes for cold water supply	m	1,000
46	Low pressure plastic pipes for cold water distribution	m	1,400
47	UPVC soil and vent pipes	m	6,200
48	White vitreous china WC suite	each	63,700
49	White vitreous china lavatory basin	each	43,900
51	Stainless steel single bowl sink and double drainer	each	61,000
Electrical work			
52	PVC insulated copper sheathed cable	m	430
53	15 amp unswitched socket outlet	each	4,700
54	Flush mounted 20 amp, 1 way light switch	each	4,050

		Unit	Rate ¥
Finishings			
55	2 coats gypsum based plaster on brick walls 20mm thick	m^2	4,500
56	White glazed tiles on plaster walls	m^2	5,950
57	Red clay quarry tiles on concrete floor	m^2	10,000
58	Cement and sand screed to concrete floors 30mm thick	m^2	1,900
59	Thermoplastic floor tiles on screed	m^2	2,350
60	Mineral fibre tiles on concealed suspension system	m^2	4,900
Glazing			
61	Glazing to wood (5mm glass)	m^2	5,500
Painting			
62	Emulsion on plaster walls	m^2	1,200
63	Oil paint on timber	m^2	1,150

Approximate estimating

The building costs per unit area given in the following page are averages incurred by building clients for typical buildings in the Tokyo area as at the first quarter 1999. They are based upon the total floor area of all storeys, measured between external walls and without deduction for internal walls. Areas in Japan are often measured in Tubo (1.818m x 1.818m).

Approximate estimating costs generally include mechanical and electrical installations but exclude furniture, loose or special equipment, and external works; they also exclude fees for professional services. The costs shown are for specifications and standards appropriate to Japan and this should be borne in mind when attempting comparisons with similarly described building types in other countries. A discussion of this issue is included in section 2. Comparative data for countries covered in this publication, including construction cost data, are presented in Part Three.

Approximate estimating costs must be treated with caution; they cannot provide more than a rough guide to the probable cost of building. All the rates in this section exclude VAT which is at 5%.

	Cost ¥ per m²	Cost ¥ per ft²
Industrial buildings		
Factories for letting (include lighting, power and heating)	231,000	21,500
Warehouses, low bay for owner occupation (including heating)	170,000	15,800
Administrative and commercial buildings		
Civic offices, fully air conditioned	312,000	29,000
Offices for letting, high rise, air conditioned	347,000	32,200
Prestige/headquarters office, high rise, air conditioned	386,000	35,900
Health and education buildings		
General hospitals (300 beds)	479,000	44,500
University (arts) buildings	360,000	33,400
Recreation and arts buildings		
Theatres (over 500 seats) including seating and stage equipment	527,000	49,000
Residential buildings		
Social/economic apartment housing, high rise (with lifts)	261,000	24,200
Private sector apartment building (luxury)	340,000	31,600
Hotel, 5 star, city centre	437,000	40,600
Hotel, 3 star, city/provincial	350,000	32,500

Regional variations

The approximate estimating costs are based on average Tokyo rates. Adjust these costs by the following factors for regional variations:

Sapporo	−6%	Kanazawa	−4%	Hiroshima	−6%
Sendai	−5%	Nagoya	−5%	Takamatsu	−6%
Niigata	−4%	Osaka	−2%	Fukuoka	−7%

Value added tax (VAT)

Value added tax (VAT) rate is currently 5%, chargeable on general building work and materials.

Exchange rates and cost and price trends

The combined effect of exchange rates and inflation on prices within a country and on price comparisons between countries is discussed in section 2.

Exchange rates

The graph below plots the movement of the Japanese yen against sterling, the ECU/Euro and the US dollar since 1990. The values used for the graph are quarterly and the method of calculating these is described and general guidance on the interpretation of the graph provided in section 2. The average exchange rate in the first quarter of 1999 was ¥ 196 to the pound sterling, ¥ 118 to the US dollar and ¥ 138 to the Euro.

The Japanese yen against sterling, the ECU/Euro, and the US dollar

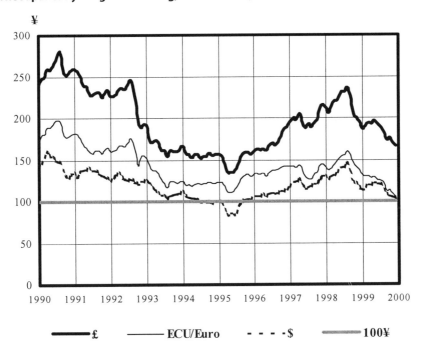

Cost and price trends

The table on the following page presents the indices for consumer prices, building material and labour costs in Japan since 1990. The indices have been rebased to 1990=100. The annual change is the percentage change between the average index of consecutive years. Notes on inflation are included in section 2.

Consumer price, building material cost and labour cost indices

Year	Consumer prices annual average	change %	Building materials costs annual average	change %	Labour costs annual average	change %
1990	100.0	3.1	100.0	3.5	100.0	3.4
1991	103.3	3.3	102.1	2.1	102.4	2.4
1992	105.1	1.7	101.5	-0.6	103.9	1.5
1993	106.4	1.2	100.7	-0.8	105.0	1.1
1994	107.1	0.7	98.1	-2.6	105.2	0.2
1995	107.0	-0.1	97.5	-0.6	105.2	0.0
1996	107.1	0.1	97.7	0.2	106.3	1.0
1997	108.9	1.7	-	-	-	-
1998	109.7	1.7	-	-	-	-

Source: International Monetary Fund, Yearbook 1998 and supplements 1999.
Statistics Bureau, Monthly Statistics.
Economic Affairs Bureau, Ministry of Construction.

Useful addresses

Government and public organizations

Ministry of Construction (MOC)
2–1–3 Kasumigaseki
Chiyoda-ku
Tokyo 100-0013
Tel: (81-3) 3580 4311
Fax: (81-3) 5251 1926

Ministry of Transport
2–1–3 Kasumigaseki
Chiyoda-ku
Tokyo 100-0013
Tel: (81-3) 3580 3111
Fax: (81-3) 3593 0474

National Land Agency
1–2–1 Kasumigaseki
Chiyoda-ku
Tokyo 100-0013
Tel: (81-3) 3593 3311
Fax: (81-3) 3501 5349

The Housing and Urban Development Corporation
1–14–6 Kundashita
Chiyoda-ku
Tokyo 102
Tel: (81-3) 32638434

Japanese Industrial Standards
4–1–24 Akasaka
Minato-ku
Tokyo 107
Tel: (81-3) 35838005
Fax: (81-3) 3586-2014

Management and Coordination Agency
The Statistics Bureau
19–1 Wakamatsu–cho
Shinjuku-ku – Tokyo
Tel: (81-3) 32021111
Fax: (81-3) 52731180

Trade and professional associations

Japan Federation of Construction Contractors
Tokyo Kensetsu Kaikan building
2–5–1 Hacchobori
Chuo-ku
Tokyo 1040032
Tel: (81-3) 35530701
Fax: (81-3) 35522360

The Associated General Constructors of Japan
Tokyo Kensetsu Kaikan buiding
2–5–1 Hacchobori
Chuo-ku
Tokyo 1040032
Tel: (81-3) 35519396
Fax: (81-3) 35553218

Association of Japanese Consulting Engineers
3–16–4 Ueno
Taito-ku
Tokyo 110
Tel: (81-3) 38398471
Fax: (81-3) 38398472

Japan Construction Consultants Association
Shin Kudan building
2-2–4 Kudan Minami
Chiyoda-ku
Tokyo 102
Tel: (81-3) 32397992
Fax: (81-3) 32391869

Planning Consultants Association of Japan
Heights New Hirakawa
2–12–18 Hirakawa-cho
Chiyoda-ku
Tokyo 1020093
Tel: (81-3) 32616058
Fax: (81-3) 32615082

The Overseas Construction Association of Japan
Nichibei building
2–24–2 Hacchobori
Chuo-ku
Tokyo 1040032
Tel: (81-3) 35531631
Fax: (81-3) 35510148

Trade and professional associations (continued)

Japan Housing Association
Kojimachi Kyodo building
3-2 Kojimachi
Chiyoda-ku
Tokyo 1020083
Tel: (81-3) 32658201
Fax: (81-3) 32658230

Japan Civil Engineering Consultants Association
Shin Kudan building
2-2-4- Kudan Minami
Chiyoda-ku
Tokyo 102
Tel: (81-3) 32397992
Fax: (81-3) 32391869

Japan Civil Engineering Contractors' Association
Tokyo Kensetsu Kaikan building
2-5-1 Hacchobori
Chuo-ku
Tokyo 1040032
Tel: (81-3) 35523201
Fax: (81-3) 35523206

Japan Institute of Architects
2–3–16 Jingumae
Shibuya-ku,
Tokyo 150
Tel: (81-3) 34087125
Fax: (81-3) 34087129

The Building Surveyors' Institute of Japan
2–26–20 Shiba
Minato-ku
Tokyo 108
Tel: (81-3) 34539591
Fax: (81-3) 34539597

Building Contractors Society
Tokyo Kensetsu Kaikan building
2-5-1 Hacchobori
Cuo-ku
Tokyo 104-0032
Tel.: (81-3) 35511118
Fax: (81-3) 35552463

Other organizations

The Building Centre of Japan
3–2–2 Toranomon
Minato-ku
Tokyo 105
Tel: (81-3) 34347155
Fax: (81-3) 34313301

Research Institute of Construction and Economy
39 Mori building
2–4–5 Azabudai
Minato-ku
Tokyo 106
Tel:(81-3) 34335011
Fax: (81-3) 34335239

Japan External Trade Organization (JETRO)
2–2–5 Toranomon
Minato-ku
Tokyo 107
Tel: (81-3) 3582 5511
Fax: (81-3) 3582 5027

Construction Research Institute
11-8 Odenma-cho Nihonbashi
Chuoo-ku
Tokyo1030011
Tel: (81-3) 36632411
Fax: (81-3) 36632417

OUR UNITY
IS YOUR STRENGTH

ARCADIS BOUW/INFRA (Grabowsky & Poort, Starke Diekstra, BV Articon, Jongen)

What you see above is a very small part of a very big picture. ARCADIS is the new Heidemij. A new whole born of a number of highly successful companies now united in an organisation of some 6,500 people working from about 200 offices. And if you've never seen or heard of us before, you're almost sure to have been touched by one of our projects.

With our united expertise, knowledge and resources and more than a hundred years' experience in project and process management we belong to the top ten multidisciplinary consulting engineering companies in the world.

The core of our activity is the design and execution of integrated solutions in environment and infrastructure, construction and real estate information. The proof of our vision is to be seen in the many thousands of projects in more than 100 countries.

Each project is part of a vision, a vision which is shared and realised responsibily. Objectives are discussed. Cooperation is sought. And the impact of our work is measured, considering the effects for today and the implications for tomorrow. This realisation that every single project, regardless of size, is ultimately part of a much bigger picture is what shapes and defines the durability of our work and delivers the greatest possible added value. Can you see it?

For more information call: +31 30 605 8605 Internet: www.arcadis.nl

Netherlands

All data relate to 1998 unless otherwise indicated.

Population

Population	15.7 million
Urban population (1997)	89%
Population under 15	18%
Population 65 and over	14%
Average annual growth rate (1985 to 1998)	0.6%

Geography

Land area	37,330 km^2
Agricultural area	59%
Capital city	Amsterdam

Economy

Monetary unit	Guilder (G)
Exchange rate (average first quarter 1999) to:	
the pound sterling	G 3.14
the US dollar	G 1.89
the Euro	G 2.20
the yen x 100	G 1.60
Average annual inflation (1995 to 1998)	2.1%
Inflation rate	1.9%
Gross Domestic Product (GDP)	G 749 billion
GDP PPP basis (1997)	$ (PPP) 343.9 billion
GDP per capita	G 47,700
GDP per capita PPP basis (1997)	$ (PPP) 22,040
Average annual real change in GDP (1995 to 1998)	3.3%
Private consumption as a proportion of GDP	60%
Public consumption as a proportion of GDP	14%
Investment as a proportion of GDP	20%

Construction

Gross value of construction output	G 78.6 billion
Gross value of construction output per capita	G 5,010
Gross value of construction output as a proportion of GDP	10.5 %
Average annual real change in gross value of construction output (1995 to 1998)	3.2%
Annual cement consumption	5.7 million tonnes

PPP Purchasing power parity

The construction industry

Construction output

The value of gross construction sector output in the Netherlands in 1998 was G 78.6 billion equivalent to or 35.4 billion ECUs (the same as January 1999 Euros).   This represents 10% of GDP.   Moreover, in addition, another G 10.2 billion is spent on services and construction by other sectors, DIY and black economy.   This constitutes around 13% of the official output for the construction sector. The breakdown by type of work in 1998 is shown below:

Output of construction sector, 1998 (current prices)

Type of work	G billion	ECUs billion	% of total
New work			
Residential building	17.6	7.9	22
Non-residential building			
Office and commercial	5.8	2.6	7
Industrial	2.5	1.1	3
Other	5.1	2.3	7
Total	13.4	6.0	17
Civil engineering	13.1	5.9	17
Total new work	44.0	19.8	56
Repair and maintenance			
Residential	14.9	6.7	19
Non-residential	16.9	7.6	22
Civil engineering	2.9	1.3	4
Total repair and maintenance	34.6	15.6	44
Total	**78.6**	**35.4**	**100**

Note: the value of the Euro in January 1999 was very similar to that of the ECU in 1998.
Source: based on Vries, O., 'The Netherlands', in Euroconstruct conference
proceedings, Prague, June 1999

The buoyant private non-residential and civil engineering sectors offset negative growth in residential construction and in pubic non-residential construction resulting in a modest increase of about 1% overall in 1998 compared with 5% in 1997.

Economic growth with low interest rates and increasing incomes are both reasons for an increase in housing demand but in 1998 output fell and the units for which building permits were given were thought to have fallen by nearly 15%

from the very high level of 1997, which suggests that output will not rise very much in the near future. A factor is the increasing price of land. Social housing in the subsidized sector is becoming expensive, subsidies have been reduced and output is falling. The proportion of flats in 1998 was about 27% of total units completed.

The strong economy, a decrease in the vacancy rate and greater profitability are stimulating the demand for office, industrial and commercial buildings. The office market is likely to focus on replacement of low-standard buildings while the industrial sector is benefitting from new projects in the chemical and machinery industries. The public non-residential sector is facing a stagnant investment. Many of the projects in health are completed or about to complete and care is shifting to external resources.

Civil engineering investment grew, but at a moderate rate. A fall in tele-communication work was significant. Only the transport sector is increasing. Private expenditure on civil projects is increasing, especially in infrastructure for public transport.

The geographical distribution of building completions in 1996 compared with that of population is shown below. The distribution of work follows roughly that of the population.

Population and construction output – 1996 regional distribution

Region	Population (%)	Construction output (%)
Groningen	3.6	2.9
Friesland	3.9	3.9
Drenthe	3.0	2.8
Overijssel	6.8	7.7
Flevoland	1.8	1.2
Gelderland	12.1	11.7
Ultrecht	6.9	7.7
Noord-Holland	15.9	14.4
Zuid-Holland	21.5	23.2
Zeeland	2.4	2.2
Noord-Brabant	14.8	16.2
Limburg	7.3	6.0
Total	**100.0**	**100.0**

Sources: Statistical Yearbook 1997, CBS Statistics Netherlands.

In the past expenditure in renovations was low, due to the abolition of subsidies for refurbishment of dwellings.   Investment is now recovering, helped by increasing private consumption and low interest rates.   In the public sector a major renovation programme for post war dwellings is proposed.   Upgrading of offices for modern technology and improvement of shopping centres is affecting renovation expenditure in the non-residential sector.

The work abroad by Dutch contractors present in the top 225 international contractors in Engineering News Record in 1997 was valued at G 3 billion or around 4% of total domestic work as follows:

Dutch construction work abroad 1997

Area	G billion	% of total
Europe	1.7	56.1
Africa	0.0	0.0
Middle East	0.8	27.5
Asia	0.2	6.3
US	0.0	0.0
Canada	0.1	1.7
Latin America	0.2	8.4
Total	**3.0**	**100.0**

Source: Engineering News Record 12 August 1998.

Characteristics and structure of the industry

Most of the technical ministries have well established in-house facilities, including design departments.   It has been estimated that about 60% of the public sector market for consultancy services is taken up by in-house offices. The largest office is in the Ministry of Public Works, which used to do all its own design and supervision of major projects but now lets about 10% to external consultants.   The Ministry of Housing will increase the work it lets to external designers up to 80%.   Other Ministries are also increasing their external work, and provincial departments are being slimmed down, giving more work to private firms.   In addition, several Dutch contractors have in-house engineering departments for turnkey projects and many suppliers also have design offices.

Private sector engineering firms have a relatively small but increasing market. Many of them are expanding into areas quite separate from engineering and are becoming risk-taking business enterprises.   Few firms in this sector employ over 500 staff but at the other extreme there are a large number of very small firms.

There are about 2,000 architects registered in the Netherlands.   The title of architect is not protected by law and there is no requirement to employ an architect on building projects, though in practice architects are used for all but the very smallest projects. Most architects are in private practice.

In 1997 there were approaching some 68,000 construction companies in the Netherlands employing over 400,000 workers; around 20 companies have more than 500 employees. It is normal practice to employ a main contractor rather than using the separate trades system.

Dutch contractors have traditionally worked overseas on building and civil engineering projects, particularly water related ones. The Netherlands have long been renowned for port and harbour engineering, dredging, land reclamation, coastal defences, waterway engineering and offshore pipeline engineering.

In 1997 there were 15 Dutch contractors in *Building* magazine's list of the 'Top 300 European Contractors' and four in the top 50. Two were in *Engineering News Record's* list of the 'Top 225 International Contractors' in 1997.

The table below shows the principal Dutch contractors and their percentage sales abroad.

Major Dutch contractors 1997

Major contractors	*Place in* Building's *Top 300 European Contractors' 1997*	*% of sales abroad*
Hbg	10	83
Volker Stevin	25	23
Ballast Nedam	29	58
NBM	31	2.3
Amstelland Tbi Holdings	51	9
Heijmans	56	9
Beheer Koop	70	54
Bam Groep	71	5.6
Boskalis Westminster	82	78
Wilma	85	–

Source: Building December 1998.

Construction companies in the Netherlands have to be licensed by the Chamber of Commerce, which requires compliance with certain educational, technical and financial standards. Most contractors are members of one or the two largest employers' organizations: *Nederlandse Aannemersbond en Patronsbond van de Bouwbedrijven in Nederland* (NAPB) or *Algemeen Verbond Ondernemers Bouwnijverheid* (AVOB). There is a federation of associations of contractors – the Association of Cooperative Price Regulating Organizations for the Construction Industry (ACPRO). Its objectives are to promote orderly competition, to prevent improper conduct in submitting tenders and to promote the establishment of economically justifiable prices. It has produced the 'Uniform Building Regulations' (UBR) (see under Contractual Arrangements). It is very usual for Dutch contractors to form joint ventures which help to protect the project in the event of one contractor failing and to spread the available workload. Usually the contractor based near the location of the project leads the team.

Clients and finance

The government directly or indirectly finances about half of all new construction work in the Netherlands but attempts are being made to obtain more private funds for infrastructure.

Some housing is built under the 'Housing Act' by municipalities or, more usually, housing associations, with a high level of state contribution to the cost of development (about 35% of new housing in the period 1971 to 1984); another part is built for sale or rent by housing associations, financial institutions, pension funds and the private sector with a moderate level of state subsidy (43% of new housing in the period 1971 to 1984); and there is housing built for sale or rent by the private sector with no subsidy for the cost of construction, though it is still eligible for mortgage tax relief (19% of new housing in the period 1971 to 1984).

Mortgages for house purchase are available from banks but they do not usually lend more than 60% to 70% of the value of the house without additional security. However the local authority may guarantee the loan and in that case loans of up to 100% of the value of the property can be obtained.

There are about 850 housing associations throughout the Netherlands, usually operating locally and with a small number of developments under their individual control.

Private non-residential building is let either by the users themselves or by property developers (*projectontwikkelaars*). The latter may be investment arms of financial institutions such as banks, insurance companies, pension funds, property companies, subsidiaries of major industrial concerns or estate agents. Public sector utilities contracts are placed by departments of Ministries or local governments.

About 75% of civil engineering – often known as GWW (ground, water and roadbuilding) – contracts are let by the public sector: central, provincial municipal or local government organizations.

Selection of design consultants

Dutch law does not prescribe any specific tender or selection procedures for consultants.

If an external design consultant is employed, he may be appointed after direct negotiation, after a call for proposals from two or more consultants or by pre-selection from a restricted number of candidates followed by a call for proposals. Negotiation with one consultant is common though other methods of appointment are being increasingly used.

Contractual arrangements

Most contracts in Holland are let to main contractors in a 'lump sum' contract and the contractors then subcontract part of the work. However demolition and piling (which is very important because of the nature of the soil), and often also mechanical and electrical engineering work, are let directly by the client. There is some nominated subcontracting. 'Management contracting' and 'separate trades contracting' are rare, although they are used where there are a number of separate specialised disciplines involved and/or where time and quality are of crucial importance.

Although there is still some open tendering for public sector work, selective tendering is steadily gaining favour and is now dominant. Normally, selective and negotiated tendering methods are used in private development. Only in exceptional circumstances will the contract not be awarded to the lowest tenderer. Clauses may be included in the contract documents to cover the reimbursement of variations in the costs of labour and materials. Where fluctuations are allowed, they are usually paid according to an official index (*Risicoregeling*).

Contracts in the Netherlands are mainly governed by the *Uniforme Administratieve Voorwaarden* (Uniform Administrative Regulations, UAV). The UAV were originally intended for government contracts but have since been generally adopted throughout the industry. The Dutch seem to have succeeded in balancing the interests of the client and the contractor and this balance is expressed in the UAV. The forms of contract in the UAV apply equally to building and civil engineering work.

The procedure is that the client asks for bids from one or more contractors – usually a single price to conform to specifications and associated drawings. When compiling a tender, the contractor prepares his own internal schedule of quantities, in operational order, from the specifications and drawings that he prices, keeping labour and materials separate. However, this document is confidential to the tenderer and does not form any part of the contract unless he is appointed, in which case he will normally make it available. Sometimes the prices (but not the quantities) will be made part of the contract for the purpose of valuing variations. A contractor who tenders makes a binding offer and cannot change his price. Most contractors who are members of a trade association follow the 'Code of Ethics' laid down in the submission of tenders. According to the procedures laid down in the 'Uniform Building Regulations' (UBR), when asked to bid for a contract the contractors will usually attend a meeting three days prior to the tender date. The 'bidding meeting' is an opportunity to compare prices and the contractors frequently declare their full bid at this stage. In fact, any contractor may withdraw his bid following the meeting. This arrangement helps to check for errors in calculations or in risk assessment and this is vital because the tender delivered to the client is binding. Under the UBR tendering costs must be borne by the client. This means that estimating costs are allocated to each project and the client has an incentive to keep the number of bidders low. It is possible that a contractor may be granted some preference at the bidding meeting for special reasons. According to the Code of Ethics only the lowest bidder may negotiate his tender with the employer, and when the tender has been submitted, another contractor, if required to submit a bid, can do it only with the approval of the lowest bidder.

Variations may be made at the client's will and the contractor has to accept them and accept any instruction from the client's agent. The contractor can refuse to carry out orders only on instructions from an arbitrator.

The UBR also deals with a range of other matters such as the correct management of applications and the financial execution of agreements. There are sanctions for infringement of the 'Code of Ethics' and the UBR. These may be imposed by the associations or by the courts.

There are no standard forms of subcontract because it has not been possible to agree one. Subcontractors are awarded the work during the pre-bid meeting but there is still controversy about the best way to agree a code of conduct. There is no 'pay-when-paid' clause. No standard form of warranty in the employer's favour exists and it is common that the employer will look for a performance bond. It is very often used to treat the subcontracting agreement in the same way as the main contracting agreement.

The settlement of disputes is governed by the UAV. Parties waive their right to intervention by the courts and agree to arbitration according to the 'Regulations of the Court of Arbitration for the Netherlands Building Industry'. Provision is made in UAV for the continuation of work pending an award.

Not all contracts are let according to the UAV, though that is the most popular. 'Frame contracts' are one alternative form. A 'frame contract' is used for large scale, ill defined and time-consuming engineering works. It permits the design to proceed simultaneously with the execution of the works and the contractor often contributes to the design process. The 'frame contract' details the terms and conditions and envisages the award of partial contracts during the course of the work to undertake various parts of the overall work. The contractor in this type of project needs to be especially reliable in terms of expertise and financial standing. The design will proceed at the same time as the realization of the works and they will adapt to each other. Client and contractors mutually agree on the choice of subcontractors. But in the case where the client chooses a subcontractor, the contractor must accept him as his subcontractor, therefore becoming responsible for his actions.

Liability and insurance

The contractor is generally responsible for the quality of materials and their fitness for purpose and also for the conformity to the specified requirements and their timely delivery, though the client is liable for 'functional' unfitness of materials prescribed by him or by a nominated subcontractor. There are, however, problems in the precise definition of 'functional'. In any case, the contractor will be responsible, at least so far as checking for defects. If there is a clear defect in design and the contractor warns of the defect but is ignored, he will not be considered responsible. This also applies to the case where warranties are given.

The contractor is responsible for work of subcontractors, including general work of nominated subcontractors. The contractor can hold the subcontractor responsible under the terms of the subcontract. The main contractor will get an extension of time in case of delay or non-execution of the works by the subcontractor but he will have to do the work himself. The client is responsible for design, including the effect of sub-surface conditions, and for equipment placed at the contractors' disposal by the client.

Liability for contractors and designers normally lasts for 10 years but may extend to 30 years. However, costs awarded against architects may amount to only half the designers' fees and it is possible to opt out of liability as part of the contract engagement.

Although insurance is not obligatory, in practice approximately 80% of new projects are covered.

Development control and standards

The 'Physical Planning Act' of 1962 (amended 1985) created a three-tier planning structure. Central government takes the key decisions; the provinces produce regional plans and the municipalities produce both structure and legally binding plans.   The municipalities have a duty to adopt their own building regulations and to issue building permits for all construction, as well as some other activities, such as tree planting and demolition.   The planning system and the building regulations system operate as one.   Planning permission can be refused only when a proposal is not in compliance with the approved plan, building regulations or permission under the 'Monuments and Historic Buildings Act'.   Any refusal must have specific stated reasons and an appeal is possible. The normal time limit for a decision is two months, though this may be extended to four months.

Building regulations are administered with planning control.   Under the 'Housing Act' the Minister of Housing and Physical Planning may issue specific regulations and the act is also the basis of municipal building by-laws.   The Society of Netherlands Municipalities drafts and keeps up to date model by-laws, though within limits these may be amended by individual municipalities.

Standards are prepared and published according to the regulations of the Netherlands Standards Institution (NNI), a private organization.   Standards have no legal status though they may be referred to in model by-laws and other documents.   A system of testing and approval exists for building products and building components on a private basis.   Several private organizations are involved in this work and they issue certificates of approval.   The use of certified products is promoted nationally but the scheme is nevertheless purely voluntary.

Construction cost data

Cost of labour

The figures opposite are typical of labour costs in the Utrecht area as at the first quarter 1999. The cost of labour indicates the cost to a contractor of employing an employee, a further description of which is given in section 2.

	Wage rate (per hour) G	Cost of labour (per hour) G	Number of hours worked per year
Site operatives			
Mason/bricklayer	18	60	1,645
Carpenter	18	60	1,645
Plumber	18	60	1,645
Electrician	18	59	1,645
Structural steel erector	17	55	1,645
HVAC installer	18	59	1,645
Skilled labour	18	59	1,645
Semi-skilled worker	17	54	1,645
Unskilled labourer	16	53	1,645
Equipment operator	17	54	1,645
Watchman/security	15	52	1,645
Site supervision			
General foreman	24	85	1,645
Trades foreman	22	74	1,645
Clerk of works	19	58	1,645
Contractors' personnel			
Site manager	26	98	1,645
Resident engineer	24	85	1,645
Resident surveyor	24	85	1,645
Junior engineer	22	72	1,645
Junior surveyor	21	63	1,645
Planner	21	63	1,645
Consultants' personnel			
Senior architect	41	268	1,707
Senior engineer	41	268	1,707
Senior surveyor	41	268	1,707
Qualified architect	32	217	1,707
Qualified engineer	32	217	1,707
Qualified surveyor	32	217	1,707

Cost of materials

The figures that follow are the costs of main construction materials, delivered to site in the Utrecht area, as incurred by contractors in the first quarter of 1999. These assume that the materials would be in quantities as required for a medium sized construction project and that the location of the works would be neither constrained nor remote. All the costs in this section exclude value added tax (VAT) which is at 17.5%.

	Unit	Cost G
Cement and aggregates		
Ordinary portland cement in 50kg bags	tonne	271
Coarse aggregates for concrete	m³	56
Fine aggregates for concrete	m³	71
Ready mixed concrete	m³	160
Ready mixed concrete	m³	165
Steel		
Mild steel reinforcement	tonne	1,670
High tensile steel reinforcement	tonne	1,670
Structural steel sections	tonne	2,680
Bricks and blocks		
Common bricks (210 x 100 x 50mm)	1,000	354
Good quality facing bricks	1,000	606
Medium quality facing bricks	1000	455
Hollow concrete blocks	1,000	2,050
Solid concrete blocks	1,000	2,190
Precast concrete cladding units with exposed aggregate finish	m²	328
Timber and insulation		
Softwood sections for carpentry	m³	808
Softwood for joinery	m³	1,160
Hardwood for joinery	m³	2,220
Exterior quality plywood	m²	35
Plywood for interior joinery	m²	26
Softwood strip flooring	m²	22
Chipboard sheet flooring	m²	7
100mm thick quilt insulation	m²	12
100mm thick rigid slab insulation	m²	28
Softwood internal door complete with frames and		
ironmongery (office)	each	606
Glass and ceramics		
Float glass (8mm)	m²	66
Sealed double glazing units	m²	167
Good quality ceramic wall tiles0	m²	51
Plaster and paint		
Plaster in 50kg bags	tonne	475
Plasterboard (15mm thick)	m²	7
Emulsion paint in 5 litre tins	litre	18
Gloss oil paint in 5 litre tins	litre	21
Tiles and paviors		
Clay floor tiles (150x 150 x 1.5mm)	m²	61
Vinyl floor tiles	m²	33

	Unit	Cost G
Precast concrete paving slabs	m²	22
Clay roof tiles	1,000	2,210
Precast concrete roof tiles	1,000	1,590
Drainage		
WC suite complete	each	353
Lavatory basin complete	each	227
100mm diameter clay drain pipes	m	15
150mm diameter cast iron drain pipes	m	59

Unit rates

The descriptions of work items below are generally shortened versions of standard descriptions listed in five languages (English, French, Italian, German and Spanish) in Appendix 3.

Where an item has a two digit reference number (e.g. 05 or 33), this relates to the full description against that number in Appendix 3. Where an item has an alphabetic suffix (e.g. 12A or 34B) this indicates that the standard description has been modified. Where a modification is major the complete modified description is included here and the standard description should be ignored.

Where a modification is minor (e.g. the insertion of a named hardwood) the shortened description has been modified here but, in general, the full description in Appendix 3 prevails.

The unit rates below are for main work items on a typical construction project in the Utrecht area in the first quarter of 1999. The rates include all necessary labour, materials and equipment, and allowances for contractor's overheads and profit (6.5%) and preliminary and general items (2%). No allowance to cover contractor's profit and attendance on specialist trades has been made (10%). All the rates in this section exclude VAT which is at 17.5%.

		Unit	Rate G
Excavation			
01	Mechanical excavation of foundation trenches	m³	13
02	Hardcore filling making up levels	m²	10
03	Earthwork support	m²	2
Concrete work			
04	Plain insitu concrete in strip foundations in trenches	m³	292
05	Reinforced insitu concrete in beds	m³	246
06	Reinforced insitu concrete in walls	m³	313

		Unit	Rate G
07	Reinforced insitu concrete in suspended floor or roof slabs	m^3	246
08	Reinforced insitu concrete in columns	m^3	338
09	Reinforced insitu concrete in isolated beams	m^3	282
10	Precast concrete slab	m^2	65

Formwork

11	Softwood formwork to concrete walls	m^2	73
12	Softwood or metal formwork to concrete columns	m^2	83
13	Softwood or metal formwork to horizontal soffits of slabs	m^2	86

Reinforcement

14	Reinforcement in concrete walls (16mm)	tonne	2,000
15	Reinforcement in suspended concrete slabs	tonne	2,100
16	Fabric reinforcement in concrete beds	tonne	6

Steelwork

| 17 | Fabricate, supply and erect steel framed structure | tonne | 3,430 |

Brickwork and blockwork

18	Precast lightweight aggregate hollow concrete block walls (100mm thick)	m^2	108
19	Solid (perforated) concrete bricks	m^2	113
20	Solid (perforated) sand lime bricks	m^2	92
21	Facing bricks	m^2	133

Roofing

22	Concrete interlocking roof tiles	m^2	56
23	Plain clay roof tiles	m^2	77
24	Fibre cement roof slates	m^2	35
25	Sawn softwood roof boarding	m^2	54
26	Particle board roof covering	m^2	41
27	3 layers glass-fibre based bitumen felt roof covering include chippings	m^2	50
28	Bitumen based mastic asphalt roof covering	m^2	39
29	Glass-fibre mat roof insulation	m^2	49
30	Loadbearing glass-fibre roof insulation	m^2	24
31	Troughed galvanised steel roof cladding	m^2	44

Woodwork and metalwork

32	Preservative treated sawn softwood 50 x 100mm	m	23
33	Preservative treated sawn softwood 50 x 150mm	m	25
34	Single glazed casement window in hardwood, size 650 x 900mm	each	410
35	Two panel glazed door in hardwood, size 850 x 2000mm	each	1,350
36	Solid core half hour fire resisting hardwood internal flush doors, size 800 x 2000mm	each	1,180
37	Aluminium double glazed window, size 1200 x 1200mm	each	1,030

		Unit	Rate G
38	Aluminium double glazed door, size 850 x 2100mm	each	1,280
39	Hardwood skirtings	m	24
40	Framed structural steelwork	tonne	4,100
41	Structural steelwork lattice roof trusses	tonne	4,360

Plumbing

42	UPVC half round eaves gutter	m	49
43	UPVC rainwater pipes with pushfit joints	m	18
44	Light gauge copper cold water tubing	m	39
45	High pressure plastic pipes for cold water supply	m	26
46	Low pressure plastic pipes for cold water distribution	m	26
47	UPVC soil and vent pipes	m	44
48	White vitreous china WC suite	each	764
49	WC basin complete	each	384
50	Glazed fireclay shower tray	each	282
51	Stainless steel single bowl sink and unit	each	292

Electrical work

52	PVC insulated copper sheathed cable	m	8
53	Socket outlet with PVC insulated copper cable	each	200
54	Flush mounted I way light switch with insulated copper cable	each	200

Finishings

55	2 coats gypsum based plaster on brick walls	m^2	35
56	White glazed tiles on plaster walls	m^2	108
57	Red clay quarry tiles on concrete floor	m^2	138
58	Cement and sand screed to concrete floors 50mm thick	m^2	16
59	Thermoplastic floor tiles on screed	m^2	50
60	Mineral fibre tiles on concealed suspension system	m^2	64

Glazing

61	Glazing to wood	m^2	97

Painting

62	Emulsion on plaster walls	m^2	17
63	Oil paint on timber	m^2	41

Approximate estimating

The building costs per unit area given on the next page are averages incurred by building clients for typical buildings in the Utrecht area at the first quarter 1999. They are based upon the total floor area of all storeys, measured between external walls and without deduction for internal walls.

Approximate estimating costs generally include mechanical and electrical installations but exclude furniture, loose or special equipment, and external works; they also exclude fees for professional services. The costs shown are for specifications and standards appropriate to the Netherlands and this should be borne in mind when attempting comparisons with similarly described building types in other countries. A discussion of this issue is included in section 2. Comparative data for countries covered in this publication, including construction cost data, are presented in Part Three.

Approximate estimating costs must be treated with caution; they cannot provide more than a rough guide to the probable cost of building. All the rates in this section exclude VAT which is at 17.5%.

	Cost G/m^2	Cost G/ft^2
Industrial buildings		
Factories for letting	1,110	103
Factories for owner occupation (light industrial use)	1,360	126
Factories for owner occupation (heavy industrial use)	1,640	152
Factory/office (high-tech) for letting (shell and core only)	2,050	190
Factory/office (high-tech) for letting (ground floor shell, first floor offices)	2,460	229
Factory/office (high-tech) for owner occupation (controlled environment, fully furnished)	2,870	267
High tech laboratory (air conditioned)	4,200	390
Warehouses, low bay (6 to 8m high) for letting (no heating)	871	81
Warehouses, low bay for owner occupation (including heating)	1,130	105
Warehouses, high bay for owner occupation (including heating)	1,440	134
Administrative and commercial buildings		
Civic offices, non air conditioned	1,740	162
Civic offices, fully air conditioned	2,510	233
Offices for letting, 5 to 10 storeys, non air conditioned	2,050	190
Offices for letting, 5 to 10 storeys, air conditioned	2,770	257
Offices for letting, high rise, air conditioned	3,130	290
Offices for owner occupation, 5 to 10 storeys, non air conditioned	2,260	210
Offices for owner occupation, 5 to 10 storeys, air conditioned	2,970	276
Offices for owner occupation, high rise, air conditioned	3,380	314
Prestige/headquarters office, 5 to 10 storeys, air conditioned	3,670	341
Prestige/headquarters office, high rise, air conditioned	4,310	400
Health and education buildings		
General hospitals (300 to 350 beds)	4,510	419
Teaching hospitals	5,130	477
Private hospitals	4,510	419
Health centres	2,360	219
Nursery schools	1,540	143

	Cost G/m²	Cost G/ft²
Primary/junior schools	1,590	148
Secondary/middle schools	1,640	152
University (arts) buildings	2,050	190
University (science) buildings	2,360	219
Management training centres	2,030	189
Recreation and arts buildings		
Theatres (over 500 seats) including seating and stage equipment	3,080	286
Theatres (less than 500 seats) including seating and stage equipment	3,590	334
Concert halls including seating and stage equipment	3,690	343
Sports halls including changing and social facilities	2,150	200
Swimming pools (international standard) including changing and social facilities	2,460	229
Swimming pools (schools standard) including changing facilities	2,410	224
Local museums including air conditioning	3,900	362
City centre/central libraries	3,590	334
Branch/local libraries	3,080	286
Residential buildings		
Social/economic single family housing (multiple units)	1,180	110
Private/mass market single family housing 2 storey detached/semidetached (multiple units)	1,230	114
Purpose designed single family housing 2 storey detached (single unit)	1,440	134
Social/economic apartment housing, low rise (no lifts)	1,130	105
Social/economic apartment housing, high rise (with lifts)	1,230	114
Private sector apartment building (standard specification)	1,440	134
Private sector apartment building (luxury)	1,540	143
Student/nurses halls of residence	950	89
Homes for the elderly (shared accommodation)	1,080	100
Homes for the elderly (self contained with shared communal facilities)	1,230	114
Hotel 5 star, city centre	3,590	333
Hotel 3 star, city/provincial	2,870	267
Motel	1,850	172

Regional variations

The approximate estimating costs are based on the Utrecht area rates. Adjust these costs by the following factors for regional variations:

West Netherlands	0%
North Netherlands	−5%
East Netherlands	−3%
South Netherlands	−5%

Value added tax (VAT)

The standard rate of VAT is currently 17.5%, chargeable on general building work and materials.

Exchange rates and cost and price trends

The combined effect of exchange rates and inflation on prices within a country and on price comparisons between countries is discussed in section 2.

Exchange rates

The graph below plots the movement of the Dutch guilder against sterling, the ECU/Euro, the US dollar and 100 Japanese yen since 1990. The values used for the graph are quarterly and the method of calculating these is described and general guidance on the interpretation of the graph provided in section 2. The average exchange rate in the first quarter of 1999 was G 3.14 to the pound sterling, G 1.89 to the US dollar and G 2.20 to the Euro.

The Dutch guilder against sterling, the ECU/Euro, the US dollar and 100 Japanese yen

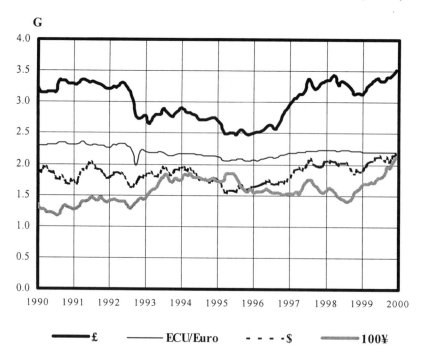

Cost and price trends

The table below presents the indices for consumer prices and residential and industrial building costs in the Netherlands since 1990. The indices have been rebased to 1990=100. The annual change is the percentage change between the average index of consecutive years. Notes on inflation are included in section 2.

The residential building cost index formerly covered new residential buildings for the social rented sector. It now , using a different method, covers the whole low rent residential building sector. It uses price index data collected by the Central Bureau of Statistics. It excludes repair and maintenance, site preparation and professional fees.

Consumer price, residential building cost and industrial building cost indices

Year	Consumer prices		Residential building costs		Industrial building costs	
	annual average	change %	annual average	change %	annual average	change %
1990	100.0	2.5	100.0	1.8	100.0	n.a.
1991	103.1	3.1	102.7	2.7	104.5	4.5
1992	106.4	3.2	108.0	5.2	106.9	2.3
1993	109.2	2.6	110.6	2.5	108.3	1.3
1994	112.2	2.7	113.3	2.4	109.3	1.0
1995	114.4	2.0	117.7	3.9	112.1	2.5
1996	116.8	2.1	123.0	4.5	114.2	1.9
1997	119.3	2.1	127	3.3	115.3	1.0
1998	121.6	1.9	129	1.6	–	–

Sources International Monetary Fund, Yearbook 1998 and supplements 1999.
European Mortgage Federation, Hypostat 1986-1996.
Central Bureau of Statistics of the Netherlands.

Useful addresses

Government and public organizations

Ministerie van Economische Zaken
Ministry of Economic Affairs
Bezuidenhoutseweg 30
Postbus 20101
2500 EC Den Haag
Tel: (31-70) 3798911
Fax: (31-70) 3474081

Ministerie van Volkshuisvesting, Ruimtelijke Ordening en Milieubeheer
Ministry of Housing, Design and the Environment
Rijnstraat 8
Postbus 20951
2500 EZ Den Haag
Tel: (31-70) 3393939

Government and public organizations (continued)

Netherlands Foreign Investment Agency
Bezuidenhoutseweg 2
Postbus 20101
2500 EC Den Haag
Tel: (31-70) 3798818
Fax: (31-70) 3796322

Normalisatie Instituut Nederlands Dutch National Standard Institution
Kalfjeslaan 2
2623 GB Delft
Tel: (31-15) 2690390
Fax: (31-15) 2690190

Centraal Bureau voor de Statistiek
National Statistics Bureau
Postbus 4000
2270 JM Voorburg
Tel: (31-70) 3373800
Fax: (31-70) 3877429

Trade and professional associations

Vereniging Groot – en tussenhdl
Wholesale Building Contractors'
Association
Meelkweg 1
3255 Te Oude Tonge
Tel: (31-187) 641930
Fax: (31-187-643054

Nederlandse Vereniging van Bouwondernemers
Dutch Association of Building Contractors
Parkweg 162
2271 AM Voorburg
Tel: (31-70) 3860204
Fax: (31-70) 3876326

Bond van Nederlandse Architecten (BAN)
Dutch Architects Association
1000 GP Amsterdam
Tel: (31-20) 5553666

Orde van Nederlandse Raadgevende Ingenieurs (ONRI)
Association of Dutch Consulting Engineers
Koninginnegracht 22
2514 AB Den Haag
Tel: (31-70) 3630756
Fax: (31-70) 3600661

Internationale Kamer van Koophandel
International Chamber of Commerce
Pr. Beatrixlaan 7
2595 Ak-s-Gravenhage
Tel: (31-70) 3819563
Fax: (31-70) 3819563

Other organizations

TNO

Research Organization
Schoemakerstraat 97
2628 Delft
Tel: (31-15) 2696900
Fax: (31-15) 2627335

Bouwcentrum EXPO

3032 AG Rotterdam
Tel: (31-10) 2433600
Fax: (31-10) 2430917

Stichting Bouwresearch

Construction research organization
Kruisplein 37
3014 DB Rotterdam
Tel: (31-10) 4117276
Fax: (31-10) 4130175

PROJECT MANAGEMENT
INTERNATIONAL
Project Management & Cost Consultancy

Project Management International provides professional project management & associated construction consultancy services. Our commitment is to provide a high quality service through a dedicated resource of highly skilled project managers and property professionals.

Services include:-

Project Management

Cost Consultancy

Project Monitoring & Loan Monitoring

Dispute Resolution

Migration Management

Management Training

Offices in:-

London	Warsaw	Paris
Project Management International 10-11 Charterhouse Square London EC1M 6EH	PMI Polska Sp.z o.o ul Wiejska 18 m 4 00-480 Warszawa	PMI Fance SARL 11 Rue La Boetie 75008 Paris
Tel : + 44 171 566 7900 Fax : + 44 171 566 7911	Tel : + 48 22 622 4441 Fax : + 48 22 629 8724	Tel : + 33 1 47 42 1905 Fax : + 33 1 47 42 2178
	e-mail : pmipw@medianet.pl	e-mail : bpmi@magic.fr

Other offices in:-
Katowice
Edinburgh
Glasgow
Manchester
Portsmouth

Poland

All data relate to 1998 unless otherwise indicated.

Population

Population	38.7 million
Urban population (1997)	64%
Population under 15	21%
Population 65 and over	12%
Average annual growth rate (1985 to 1998)	0.3%

Geography

Land area	312,683 km^2
Agricultural area	61%
Capital city	Warsaw

Economy

Monetary unit	Zloty (Zl)
Exchange rate (average first quarter 1999) to:	
the pound sterling	Zl 5.82
the US dollar	Zl 3.47
the Euro	Zl 4.10
the yen x 100	Zl 2.94
Average annual inflation (1995 to 1998)	15.9%
Inflation rate	11.7%
Gross Domestic Product (GDP)	Zl 536 billion
GDP PPP basis (1997)	(PPP) 280.7 billion
GDP per capita	Zl 13,860
GDP per capita PPP basis (1997)	$ (PPP) 7,262
Average annual real change in GDP (1995 to 1998)	4.8%
Private consumption as a proportion of GDP (including NPISH*)	65%
Public consumption as a proportion of GDP	17.0%
Investment as a proportion of GDP	21.0

Construction

Gross value of construction output	Zl 57.5 billion
Gross value of construction output per capita	Zl 1,487
Gross value of construction output as a proportion of GDP	10.7%
Average annual real change in gross value of construction output (1995 to 1998)	9.4%
Annual cement production	13.0 million tonnes

PPP Purchasing power parity
*Non-profit institutions serving households.

The construction industry

Construction output

The value of the gross output of the construction sector in Poland in 1998 was Zl 57.5 billion equivalent to 14.4 billion ECUs (nearly the same in January 1999 Euros). This represents 10.7% of GDP. In addition it is estimated that the construction work undertaken by other sectors, by the black economy and DIY was valued at Zl 12.6 billion, or 22% of construction sector output. The breakdown of work, including the informal output, is shown in the table below.

Output of construction, 1998 (current prices)

Type of work	Zl billion	ECUs billion	% of total
New work			
Residential building	7.6	1.9	11
Non-residential building			
Office and commercial	8.1	2.0	12
Industrial	13.2	3.3	19
Other	5.3	1.3	8
Total	26.6	6.7	38
Civil engineering	13.6	3.4	19
Total new work	48.0	12.0	68
Repair and maintenance			
Residential	7.3	1.8	10
Non-residential	10.2	2.6	15
Civil engineering	4.6	1.2	7
Total repair and maintenance	22.2	5.6	32
Total	**70.1**	**17.6**	**100**

Note: the value of the Euro in January 1999 was very similar to that of the ECU in 1998.
Source: based on Sochacki, M., 'Poland' in Euroconstruct conference
          proceedings, Prague, June 1999.

Poland and the construction industry in Poland have developed remarkably well in the transition to a market economy and growth seems likely to continue, although perhaps at a slower rate – already evident in 1998 when the rate of growth fell to a high 11% compared with heady 13% in 1997. Such high rates are difficult to maintain.

Residential building increased in 1997 and 1998 and in 1998 the number of dwellings completed is thought to have reached 75,000. Of these, according to the Central Statistical Office, nearly half were private and this proportion is growing year by year. In addition, those built with public funds for sale or rent have increased to about 10% of the total. However, co-operative and municipal housing is falling in importance. Co-operative housing is down to about a third of the total. Difficulties in the housing sector include high rates of interest, high prices in relation to income and reduced subsidies. There are problems with obtaining mortgages and loans but these are diminishing. An increase in housing construction is expected as these factors improve. It has been estimated that about a million new dwellings are needed – new residential construction as a proportion of the total construction output is one of the lowest in Europe. About half the construction of dwellings is in flats.

Non residential construction is growing fast, although growth in 1998 at 13.5% was less than that in 1997 of over 20%. The highest increases have been in office and commercial buildings, especially in the cities, with industrial buildings also rising. Foreign investors are putting money into commercial and industrial property, especially shops and tourist facilities. Public sector non-residential building construction is limited by the controls on public expenditure, although the need is seen to exist. The proportion of total construction output accounted for by new non-residential construction is very high compared to other European countries.

Civil engineering has generally been increasing but the rate of growth slowed down in 1998. Government expenditure in this area has increased and is expected to remain positive. The need to meet standards of environmental quality is boosting work in this sector. Repair and maintenance is 32% of all work which is high in relation to what it was in Poland as a planned economy. The needs are there but finance is a serious problem. In the private commercial and office sector and in industrial buildings renovation and rehabilitation work has been important. It has been estimated that about 300,000 existing dwellings need refurbishment.

The geographical distribution of population in 1996 (which is often a good indicator of location of construction output) is shown on the next page.

Characteristics and structure of the industry

The relative importance of the State and private sectors within the construction industry in Poland has changed dramatically, as is shown by the second table overleaf.

Population, 1996 regional distribution

Provinces in the region	Population (%)
Katowice, Bielsko, Czestochwa	13.7
Bydgoszcz, Torun, Wloclawek	7.5
Wloclaw, Opole, Jelenia Góra, Legnica	10.9
Kraków, Nowy Sacz	7.8
Gdansk	5.1
Bialystok, Suwalki, Olsztyn, Lomza, Ostroleka	9.2
Warsaw	11.9
Lódz, Piotrków, Sieradz	6.5
Koszalin, Slupsk	3.4
Pozna·, Zielona Góra, Gorzów, Konin, Pila, Leszno	18.8
Szczecin	5.1
Total	**100.0**

Source: Polish Agency for Foreign Investment, London 1998.

Relative importance of various types of construction firms (no. of firms)

	end 1995	end 1996	% change
State owned	789	724	-8.2
Communal	106	106	0.0
Co-operative	1,034	913	-11.7
Civil partnerships	17,387	19,085	10.0
Joint stock	682	772	13.2
limited liability	12,834	13,745	7.1
Other companies (e.g. limited, general)	166	191	15.1
Total	**32,998**	**35,536**	**7.7**
sole traders	169,464	197,426	16.5

Source: Polish Agency for Foreign Investment, London 1998.

In 1997 and 1998 the trend continued to private sector firms with about 94% now in private hands. The Polish construction industry has been one of the most rapidly privatized industries in the country. Tax incentives and exemptions along with more relaxed regulation benefited the private sector. Setting up private companies is easier and so is the liquidation or privatization of state-owned companies. The share of the private sector, when measured in terms of revenues and employment, put the construction industry immediately after agriculture and internal trade. The share of the private sector was smaller when measured in terms of investment, assets, exports, research or technological innovation. The construction industry also ranked third after banking and trade in the number of construction firms quoted in the Warsaw Stock Exchange. Motostal-Export and Exbud are the largest. Since 1995 there has been consolidation in the industry. Many companies have undertaken restructuring programmes in order to reduce costs and improve quality. Greater consolidation will occur as more foreign companies enter the market.

Foreign investment is growing, because of low production costs, enormous market potential and geographical location advantages. The Government promotes foreign investment and grants particular advantages to firms that produce for export and introduce new technology. Foreign contractors use Polish sub-contractors and materials.

The economic changes of this decade led to a fall in employment by around 23% from 1993 to 1996 in the construction sector. Nevertheless, at the end of 1996 the employment level was around 557,000 workers in the industry, although illegal employment is estimated at around 30% of the previous figure. The result of the decrease in employment was a substantial increase in productivity. Employment decreased in public companies and increased in private companies where the share of total employment was at around 85% of total employment.

The public sector had in the past been dominated by large companies employing 500 to 1,000 persons or even more. These companies were inflexible and monopolized both the material and component production and construction. They mainly used large prefabricated components which they produced and they often owned transport and assembly organizations, as well as auxiliary operating units, e.g. hotels and other services. The status of large companies was reinforced in 1996 as a result of changes in legislation and regulation made in 1994. Public tenders require deposits as a financial requirement to bid for work and this has a negative effect on small firms that cannot enter some sectors. At the same time, the presence of some big firms is consistent with the trend in developed countries.

Private companies in the past were mainly fairly small and a private establishment on average employed only two persons. They were engaged in building repair work and also in finishing works on newly assembled housing. It is believed that once the shortage of capital is reduced and the construction output grows, some will develop into medium sized firms.

Polish contractors did not feature in 1997 in either *Building* magazine's 'Top 300 European Contractors' or *Engineering News Record's* 'Top 225 International Contractors'.

The following table shows the structure of the building industry in 1996 by size of output, size of labour force and by number of firms.

Structure of the building industry in 1996 (percentages)

	Production %	Employees %	Proportion of firms %
Small firms			
less than 20 employees	44	38	94
Medium size firms			
20 to 50 employees	13	9	3
Large firms			
over 50 employees	44	53	3
Total	**100**	**100**	**100**

Source: GUS – Polish Central Statistical Office.

Many of the large construction firms are involved in mechanical and electrical works as well as building or civil construction.

The main institution regulating the construction sector is the Ministry of Physical Planning and Construction. It is concerned with physical planning, residential and municipal management, management of landlord matters, cartography, technical construction matters and regulations concerning construction and building materials.

Clients and finance

Housing cooperatives have been the main investors in housing. They have a monopolistic and compulsory character because for the majority of people from low-income groups they offer the only way of having a flat of their own. The difference between the average family income and the cost of a small family flat means that they are not able to meet the cost of a mortgage. Most co-operatives are large, with more than 1,000 dwellings and they have large reserves of land.

Investors' own resources are the most important source of finance for building. The banks also offer investment credits, mortgages and commercial credits. The frail mortgage and legal system together with inflation, slow the development of the market for bank loans. If inflation decreases and the economic and legal situation improves, banks will be able to lead the market for credit. Another important source of building finance is foreign capital. Official estimates show that capital for building investment in 1995 came from the following sources: own investor's capital 65%, state budget 4%, bank credit 8%, foreign capital 11%, other capital 12%.

According to the Law of Support for Housing Construction new forms of saving for housing will enable savers to obtain cheap mortgage credit. Moreover, the Housing Stimulation Act allows the establishment of Social Housing Societies to build flats for rent at moderate rates. The National Economy Bank has established a National Housing Fund to give preferential credits to the Societies and Housing Cooperatives for the building of dwellings and for tenants' housing cooperatives. A law of 1997, effective from January 1998, paved the way for the creation of mortgage banks. The first one has already been established with the involvement of a German bank as well as Polish banks.

A number of Polish developers are now in existence. The vast majority have only a small capital and tend to avoid excessive risks by sharing investments. A number of key stage payments are made by the client following an initial non-refundable down payment. Some developers provide credit. Developers may act for the smaller housing cooperatives. They concentrate their activities on single family housing but also build multi-storey blocks in towns. Most of the capital invested in the office market is from foreign investors.

Selection of design consultants

The very big projects are usually undertaken under the direction of construction managers or management contractors. Clients may hire architects, project managers or civil engineers as advisers on the feasibility of projects, planning and supervision work. Consultants need a licence. There are different licences for design and design verification, for supervision of work on behalf of the client and for construction management.

At present only licences issued by a Polish authority can be used. The applicant must pass a State exam.

Contractual arrangements

The process of adapting the legal regulations for building to the new market conditions started at the end of 1994. Building projects can begin only if a Building Permit has been granted. Structures must be built in accordance with the law and current building knowledge. They also need to be safe and energy efficient. It is also very important that building activity complies with the requirements of general standards and building laws and conforms to EU standards and the directives of the EU Council. The State regulates the quality of design and the choice of building materials.

The technical, spatial and administrative aspects of building design and construction, as well as the maintenance and demolition of structures are regulated by the 'Building Law Act'. There are also regulations concerning contracts and the execution and acceptance of works.

The State promotes efficient construction activity in the sense of profitability through open market competition. Open or restricted tenders are being increasingly used. The investor makes a cost estimate of the project that is based on norms. At the same time the contractor also estimates cost on the basis of his own method and adds a profit. Negotiation then takes place. There is some warranty provisions in the contract for physical defects with duration of one to three years. Bankers' guarantees are recommended. The investor is free to choose his funding and to select sub-contractors.

Almost all forms of contracts exist. For big State projects or World Bank financed projects the most common method of procurement is a lump-sum contract which can be priced either in Polish Zloty or Euros. The new Procurement Law lays down tender procedures for public projects.

There are project management services available for those organizations that do not have their own in-house facilities.

Development control and standards

Physical planning is regulated by the 'Physical Planning Act 1984'. Building construction is regulated by the 'Building Act 1974'. There are obligatory technical standards for construction which are established by the Polish Committee for Standardization, Measurement and Quality. The main technical standards are *Poloskie Normy* (PNB), branch standards and factory or local standards.

Presently, work is under way to align Polish standards with the European EUROCODE (EC).

Quality standards for construction works are described in the 'Technical requirements for the execution and acceptance of objects'. A warranty period of 3 to 5 years is provided to promote good quality.

The institution which has the task to monitor the observance of the Building Law is the Main Office of Building Supervision (GUNB) which cooperates with the State Sanitary Inspectorate and the State Fire Service. Among their basic tasks are:

- building and engineering supervision, including safety for people and property at all stages of the process, according to the 'Act of 24 October 1974 – Building Code' and the 'Decree of the Council of Ministers' of 13 March 1975 on building and engineering supervision.

- building and town planning supervision, including compatibility of developments with local spatial plans, environmental controls and protection, according to the 'Decree of the Minister of Local Industry and Environment Preservation' of 20 February 1975 on building and town planning supervision.

In order to be admitted to the EU Poland needs to adjust Polish standards and regulations to the EU building standards and regulations. Three EU directives are very important: the first one refers to the task of integrating laws and regulations about building materials; the second is about conforming to safety and hygiene regulations at construction sites; the third one refers at the mutual acknowledgement of diplomas, certificates and, in general, any document regarding professional qualification and to the freedom to provide professional services.

The Polish Law on Procurement, in its late 1995 amended version, takes full account of EU standards.

Construction cost data

Cost of materials

The figures that follow are the costs of main construction materials, delivered to site in the Warsaw area, as incurred by contractors in the first quarter of 1999. These assume that the materials would be in quantities as required for a medium sized construction project and that the location of the works would be neither constrained nor remote. All the costs in this section exclude value added tax (VAT) which is charged at 7% generally or 22% on certain materials.

	Unit	Cost Zl
Cement and aggregates		
Ordinary portland cement in 50kg bags	tonne	155
Coarse aggregates for concrete	m^3	35
Fine aggregates for concrete	m^3	45
Ready mixed concrete (B40)	m^3	170
Ready mixed concrete (B20)	m^3	140
Steel		
Mild steel reinforcement	tonne	1,250
High tensile steel reinforcement	tonne	1,280
Structural steel sections	tonne	2,000

	Unit	Cost Zl
Bricks and blocks		
Common bricks (250 x 120 x 65mm)	1,000	500
Good quality facing bricks (250 x 120 x 65mm)	1,000	2,150
Medium quality facing bricks	1,000	1,930
Hollow concrete blocks (88 x 220 x 288mm)	1,000	890
Solid concrete blocks (590 x 240 x 240mm)	1,000	3,720
Timber and insulation		
Softwood sections for carpentry (seasoned)	m²	650
Softwood for joinery	m³	590
Hardwood for joinery (oak)	m³	1,800
Exterior quality plywood	m²	50
Plywood for interior joinery	m²	30
Softwood strip flooring (19mm)	m²	12
Chipboard sheet flooring (mm)	m²	9
200mm thick quilt insulation	m²	12
Softwood internal door complete with frames and ironmongery	each	600
Glass and ceramics		
Float glass (2.5mm)	m²	45
Good quality ceramic wall tiles (100 x 100mm)	m²	35
Plaster and paint		
Plaster in 50kg bags	tonne	130
Plasterboard (12mm thick)	m²	7
Emulsion paint in 5 litre tins	litre	25
Gloss oil paint in 5 litre tins	litre	60
Tiles and paviors		
Clay floor tiles (300 x 300 x 9mm)	m²	31
Vinyl floor tiles (300 x 300 x 3mm)	m²	38
Clay roof tiles	1,000	1,160
Precast concrete roof tiles	m²	27
Drainage		
WC suite complete	each	430
Lavatory basin complete	each	190
150mm diameter cast iron drain pipes	m	58

Unit rates

The descriptions of work items below are generally shortened versions of standard descriptions listed in five languages (English, French, Italian, German and Spanish) in Appendix 3.

Where an item has a two digit reference number (e.g. 05 or 33), this relates to the full description against that number in Appendix 3. Where an item has an alphabetic suffix (e.g. 12A or 34B) this indicates that the standard description has been modified. Where a modification is major the complete modified description is included here and the standard description should be ignored.

Where a modification is minor (e.g. the insertion of a named hardwood) the shortened description has been modified here but, in general, the full description in Appendix 3 prevails.

The unit rates below are for main work items on a typical construction project in the Warsaw area in the first quarter of 1999. The rates include all necessary labour, materials, equipment and, where appropriate, allowances for contractor's overheads and profit, preliminary and general items and contractor's profit and attendance on specialist trades.

		Unit	Rate Zl
Excavation			
01	Mechanical excavation of foundation trenches	m^3	80
02	Hardcore filling making up levels	m^2	50
03	Earthwork support	m^2	12
Concrete work			
04	Plain insitu concrete in strip foundations in trenches	m^3	279
05	Reinforced insitu concrete in beds	m^3	290
06	Reinforced insitu concrete in walls	m^3	315
07	Reinforced insitu concrete in suspended floor or roof slabs	m^3	350
08	Reinforced insitu concrete in columns	m^3	384
09	Reinforced insitu concrete in isolated beams	m^3	370
10	Precast concrete slab	m^2	15
Formwork			
11	Softwood formwork to concrete walls	m^2	36
12	Softwood formwork to concrete columns	m^2	45
13	Softwood or metal formwork to horizontal soffits of slabs	m^2	53
Reinforcement			
14	Reinforcement in concrete walls (16mm)	tonne	2,800
15	Reinforcement in suspended concrete slabs	tonne	2,800
16	Fabric reinforcement in concrete beds	m^2	15
Steelwork			
17	Fabricate, supply and erect steel framed structure	tonne	10,000

		Unit	Rate Zl
Brickwork and blockwork			
18	Precast lightweight aggregate hollow concrete block walls	m^2	42
19	Solid (perforated) concrete bricks	m^2	53
21	Facing bricks	m^2	130
Roofing			
22	Concrete interlocking roof tiles	m^2	115
23	Plain clay roof tiles	m^2	195
25	Sawn softwood roof boarding	m^2	50
27	3 layers glass-fibre based bitumen felt roof covering	m^2	92
28	Bitumen based mastic asphalt roof covering	m^2	110
29	Glass-fibre mat roof insulation	m^2	25
30	Rigid sheet loadbearing roof insulation 75mm thick	m^2	16
31	Troughed galvanised steel roof cladding	m^2	65
Woodwork and metalwork			
32	Preservative treated sawn softwood 50 x 100mm	m	8
33	Preservative treated sawn softwood 50 x 150mm	m	13
34	Single glazed casement window in hardwood, size 650 x 900mm	each	250
35	Two panel glazed door in hardwood, size 850 x 2000mm	each	1,500
36	Solid core half hour fire resisting hardwood internal flush doors, size 800 x 2000mm	each	2,900
37	Aluminium double glazed window, size 1200 x 1200mm	each	1,300
38	Aluminium double glazed door, size 850 x 2100mm	each	1,500
39	Hardwood (Oak) skirtings	m	100
40	Framed structural steelwork in universal joist sections	tonne	8,600
41	Structural steelwork lattice roof trusses	tonne	10,800
Plumbing			
42	UPVC half round eaves gutter	m	36
43	UPVC rainwater pipes	m	45
44	Light gauge copper cold water tubing	m	24
45	High pressure plastic pipes for cold water supply	m	18
47	UPVC soil and vent pipes	m	40
48	White vitreous china WC suite	each	780
49	White vitreous china lavatory basin	each	450
50	Steel enameled shower tray	each	495
51	Stainless steel single bowl sink and double drainer	each	280

		Unit	Rate Zl
Electrical work			
52A	PVC insulated and PVC sheathed copper cable core and earth	m	6
53	13 amp unswitched socket outlet	each	40
54	Flush mounted 20 amp, 1 way light switch	each	38
Finishings			
55	2 coats gypsum based plaster on brick walls	m^2	35
56	White glazed tiles on plaster walls 150 x 150mm	m^2	95
57	Medium-high quality floor tiles 3000 x 300mm	m^2	105
58	Cement and sand screed to concrete floors 50mm thick	m^2	30
59	Thermoplastic floor tiles on screed	m^2	80
60	Mineral fibre tiles on concealed suspension system	m^2	95
Glazing			
61	Glazing to wood	m^2	48
Painting			
62	Emulsion on plaster walls	m^2	12
63	Oil paint on timber	m^2	18

Approximate estimating

The building costs per unit area given on the next page are averages incurred by building clients for typical buildings in the Warsaw area as at the first quarter 1999. They are based upon the total floor area of all storeys, measured between external walls and without deduction for internal walls.

Approximate estimating costs generally include mechanical and electrical installations but exclude furniture, loose or special equipment, and external works; they also exclude fees for professional services. The costs shown are for specifications and standards appropriate to Poland and this should be borne in mind when attempting comparisons with similarly described building types in other countries. A discussion of this issue is included in section 2. Comparative data for countries covered in this publication, including construction cost data, are presented in Part Three.

Approximate estimating costs must be treated with caution; they cannot provide more than a rough guide to the probable cost of building. All the rates in this section exclude VAT which is at 7%.

	Cost Zl per m²	Cost Zl per ft²
Industrial buildings		
Distribution centre (11,000 m²)	1,740	161
Warehouses, low bay for owner occupation (including heating)	2,330	216
Administrative and commercial buildings		
Offices for letting, 5 to 10 storeys, non air conditioned	3,400	316
Offices for letting, high rise, air conditioned (6 storey)	3,710	345
Recreation and arts buildings		
Theatres (over 500 seats) including seating and stage equipment	4,160	387
Sports halls including changing and social facilities	2,950	274
Swimming pools (international standard) including changing facilities	1,910	177
Residential buildings		
Social/economic single family housing (multiple units)	1,220	113
Purpose designed single family housing 2 storey detached (single unit)	1,630	152
Private sector apartment building (standard specification)	2,430	226
Private sector apartment building (luxury)	2,640	245
Motel	1,460	135
Others		
Drive thru fast food (excl. fixtures fittings, equipment incl. external works)	4,690	435
Supermarket (4,500 to 5,000 m²)	2,670	248
Supermarket (11,000 to 12,000 m²)	2,290	213
Hypermarket (20,000 m²)	2,120	197

Regional variations

The approximate estimating costs are based on average Warsaw rates. Adjust these costs by the following factors for regional variations:

Southern Poland – Krakow	–5%
North – Gdansk	–4%
West – Poznan	–4%
East – Bydgoszczl	–6%

Value added tax (VAT)

Value added tax is levied on most materials at, 7% but on selected items at 22%.

Exchange rates and cost and price trends

The combined effect of exchange rates and inflation on prices within a country and on price comparisons between countries is discussed in section 2.

Exchange rates

The graph below plots the movement of the Polish zloty against sterling, the ECU/Euro, the US dollar and 100 Japanese yen since 1995 the date when the zloty was substantially devalued.. The values used for the graph are quarterly and the method of calculating these is described and general guidance on the interpretation of the graph provided in section 2. The average exchange rate in the first quarter of 1999 was Zl 5.82 to the pound sterling, Zl 3.47 to the US dollar and Zl 4.10 to the Euro.

The Polish zloty against sterling, the ECU/Euro, the US dollar and 100 Japanese yen

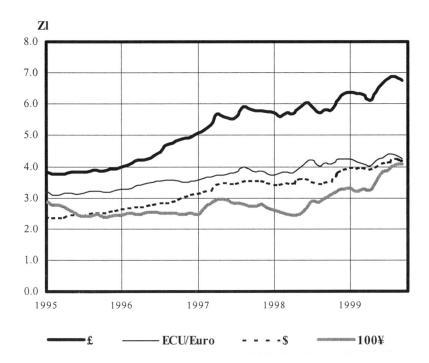

Cost and price trends

The table below presents the indices for consumer prices in Poland since 1990. The consumer price index has been rebased to 1990=100. The annual change is the percentage change between the average index of consecutive years. Notes on inflation are included in section 2.

Consumer price index

	Consumer prices	
Year	annual average	change %
1990	100.0	365.1
1991	176.7	76.7
1992	256.8	45.3
1993	351.5	36.9
1994	468.4	33.3
1995	593.9	26.8
1996	713.9	20.2
1997	827.3	15.9
1998	924.1	11.7

Source: International Monetary Fund
Yearbook 1998 and supplements 1999.

Useful addresses

Government and public organizations

Ministry of Internal Affairs and Administration
Batorego 5
02-591 Warszawa
Tel: (48-22) 6210251
Fax: (48-22) 497494

Polish Agency for Foreign Investment
Roz 2
00-559 Warszawa
Tel: (48-22) 6216261
Fax: (48-22) 6218427

Office of Housing and Urban Development
Wspólna 2-4
00-926 Warszawa
Tel: (48-22) 6618111
Fax: (48-22) 6295389

Department of Investor Servicing
Roz 2
00-559 Warszawa
Tel: (48-22) 6210706
Fax: (48-22) 6226169

Polish Centre for Testing and Certification
Klobucka 23A
02-699 Warszawa
Tel: (48-22) 430595

Central Office of Measures
Elektoralna 2
00-950 Warszawa
Tel: (48-22) 200241
Fax: (48-22) 208378

The Main Office of Building Supervision
Krucza 38-42
00-512 Warszawa
Tel: (48-22) 6618010
Fax: (48-22) 6618142

The Office for Public Procurement
Szucha 2-4
00-582 Warszawa
Tel: (48-22) 6946750
Fax: (48-22) 6946206

Institute of Construction Engineering
Filtrowa 1
00-950 Warszawa
Tel: (48-22) 250471
Fax: (48-22) 251303

State Hygiene Institute
Chocimska 24
00-791 Warszawa
Tel: (48-22) 497814
Fax: (48-22) 497814

Trade and professional associations

Polish Housing Construction Association
Chmielna 54-57
80-748 Gdansk
Tel: (48-58) 316851
Fax: (48-22) 314217

The National Construction Chamber
Zielna 49
00-108 Warszawa
Tel: (48-22) 6207082
Fax: (48-22) 6242172

The Chamber of Construction Designing and Consulting
Novogrodzka 12
00-511 Warszawa
Tel: (48-22) 6298039
Fax: (48-22) 6211572

Chamber of Commerce of the Building Industry
Wspólna 2
00-926 Warszawa
Tel: (48-22) 6283528
Fax: (48-22) 6283528

The Corporation of Building
Contractors (UNIBUD)
Nowogrodzka 21
00-950 Warszawa
Tel: (48-22) 298213
Fax: (48-22) 298081

Chamber of Architectural Designing
Foksal 2
00-366 Warszawa
Tel: (48-22) 6200616
Fax: (48-22) 6241251

Other organizations

Permanent Building Exhibition
Bartycka 26
00-716 Warszawa
Tel: (48-22) 404662
Fax: (48-22) 404662

Housing Research Institute
Filtrowa 1
00-925 Warszawa
Tel: (48-22) 250683
Fax: (48-22) 250953

Portugal

All data relate to 1998 unless otherwise indicated.

Population
Population	9.9 million
Urban population (1997)	37%
Population under 15	17%
Population 65 and over	15%
Average annual growth rate (1985 to 1998)	0.3%

Geography
Land area	92,391 km^2
Agricultural area	44%
Capital city	Lisbon

Economy
Monetary unit	Escudo (Esc)
Exchange rate (average first quarter 1999) to:	
the pound sterling	Esc 285.80
the US dollar	Esc 171.60
the Euro	Esc 200.50
the yen x 100	Esc 145.20
Average annual inflation (1995 to 1998)	2.7%
Inflation rate	2.8%
Gross Domestic Product (GDP)	Esc 19,370 billion
GDP PPP basis (1997)	$ (PPP) 149.5 billion
GDP per capita	Esc 1,951,000
GDP per capita PPP basis (1997)	$ (PPP) 15,055
Average annual real change in GDP (1995 to 1998)	2.4%
Private consumption as a proportion of GDP	70%
Public consumption as a proportion of GDP	21%
Investment as a proportion of GDP	27%

Construction
Gross value of construction output	Esc 2,714 billion
Gross value of construction output per capita	Esc 274,100
Gross value of construction output as a proportion of GDP	14.0%
Average annual real change in gross value of construction output (1995 to 1998)	7.2%
Annual cement consumption	9.4 million tonnes

PPP Purchasing power parity

The construction industry

Construction output

The value of the gross output of the construction sector in Portugal in 1998 was
Esc 2,714 billion equivalent to 13.5 billion ECUs (nearly the same as January
1999 Euros). This represents 14.0% of GDP. In addition it is estimated that
Esc 298 billion of output or an extra 11% are undertaken by other sectors, the
black economy and DIY. The breakdown of output is shown in the table below:

Output of construction sector 1998 (current prices)

Type of work	Esc billion	ECUs billion	% of total
New work			
Residential building	849	4.2	31
Non-residential building			
Office and commercial	241	1.2	9
Industrial	98	0.5	4
Other	357	1.8	13
Total	697	3.5	26
Civil engineering	1,005	5.0	37
Total new work	2,550	12.7	94
Repair and maintenance			
Residential	59	0.3	2
Non-residential	51	0.3	2
Civil engineering	53	0.3	2
Total repair and maintenance	164	0.8	6
Total	**2,714**	**13.5**	**100**

Note: the value of the Euro in January 1999 was veru similar to that of the ECU in 1998.
Source: based on Afonso, F.P., 'Portugal' in Euroconstruct conference proceedings,
        Prague, June 1999.

The proportion of total construction output in renovation work is extremely low
compared to any country in the world and certainly any country in Europe. A
part – possibly a large part – of the 11% additional work in black market etc.
may be repair and maintenance but even this would not bring the total to usual
levels. It may be that there is significant under-counting, due to deliberate
avoidance of VAT. The high level of new construction pushes down the
percentage of repair and maintenance.

1997 was a boom year for construction with an increase in output of nearly 13%. 1998 output increased by just under 5%. All sectors for new work followed a similar pattern of large increases in 1997 followed by a moderate one in 1998. Renovation increased marginally in building for 1998 but decreased in civil engineering.

Residential building construction is very buoyant. In the three years 1995 to 1998, according to the Central Bank, the growth in residential mortgage lending was 24%, 27% and 35%. Reasons include a fall in interest rates, increases in incomes and a reduction of unemployment. If these continue, residential building may well continue to grow in the short term, a view which is also supported by the continuing increase in residential building permits. An increase in house prices may have a dampening effect. About two thirds of units constructed are in flats. The market of private housing for sale is very healthy but the rental market is less bouyant

Non-residential construction in 1998 barely changed though the growth in 1997 was 12%. The same pattern emerges in the private sector and in the public sector schools, universities and hospital construction. Public sector investment has been hit by the Maastricht criteria. The reasons for the fall in the private sector are less obvious in view of favourable interest rates and a satisfactory growth in the economy, but no early rise can be expected.

The civil engineering market, being still largely publicly funded, is also affected by constraints on the economy as a result of joining the EU. However, the EU itself funds much of the infrastructure construction. Privately funded projects, especially for road construction, environmental infrastructure are expected to increase.

Repair and maintenance is, as already stated, very low. In housing this has partly been due to rent control over 40 years and continuing low rents. It has been estimated that a third of housing needs renovation. It may be that easing of the pressure of new construction will permit an increase in housing repair and maintenance.

In non-residential buildings also there is a need for renovation both in the private and public sector. It seems that public renovation, especially historical buildings in the cities, increased which caused an overall increase in 1998. The infrastructure of schools, universities and hospitals badly needs attention and it seems that the public sector is increasing repair and maintenance expenditure. Private sector institutions and companies are increasingly buying old buildings, for renovation as offices, rather than building new. It seems not all such activity shows in the figures.

Civil engineering infrastructure also needs maintenance and budgets may allow this if new construction expenditure does not increase so much. Overall the outlook is for an increase in repair and maintenance.   It could hardly be permitted to fall from such very low levels.

The geographical distribution of population and dwelling completions in 1992 is shown below.   The figures show a spread roughly related to population. Although these figures are old they are probably a good indication of current geographical spread.

Population and dwelling completions 1992 regional distribution

Regions	Population (%)	Dwelling completions (%)
Aviero	6.7	7.6
Beja	1.7	0.9
Braga	7.6	7.0
Bragança	1,6	1.4
Castelo Branco	2.2	2.0
Coimbra	4.3	4.3
Évora	1.8	1.2
Faro	3.5	8.7
Guardia	1.9	2.1
Leiria	4.3	5.9
Lisboa	20.8	18.9
Portalegre	1.4	0.7
Porto	16.7	16.1
Santarém	4.5	4.2
Setúbal	7.3	7.7
Viana do Castelo	2.5	2.6
Vila Real	2.4	2.3
Viseu	4.1	3.9
Açores	2.4	0.8
Madeira	2.6	1.7
Total	**100.0**	**100.0**

Sources: Anuário Estatístico de Portugal 1993, Instituto Nacional de Estatística.

Characteristics and structure of the industry

Both the 'main contractor' and the 'separate trades' system's operate in Portugal. The contractor usually undertakes detailed design.

Engineers are dominant in the design of construction projects and there are about 15 engineers for every one architect.   Engineers' education is more broadly based than in the UK and engineers hold many of the key posts in government and in private industry. There are two types of engineers: those with a five year university degree and those with a three year technical college qualification, usually referred to as technical engineers. Engineers are involved not only in design, but also in site and project management and administration.

Architects have improved their status and influence in recent years. The Association of Architects is an organization to which, by law, all Portuguese architects working in Portugal must belong. Non-resident European Community architects have to register with the Association to undertake work in Portugal but need not be members. Most architects practice as individuals and there are few large practices. A law of 1988 redefines the lines of professional competence between the architect and the engineer but, since the allocation is not mandatory, architects have no monopoly over the design of buildings. The architect also has no role in project supervision.

Estimates of the number of contractors vary from 20,000 to 30,000 firms. Employment was estimated to be around 350,000 persons in 1996. The industry is made up of a large number of small and medium sized enterprises. There are few large firms but they do a relatively high proportion of the work, especially in the civil engineering sector. Around two-thirds of Portuguese companies employ less than 5 workers and around 7% employ more than 20 workers. There are seven Portuguese companies in *Building* magazine's list of 'Top 300 European Contractors' for 1997 (see table below). 'Engil' is the only Portuguese contractor listed in *Engineering News Record's* list of the 'Top 225 International Contractors' for 1997, at rank 182.

Major Portuguese contractors, 1997

Major contractors	*Place in* Building's 'Top 300 European Contractors'
Somague	91
Soares da Costa	119
Teixeira Duarte	133
Engil	151
Mota and Companhia	156
Bento Pedroso Construcoes	193
Edifer	215

Source: Building December 1998.

Most work abroad is undertaken in former Portuguese colonies. All contractors need a licence to operate and the category of work for which the licence is given depends on various factors, including the financial and technical capacity of the company in that field.

Clients and finance

The public sector in Portugal benefits from considerable financial support from the European Union (Structural Funds from the European Regional Development Fund, the European Social Fund and the European Agricultural Guidance and Guarantee Fund). A large proportion of the share destined for construction is for civil engineering projects.

New property investment companies (*Sociedade gestora de investimento imobiliario – SGILs*) have been established to invest in construction and these receive major tax benefits. The three institutions which specialize in property financing are the General Savings Bank (*Caixa General dos Depositos – CGD*) for civil construction and owner occupied housing, the Real Estate Credit Bank (*Credito Predial Portugues – CPP*) for mortgage finance, and the National Development Bank (*Banco Nacional de Fomento* – BNF) for development projects.

Contractual arrangements

Residential output accounts for a very significant share of construction work in Portugal. This factor, along with a significant presence of foreign contractors and developers, has created a split between the north and the south of the country. Big contractors in the North usually do their work under a lump sum contracting arrangement, with the client using an architect for the design. Most of the work is done by contractors in-house and subcontracting is used only for specialized trades. In the residential sector in the South, however, most of the work is carried out on a design and build basis.

Construction contracts refer to the Civil Code along with the legislation governing public works contracts for procedures that are not provided for in the Civil Code. The legislation also provides for control of the work at different stages, supervision of the quality of work, checks on the conformity to law of contractual provisions, assessment sanction and surveys of provisional work, final work and performance during the guarantee period.

A law of 1986 regulates public procurement procedures for construction by what is known as the *régime juridico*. The *régime* is a document of 236 articles which constitutes a legal framework for contracts and of which certain clauses will become embedded in the contract unless specifically excluded. It also determines procedures and payment methods.

The normal contractor selection method is 'open' so long as firms have the appropriate category of licence to undertake the work. Choice is then usually based on price. There is an increasing practice of negotiation and undertaking building contracts with specific clauses.

The building process is agreed by both designers and contractors but is not formally regulated. Several stages are needed in the building process: planning permission, building permit, procurement, acceptance and post-construction liability.

Public procurement procedures are determined by a statute that provides the legal framework for contracts. 'Normal' procedures pertain to simple projects (like the EU 'open procedure'). Moreover, there is special legislation for the construction of large and/or complex projects, for which special 'prequalification' is required (like the EU 'restricted procedure'). Requirements include technical and financial capabilities, track record on projects and holdings of equipment. About five contractors or consortia are invited and given final drawings and documentation for the tender process. Proposals must follow the rules in 'decree 348-A/86' and in '*Portarais* 6QS-A-B-C/86'. Each stage of the process is separately considered and the contractor must submit documents in separate envelopes as follows:

- qualifications, financial capacity and proof of payment of taxes
- specifications of materials and equipment in accordance with tender documents
- cost and price data including unit prices of materials, catalogues and specifications of equipment or systems to be used
- documents setting out how the contract will be carried out including, for example, the programme, contract period, payment revision procedures, bonuses and penalties and defects liability period.

The lowest price is not the sole selection criterion; all the other factors are included in the final analysis.

The three most popular contractual methods for both evaluating the cost of work and making payment to contractors are:

- Total 'lump sum' (fixed fee) contracts. Payment is in accordance with a schedule and programme of works presented with the proposal.
- Unit rates and quantities contracts. The evaluation of quantities and progress of works is the responsibility of the client or his team known as *fiscals*. Payment is made monthly based on unit prices applied to the quantity of work done.
- 'Cost plus'. The quantities of works may be defined by the client at the time of signing the contract or gradually with the evolution of the project. An agreed management margin is added to the actual cost of the work by the contractor. This type of contract needs highly qualified cost-control personnel on both sides to operate efficiently.

Many private contracts follow the public system using either a main contractor or the separate trades' system. However, in Portugal there is great flexibility in the way contracts are organized and it is difficult to determine which is the prevalent method. No standard forms of contract for private works exist. Typical provisions include protection for a contractor against increases in the contract sum where the works are delayed for reasons not dependent on the contractor's will. The payment structure implies that the contractor submits an invoice (with a record of the previous month measured work) and the client must submit any objection within five days of the date of the invoice and must pay three days thereafter any amount not in dispute. The client pays interest on overdue payments.

Liability and insurance

The 'Civil Code' established in 1867 and revised for construction in 1966 and 1977 deals with civil liability for construction for all parties to the process. Specific articles are 1207 and 1230. In theory a builder can be held responsible for breaches under the building contract and the 'Civil Code', though the courts have yet to exercise such powers outside the contractual terms.

Post-construction liability for private works under the 'Civil Code' extends to five years. Post-construction liability for public works is regulated by the *régime juridico* and lasts for two years, though this is usually reduced in the contract to one year.

All post-construction liability rests with the contractor, who normally undertakes the detailed design. There are no specific laws dealing with the engineer's or architect's liability. If the fault is a major design fault, it is up to the courts to determine liability. However, the Portuguese legal system is very slow; it may well take over 10 years for a case to go through the system. The danger of an architect, engineer, or indeed a contractor, being taken to court is slight. There has been an increased tendency to use the arbitration process because of the slowness of the court system.

Contractors usually insure themselves with an all-risks policy and a latent defects policy. Some professionals take out indemnity insurance for all their work. However, as their risks are small, this is not very common.

Development control and standards

The relevant basic legal text on planning for private work is 'decree 166/70' of 15 April 1970. Public works need no formal permission from local authorities. It is recommended that plans should be produced by local authorities, ranging from general strategic development plans to detailed plans for a certain area. However, under the Portuguese constitution, municipalities are autonomous organizations and if they have not drawn up any plans, the granting of permission is done on an *ad hoc* basis. Before applying for permission, any prospective developer can ask for a description of the type of work which may be undertaken on a specific piece of land.

Planning permission is compulsory for almost all kinds of private construction works, including extensions, restoration, rebuilding or demolition of buildings. There exist exceptions for only minor maintenance or conversion works. The law that regulates the procedure for building permission, and provides a guideline of the basis on which permission is obtained is applicable only to private works because public works do not need approval from the municipal authorities. Project proposals are first presented to the local authority and are then assessed by the planning architects, technical services and health departments. The proposal is then returned with any requirements that must be met to obtain a licence. The architect has to re-submit his proposal with a declaration of conformity to the general and specific norms of construction and appropriate regulations as stated by the law, as well as with the requirements by the local authority. If these conditions are satisfied, a building permit can be granted.

There are different procedures for small projects, large projects and tourist developments. Applications for small and large project are submitted to municipalities, whereas applications for tourist developments are submitted to the Ministry of Tourism. For large projects in particular, the procedure involves submission to, or consultation with, up to eighteen government or sub-government departments. Obtaining planning permission is a lengthy, bureaucratic and uncertain process.

The main objectives of technical control legislation in Portugal are health and safety. The building regulations are set out in 'ministerial decree 718/87' of 21 August 1987. They generally do not exercise much control of building design, choice of materials or fire resistance, except in Lisbon and Oporto and on very special projects.

While the Ministry of Public Works has overall responsibility for building standards and norms, it is the Commission for Revision and Establishment of Technical Regulations, together with the Superior Council for Public Works, which produces the regulations. The *Laboratorio Nacional de Engenharia Civil* (LNEC) is responsible for the approval and classification of building materials and components, plus quality control. In reality, however, few products are tested and approved, partly because the process takes so long. Moreover, LNEC is mainly interested in civil engineering research, and not in the approval of building materials. Standards set are usually European Community standards adopted by Portugal but they are not always observed and some materials are of poor quality. Regulations are, in theory, enforced by the municipalities, but in practice the law requires only that the designer declares that he has conformed to the construction norms and observed all relevant regulations and any conditions laid down by the municipality. The site is inspected before foundations are laid and again after the foundations have been built. On completion, an occupation licence is granted if all regulations have been observed.

Construction cost data

Cost of materials

The figures that follow are the costs of main construction materials, delivered to site in the Lisbon area, as incurred by contractors in the first quarter of 1999. These assume that the materials would be in quantities as required for a medium sized construction project and that the location of the works would be neither constrained nor remote. All the costs in this section exclude value added tax (VAT) which is at 17%.

	Unit	Cost Esc
Cement and aggregates		
Ordinary portland cement in 50kg bags	tonne	15,300
Coarse aggregates for concrete	m^3	2,130
Fine aggregates for concrete	m^3	1,960
Ready mixed concrete	m^3	11,500
Ready mixed concrete	m^3	13,500
Steel		
Mild steel reinforcement	tonne	71,200
Structural steel sections	tonne	81,700
Bricks and blocks		
Common bricks	1,000	43,100
Good quality facing bricks	1,000	85,100
Hollow concrete blocks	no.	135
Solid concrete blocks	no.	160
Precast concrete cladding units with exposed aggregate finish	m^2	2,400
Glass and ceramics		
Float glass (4mm thick)	m^2	3,000
Sealed double glazing units	m^2	8,050
Good quality ceramic wall tiles	m^2	2,270

	Unit	Cost Esc
Plaster and paint		
Plaster in 50kg bags	tonne	28,900
Emulsion paint in 5 litre tins	litre	1,150
Gloss oil paint in 5 litre tins	litre	1,730
Tiles and paviors		
Clay floor tiles	m^2	1,380
Vinyl floor tiles	m^2	3,620
Precast concrete roof tiles	1,000	81,700
Clay roof tiles	1,000	80,500
Plumbing		
Lavatory basin complete	no.	20,700
100mm diameter clay drain pipes	m	2,300

Unit rates

The descriptions of work items following are generally shortened versions of standard descriptions listed in five languages (English, French, Italian, German and Spanish) in Appendix 3.

Where an item has a two digit reference number (e.g. 05 or 33), this relates to the full description against that number in Appendix 3. Where an item has an alphabetic suffix (e.g. 12A or 34B) this indicates that the standard description has been modified. Where a modification is major the complete modified description is included here and the standard description should be ignored.

Where a modification is minor (e.g. the insertion of a named hardwood) the shortened description has been modified here but, in general, the full description in Appendix 3 prevails.

The unit rates below are for main work items on a typical construction project in the Lisbon area in the first quarter of 1999. The rates include all necessary labour, materials and equipment, contractor's overheads and profit (10%), preliminary and general items (15%) and contractor's profit and attendance on specialist trades (5%). All the rates in this section exclude VAT which is at 17%.

		Unit	Rate Esc
Excavation			
01	Mechanical excavation of foundation trenches	m^3	1,820
02	Hardcore filling making up levels	m^2	780
03	Earthwork support	m^2	1,040

	Unit	Rate Esc
Concrete work		
04 Plain insitu concrete in strip foundations in trenches	m^3	14,800
05 Reinforced insitu concrete in beds	m^3	15,600
06 Reinforced insitu concrete in walls	m^3	14,800
07 Reinforced insitu concrete in suspended floor or roof slabs	m^3	16,900
08 Reinforced insitu concrete in columns	m^3	14,800
09 Reinforced insitu concrete in isolated beams	m^3	16,900
10 Precast concrete slab	m^2	10,400
Formwork		
11 Softwood formwork to concrete walls	m^2	2,340
12 Softwood or metal formwork to concrete columns	m^2	2,340
13 Softwood or metal formwork to horizontal soffits of slabs	m^2	2,210
Reinforcement		
14 Reinforcement in concrete walls (16mm)	kg	130
15 Reinforcement in suspended concrete slabs	kg	125
16 Fabric reinforcement in concrete beds	m^2	325
Brickwork and blockwork		
18 Precast lightweight aggregate hollow concrete block walls (100mm thick)	m^2	3,450
19 Solid (perforated) concrete bricks (100mm thick)	m^2	3,320
20 Solid perforated sand lime bricks	m^2	3,500
21 Facing bricks	m^2	3,750
22 Concrete interlocking roof tiles	m^2	4,550
23 Plain clay roof tiles	m^2	5,850
24 Fibre cement roof tiles	m^2	3,900
26 Particle board roof coverings	m^2	3,900
Roofing		
27 3 layers glass-fibre based bitumen felt roof covering include chippings	m^2	4,160
28 Bitumen based mastic asphalt roof covering	m^2	2,860
29 Glass-fibre mat roof insulation	m^2	3,250
30 Load bearing glass-fibre roof insulation	m^2	1,560
31 Galvanised steel roof cladding	m^2	5,460
Woodwork and metalwork		
36 Solid core half hour fire resisting flush internal door	each	104,000
Plumbing		
42 UPVC half round eaves gutter	m	4,490
43 UPVC rainwater pipes with pushfit joints	m	2,920
44 Light gauge copper cold water tubing	m	7,150
45 High pressure UPVC pipes for cold water supply	m	3,380
46 Low pressure UPVC pipes for cold water distribution	m	4,160

		Unit	Rate Esc
47	UPVC soil and vent pipes	m	3,250
48	China WC suite	each	23,400
49	Lavatory basin	each	15,600
50	Glazed fireclay shower tray	each	15,600
51	Stainless steel single bowl sink	each	32,500
Electrical work			
52	PVC insulated copper sheathed cable	m	402
53	Socket outlet and copper cable	each	4,520
54	Lightswitch and copper cable	each	4,070
Finishings			
55	2 coats gypsum based plaster on brick walls	m²	1,950
56	White glazed tiles on plaster walls	m²	3,900
57	Red clay quarry tiles on concrete floor	m²	3,640
58	Cement and sand screed to concrete floors 50mm thick	m²	715
59	Thermoplastic floor tiles on screed	m²	3,380
60	Mineral fibre tiles on concealed suspension system	m²	5,850
Glazing			
61	Glazing to wood	m²	3,900
Painting			
62	Emulsion on plaster walls	m²	520
63	Oil paint on timber	m²	3,900

Approximate estimating

The building costs per unit area given overleaf are averages incurred by building clients for typical buildings in the Lisbon area as at the first quarter 1999. They are based upon the total floor area of all storeys, measured between external walls and without deduction for internal walls.

Approximate estimating costs generally include mechanical and electrical installations but exclude furniture, loose or special equipment, and external works; they also exclude fees for professional services. The costs shown are for specifications and standards appropriate to Portugal and this should be borne in mind when attempting comparisons with similarly described building types in other countries. A discussion of this issue is included in section 2. Comparative data for countries covered in this publication, including construction cost data, are presented in Part Three.

Approximate estimating costs must be treated with caution; they cannot provide more than a rough guide to the probable cost of building. All the rates in this section exclude VAT which is at 17%.

	Cost Esc/m²	Cost Esc/ft²
Industrial buildings		
Factories for letting	45,000	4,180
Factories for owner occupation (light industrial use)	45,000	4,180
Factories for owner occupation (heavy industrial use)	55,000	5,110
Factory/office (high-tech) for letting (shell and core)	60,000	5,570
Factory/office (high-tech) for letting ground floor shell, first floor offices)	60,000	5,570
Factory/office (high-tech) for owner occupation (controlled environment, fully furnished)	70,000	6,500
High tech laboratory workshop centres	90,000	8,360
Warehouses, low bay (6 to 8m high) for letting (no heating)	45,000	4,180
Warehouses, low bay for owner occupation (including heating)	50,000	4,650
Warehouses, high for owner occupation (including heating)	60,000	5,570
Cold stores/refrigerated areas	90,000	8,360
Administrative and commercial buildings		
Civic offices, non air conditioned	90,000	8,360
Civic offfices, fully air conditioned	120,000	11,100
Offices for letting, 5 to 10 storeys, non air conditioned	80,000	7,430
Offices for letting, 5 to 10 storeys, air conditioned	110,000	10,200
Offices for letting, high rise, air conditioned	90,000	8,360
Offices for owner occupation, 5 to 10 storeys, air conditioned	110,000	10,200
Offices for owner occupation, 5 to 10 storeys, non air conditioned	80,000	7,430
Offices for owner occupation, high rise, air conditioned	90,000	8,360
Prestige/headquarters office, 5 to 10 storeys, air conditioned	150,000	13,900
Prestige/headquarters office, high rise, air conditioned	100,000	9,290
Health and education buildings		
General hospitals	140,000	13,000
Teaching hospitals	140,000	13,000
Private hospitals	140,000	13,000
Health centres	120,000	11,100
Nursery schools	120,000	11,100
Primary/junior schools	120,000	11,100
Secondary/middle schools	110,000	10,200
University (arts) buildings	130,000	12,100
University (science) buildings	130,000	12,100
Residential buildings		
Social/economic single family housing	60,000	5,570

	Cost Esc/m²	Cost Esc/ft²
Private/mass market single family housing 2 storey detached/semidetached (multiple units)	70,000	6,500
Purpose designed single family housing single unit	80,000	7,430
Social/economic single apartment housing low rise	55,000	5,110
Social/economic single apartment housing high rise	60,000	5,570
Private sector apartment building (standard specification)	90,000	8,360
Private sector apartment building (luxury)	140,000	13,000
Student/nurses halls of residence	95,000	8,830
Homes for the elderly (shared accommodation)	100,000	9,290
Homes for the elderly (self contained)	120,000	11,100
Hotel 5 star city centre	180,000	16,700
Hotel 3 star city/provincial	120,000	11,100
Motel	100,000	9,290

Value added tax (VAT)

The standard rate of value added tax (VAT) is currently 17%, chargeable on general building work and materials.

Exchange rates and cost and price trends

The combined effect of exchange rates and inflation on prices within a country and on price comparisons between countries is discussed in section 2.

Exchange rates

The graph on the following page plots the movement of the Portuguese escudo against sterling, the ECU/Euro, the US dollar and 100 Japanese yen since 1990. The values used for the graph are quarterly and the method of calculating these is described and general guidance on the interpretation of the graph provided in section 2. The average exchange rate in the first quarter of 1999 was Esc 286 to the pound sterling, Esc 172 to the US dollar and Esc 200 to the Euro.

The Portuguese escudo against sterling, the ECU/Euro, the US dollar and 100 Japanese yen

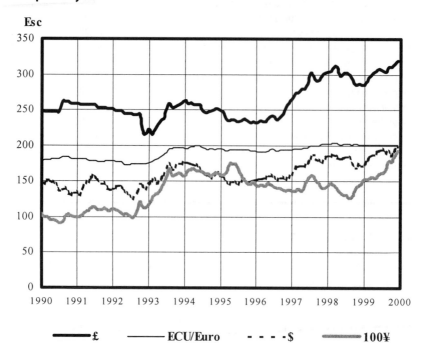

Cost and price trends

The table overleaf presents the indices for consumer prices and residential multiple dwelling costs in Portugal since 1990. The indices have been rebased to 1990=100. The annual change is the percentage change between the average index of consecutive years. Notes on inflation are included in section 2. It is noteworthy that dwelling costs have risen faster than consumer prices.

Consumer price and residential multiple dwelling cost indices

Year	Consumer prices annual average	change %	Residential multiple dwelling costs annual average	change %
1990	100.0	13.4	100.0	
1991	111.4	11.4	112.9	12.9
1992	121.3	8.9	124.4	10.2
1993	129.6	6.8	131.0	5.3
1994	135.9	4.9	139.6	6.6
1995	141.5	4.1	148.5	6.4
1996	145.5	2.8	154.9	4.3
1997	148.3	1.9	161.9	4.3
1998	152.4	2.8	–	–

Sources: International Monetary Fund, Yearbook 1998 and supplements 1999.
OECD, Main Economic Indicators.

Useful addresses

Government and public organizations

Ministério do Equipamento, do
Planeamento e daAdministraçao do
Território
Ministry of Planning
R. de São Mamede (ao Caldas) 21
1110 Lisboa
Tel: (351-1) 8861119
Fax: (351-1) 8863827

Ministério do Equipamento, do
Planeamento e daAdministraçao do
Território
Secretary of State for Public Works
Ministry of Planning
R. de São Mamede (ao Caldas) 21
1110 Lisboa
Tel: (351-1) 8861119
Fax: (351-1) 8862316

Instituto Nacional de Estatistica
National Statistical Institute
Avenida António José Almeida
1000 Lisboa
Tel: (351-1) 8470050
Fax: (351-1) 8489480

Trade and professional associations

Associaçao dos Industriais da Construçao Civil e Obras Públicas do Norte (AICCOPN)
Constractor's Assocation – North
Rua Alvares Cabral 306
4050 Porto
Tel: (351-2) 2080408
Fax: (351-2) 2081644

Associaçao de Empresas de Construçao e Obras Públicas (AECOPS)
Contractor's Association
Rua Suque de Palmela 20
1250 Lisboa
Tel: (351-1) 3110200
Fax: (351-1) 3562816

Associaçao dos Arquitectos Portuguese
Association of Portuguese Architects
Travessa Carvalho
12000 Lisboa
Tel: (351-1) 3432454
Fax: (351-1) 3432450

Associaçao Portuguesa de Projectistas e Consultores (APC)
Portuguese Consultants Association
Av. António Augusta Aguiar 126,7°
1050 Lisboa
Tel: (351-1) 3140476
Fax: (351-1) 3150413

Ordem dos Engenheiros
The 'Order' of Engineers
Av. António Augusta Aguiar 3–D
1050 Lisboa
Tel: (351-1) 3562438
Fax: (351-1) 3524632

Other organizations

Laboratório Nacional de Engenharia Civil (LNEC)
National Civil Engineering Laboratory
Avenue do Brasil 101
1700 Lisboa
Tel: (351-1) 8482131
Fax: (351-1) 8497660

Instituto Portugues da Qualidade
Portuguese Quality Assurance Institute
Rua C, Avenida Três Vales
2825 Monte da Caparica
Tel: (351-1) 2948100
Fax: (351-1) 2948101

ÚEOS – Komercia, a.s., Ružová dolina 27, 824 69 Bratislava, Slovak Republic

Tel: 00421 7 5823 7111   Fax: 00421 7 5341 1603

ÚEOS – Komercia, a.s. (joint stock company) is a private research and consultancy company, established in 1992 by transformation of former Ustav ckonomiky a organizacie stavebnictva, Bratislava (Institute of Building Economics and Organisation) founded in 1963.

Basic fields of company activities are as follows:

☞ applied economical research and development,

☞ entrepreneurial and economic advisory,

☞ monitoring and field research,

☞ evaluation of companies assets, real estate,

☞ public procurement,

☞ classification of building works,

☞ technical assistance,

☞ development of economic and calculation software,

☞ organisation of seminars, courses and further professional training,

☞ commercial, intermediate and publishing activities.

ÚEOS – Komercia, a.s, services are oriented towards:

● central and regional administration, municipalities, etc.,

● enterprises (small, medium and large),

● entrepreneurial associations and further similar groups,

● research and development organisations,

● foreign firms and institutions.

ÚEOS – Komercia, a.s. is a renowned and widely known company with special strength in development of market strategies, public procurement and in economical consulting. The company belongs to a group of most qualified and experienced companies in the Slovak Republic

Slovak Republic

All data relate to 1998 unless otherwise indicated.

Population

Population	5.4 million
Urban population (1997)	60%
Population under 15	21%
Population 65 and over	11%
Average annual growth rate (1995 to 1997)	0.2%

Geography

Land area	48,845 km²
Agricultural area	51%
Capital city	Bratislava

Economy

Monetary unit	Slovak Koruna (Sk)
Exchange rate (average first quarter 1999) to:	
the pound sterling	Sk 60.24
the US dollar	Sk 35.93
the Euro	Sk 42.41
the yen x 100	Sk 30.40
Average annual inflation (1995 to 1998)	6.2%
Inflation rate	6.7%
Gross Domestic Product (GDP)	Sk 718 billion
GDP PPP basis (1997)	$ (PPP) 46.3 billion
GDP per capita	Sk 133,500
GDP per capita PPP basis (1997)	$ (PPP) 8,605
Average annual real change in GDP (1995 to 1998)	5.8%
Private consumption as a proportion of GDP	49%
Public consumption as a proportion of GDP	22%
Investment as a proportion of GDP	39%

Construction

Gross value of construction output	Sk 63.7 billion
Gross value of construction output per capita	Sk 11,830
Gross value of construction output as a proportion of GDP	8.9%
Average annual real change in gross value of construction output (1995 to 1998)	4.7%
Annual cement consumption	1.5 million tonnes

PPP Purchasing power parity

The construction industry

Construction output

The value of the gross output of the construction sector in the Slovak Republic in 1998 was Sk 63.7 billion, equivalent to 1.6 billion ECUs (the ECU in 1998 was equivalent in Slovak Korunas to 104% of the January 1999 value of the Euro). This represents 8.9% of GDP. In addition it is estimated that there was a further Sk 20.4 billion (32%) of construction work undertaken by other sectors, the black economy and DIY. The breakdown of the output of the sector is shown below:

Output of construction sector, 1998 (current prices)

Type of work	Sk billion	ECUs billion	% of total
New work			
Residential building	9.0	0.22	14
Non-residential building	25.7	0.63	40
Civil engineering	18.4	0.45	29
Total new work	53.1	1.30	83
Repair and maintenance			
Residential	1.6	0.04	3
Non-residential	5.7	0.15	10
Civil engineering	3.1	0.08	5
Total repair and maintenance	10.1	0.27	17
Total	**63.7**	**1.56**	**100**

Note: the value of the ECU in 1998 in terms of Slovak currency was 4% more
        than the value of the Euro in January 1999.
Source: based on Kucera, T. 'Slovak Republic' in Euroconstruct conference
        proceedings, Prague, June 1999.

The economy of the Slovak Republic is in some difficulties due to the continuing inheritance of the planned economy and failure to adjust satisfactorily to a market economy. Unemployment has risen and the growth of GDP has slowed. It is not, therefore, surprising that construction output in 1998 fell by about 4% after rises approaching 10% per annum for the previous two years. Until the problems of the economy are to some extent corrected the fall may continue.

The largest falls have been in non-residential construction and in civil engineering, both of which declined by between 4 and 6% on 1997 levels. Residential construction which is a very low percentage of total output, maintained an increase amounting in 1998 to 11%. Renovation work which is about 17% of all work followed a similar pattern, falling in non-residential buildings and civil engineering but keeping up in the residential sector.

There is a great shortage of housing and the dwelling stock is increasing only very slowly. Completions in 1998 were expected to be about 8,200 dwellings of which flats accounted for nearly 60%. There are several problems in increasing the construction of housing: the high rate of interest, shortage of finance and high price increases for land.

Within the non-residential building sector, office and commercial new buildings and repair and maintenance account for 22%, industrial building for 37%, schools, universities and hospitals for 15% and miscellaneous for 26%. In total it has a high proportion of total output. The falls in 1998 have occurred mainly in the privately financed work (where foreign finance is important). With a budget deficit it cannot be expected that government spending will be unscathed. Several projects have not been completed. Civil engineering work which is financed mainly from public funds has already fallen.

Characteristics and structure of the industry

In 1989 the construction sector was employing around 10.3% of all employees in the economy. By 1996 the proportion had decreased to 6.9%.

In 1989 there were 231 companies in the construction sector in the Slovak Republic, if self-employed people are included. 82 companies had more than 100 employees. The average size of a company was around 2,500 employees. At the end of 1994, there were 47,666 construction companies registered, of which 44,584 were self-employed persons and 3,082 were legal entities. The changes in the size of construction firms are shown in the table below:

Size of building firms in the Slovak Republic, 1994 and 1996 (percentages)

	Self-employed persons	1-10	11-24	25-99	100-499	Over 500	Total
1994	93.5	9.1	1.1	0.8	0.5	0.1	**100.0**
1996	87.9	7.7	2.2	1.5	0.6	0.1	**100.0**

Source: ÚEOS-Komercia 1998.

After 1990 the ownership structure of construction companies has changed substantially. Before 1990, all construction companies belonged to the State. As an effect of the current privatization process, at the end of the year 1997, more than 80% of all construction companies were in private hands, as it is shown in the following table:

Construction companies according to ownership in 1993 and 1997 (percentages)

	1993	1997
Private (home)	66.8	83.3
Co-operative	5.4	5.2
State-owned	14.0	3.3
Community owned (communal)	2.0	0.1
Foreign	1.7	1.2
Mixed	4.2	1.4
Overall total	**100.0**	**100.0**

Source: ÚEOS-Komercia 1998.

The first phase of privatization of construction companies is Slovakia was mainly by means of vouchers, given to the population including workers, to purchase shares. The second phase used other methods such as public auction, competitive bids or direct sale to predetermined owners.

Design offices in the government sector are large organizations of several hundred persons, sometimes connected to large contracting organizations but often independent. In addition, there are architects working on their own account on design projects. The number of independent architects tends to rise while the size of public design offices is reducing. The qualifications of designers are not controlled but there is a central register of those authorized to undertake design work. There are two types of architect: one is trained only in the beaux arts tradition, while the other is part engineer and part architect with the title of engineer/architect.

Clients and finance

Investment in the Slovak Republic fell from 1993 to 1994, increased from 1995 to 1997 but fell in 1998.

The social sector – housing, education and health – is receiving a decreasing share of investment. A major problem is to obtain finance for housing. There is a construction savings scheme subsidized by the state by a bonus related to the amount the families have saved. A state housing development fund is not working very well and, because of shortage of funds, dwellings are being left unfinished. The mortgage system is in its infancy.

Construction cost data

Cost of labour

The figures below are typical of labour costs in the Bratislava area as at the first quarter 1999. The wage rate is the basis of an employee's income, while the cost of labour indicates the cost to a contractor of employing that employee. The difference between the two covers a variety of mandatory and voluntary contributions - a list of items which could be included is given in section 2.

	Wage rate (per month) Sk	Cost of labour (per month) Sk	Number of hours worked per year
Site operatives			
Mason/bricklayer	10,100	24,600	1,870
Carpenter	10,100	24,600	1,870
Plumber	11,200	34,200	1,870
Electrician	11,200	34,200	1,870
Structural steel erector	11,200	34,200	1,870
HVAC installer	11,200	34,200	1,870
Skilled tradesman	13,000	39,700	1,870
Semi-skilled worker	11,000	33,400	1,870
Unskilled labourer	6,500	18,200	1,870
Equipment operator	11,200	34,200	1,870
Watchman/security	7,000	17,500	1,870
Site supervision			
General foreman	12,000	–	–
Trades foreman	11,000	–	–
Clerk of works	9,000	–	–
Contractors' personnel			
Site manager	18,000	–	–
Resident engineer	15,000	–	–
Resident surveyor	10,000	–	–
Junior engineer	8,000	–	–
Junior surveyor	8,000	–	–
Planner	9,500	–	–

Cost of materials

The figures that follow are the costs of main construction materials, delivered to site in the Bratislava area, as incurred by contractors in the first quarter of 1999. These assume that the materials would be in quantities as required for a medium sized construction project and that the location of the works would be neither constrained nor remote. All the costs in this section exclude value added tax (VAT) which is at 23%.

	Unit	Cost Sk
Cement and aggregates		
Ordinary portland cement in 50kg bags	tonne	1,600
Coarse aggregates for concrete	m³	320
Fine aggregates for concrete	m³	380
Ready mix concrete B10	m³	1,050
Ready mix concrete B35	m³	1,380
Steel		
Mild steel reinforcement	tonne	19,000
High tensile steel reinforcement	tonne	24,000
Structural steel sections	tonne	27,000
Bricks and blocks		
Common bricks (29 x 14 x 6.5)	1,000	5,800
Good quality facing bricks (29 x 14 x 6.5)	1,000	6,800
Medium quality facing bricks	1,000	4,500
Precast concrete cladding units	m³	6,000
Timber and insulation		
Softwood sections for carpentry	m³	5,800
Softwood for joinery	m³	7,400
Hardwood for joinery	m³	9,600
Exterior quality plywood (6mm)	m²	450
Plywood for interior joinery (5mm)	m²	420
Softwood strip flooring (26mm)	m²	600
Chipboard sheet flooring	m²	420
100mm thick quilt insulation	m²	240
100mm thick rigid slab insulation	m²	470
Softwood internal door complete with frames and ironmongery	each	3,080
Glass and ceramics		
Float glass (4mm)	m²	190
Sealed double glazing units (300 x 300mm)	m²	320
Good quality ceramic wall tiles (150 x 150mm)	m²	440

	Unit	Cost Sk
Glass and ceramics		
Float glass (4mm)	m²	190
Sealed double glazing units (300 x 300mm)	m²	320
Good quality ceramic wall tiles (150 x 150mm)	m²	440
Plaster and paint		
Plaster in 50kg bags	yonne	3,900
Emulsion paint in 5 litre tins	litre	150
Gloss oil paint in 5 litre tins	litre	350
Tiles and paviors		
Clay floor tiles (150 x 150 x 2.5mm)	m²	280
Vinyl floor tiles (300 x 310 x 3mm)	m²	90
Precast concrete paving slabs	m²	320
Clay roof tiles	1,000	10,200
Precast concrete roof tiles	1,000	13,700
Plumbing		
WC suite complete	no.	2,000
Lavatory basin complete	no.	1,600
100mm diameter clay drain pipes	m	20
150mm diameter cast iron drain pipes	m	320

Unit rates

The descriptions of work items overleaf are generally shortened versions of standard descriptions listed in five languages (English, French, Italian, German and Spanish) in Appendix 3.

Where an item has a two digit reference number (e.g. 05 or 33), this relates to the full description against that number in Appendix 3. Where an item has an alphabetic suffix (e.g. 12A or 34B) this indicates that the standard description has been modified. Where a modification is major the complete modified description is included here and the standard description should be ignored.

Where a modification is minor (e.g. the insertion of a named hardwood) the shortened description has been modified here but, in general, the full description in Appendix 3 prevails.

The unit rates following are for main work items on a typical construction project in the Bratislava area in the first quarter of 1999. The rates include all necessary labour, materials and equipment, and allowances for contractor's overheads and profit (7%). No allowance has been made to cover preliminary and general items (6%) or contractor's profit and attendance on specialist trades (6%). All the rates in this section exclude VAT which is at 23%.

		Unit	*Rate Sk*
Excavation			
01	Mechanical excavation of foundation trenches	m³	210
02	Hardcore filling in making up levels	m³	85
03	Earthwork support	m²	48
Concrete work			
04	Plain insitu concrete in strip foundations in trenches	m³	1,810
05	Reinforced insitu concrete beds	m³	2,120
06	Reinforced insitu concrete in walls 300mm thick	m³	2,320
07	Reinforced insitu concrete in suspended floor	m³	2,100
08	Reinforced insitu concrete in columns0	m³	2,500
Formwork			
09	Reinforced insitu concrete in isolated beams	m²	2,500
10	Precast concrete slab (inc. reinforcement)	m'	7,300
11	Softwood formwork to concrete walls	m²	230
12	Softwood or metal formwork to concrete columns	m²	510
13	Softwood or metal formwork to horizontal soffits of slabs	m²	430
Reinforcement			
14	Reinforcement in concrete walls (16mm)	tonne	24,500
15	Reinforcement in suspended concrete slabs	tonne	26,500
16	Fabric reinforcement in concrete beds	tonne	24,800
Steelwork			
17	Fabricate, supply and erect steel framed structure	tonne	52,000
Brickwork and blockwork			
18	Precast lightweight aggregate hollow concrete block walls (200mm thick)	m³	5,500
19	Solid perforated clay common bricks	m²	560
20	Solid perforated sand lime bricks	m²	620
21	Facing bricks	m²	810
Roofing			
22	Concrete interlocking roof tiles	m²	560
23	Plain clay roof tiles	m²	450
24	Fibre cement roof slates	m²	460
25	Sawn softwood roof boarding	m²	350
26	Particle board roof coverings	m²	400
27	Bitumen felt roof covering	m²	320
28	Bitumen based mastic asphalt roof covering	m²	318
29	Glass-fibre mat roof insulation	m²	342
30	Loadbearing glass-fibre roof insulation	m²	304
31	Galvanised steel roof cladding	m²	392

		Unit	Rate Sk
Woodwork and Metalwork			
32	Presevative treated sawn softwood, framed in partitions	m	28
33	Presevative treated sawn softwood, pitched roof	m	42
34	Single glazed casement window in hardwood	each	2,420
35	Two panel door with panels open for glass in hardwood	each	4,200
36	Solid core half hour fire resisting hardwood	each	12,500
37	Aluminium double glazed window	each	9,360
38	Aluminium double glazed doorset and hardwood frame	each	11,800
39	Hardwood skirtings	m	85
40	Framed structural steelwork in universal joist sections	tonne	62,000
41	Structural steelwork lattice roof trusses	tonne	55,100
Plumbing			
42	UPVC half round eaves gutter	m	110
43	UPVC rainwater pipes with pushfit joints	m	125
44	Light gauge copper cold water tubing	m	285
45	High pressure UPVC cold water supply	m	65
46	Low pressure UPVC cold water distribution	m	60
47	UPVC soil and vent pipes	m	44
48	White vitreous china WC suite	each	2,800
49	White vitreous china lavatory basin	each	1,820
50	Glazed fireclay shower tray	each	2,420
51	Stainless steel single bowl sink and double drainer	each	3,500
Electrical work			
52	PVC insulated and copper sheathed cable	m	18
53	13 amp unswitched socket outlet	each	80
54	Flush mounted 20 amp, 1 way light switch	each	75
Finishings			
55	2 coats gypsum based plaster on brick walls	m^2	220
56	White glazed tiles on plaster walls	m^2	550
57	Red clay quarry tiles on concrete floor	m^2	400
58	Cement and sand screed to concrete floors 50mm thick	m^2	120
59	Thermoplastic floor tiles on screed	m^2	331
60	Mineral fibre tiles on concealed suspension system	m^2	750
Glazing			
61	Glazing to wood	m^2	340
Painting			
62	Emulsion on plaster walls	m^2	28
63	Oil paint on timber	m^2	140

Approximate estimating

The building costs per unit area given opposite are averages incurred by building clients for typical buildings in the Bratislava area as at the first quarter 1999. They are based upon the total floor area of all storeys, measured between external walls and without deduction for internal walls.

Approximate estimating costs generally include mechanical and electrical installations and external works but exclude furniture, loose or special equipment; they also exclude fees for professional services. The costs shown are for specifications and standards appropriate to the Slovak Republic and this should be borne in mind when attempting comparisons with similarly described building types in other countries. A discussion of this issue is included in section 2. Comparative data for countries covered in this publication, including construction cost data, are presented in Part Three.

Approximate estimating costs must be treated with caution; they cannot provide more than a rough guide to the probable cost of building. All the rates in this section exclude VAT which is at 6%.

	Cost Sk per m^2	Cost Sk per ft^2
Industrial buildings		
Warehouses low bay (6-8m high) for letting (no heating)	13,400	1,240
Warehouses low bay for owner occupation (including heating)	15,800	1,470
Administrative and commercial buildings		
Civic offices , non air conditioned	16,000	1,490
Civic offices , air conditioned	22,600	2,100
Offices, for owner occupation, 5 to 10 storeys, non air conditioned	19,500	1,810
Residential buildings		
Social/economic single family housing	12,100	1,120
Private/mass market single family housing	16,100	1,500
Purpose designed single family housing	15,400	1,430
Social/economic apartment housing, low rise (no lifts)	10,700	994
Social/economic apartment housing, high rise (with lifts)	9,820	912
Private sector apartment building (standard specification)	13,500	1,250
Private sector apartment building (luxury)	15,500	1,440
Homes for the elderly (self contained)	13,200	1,230
Other building types		
Collective multiple garage	4.420	411
Church	12,100	1,120
Superstore	20,600	1,910
Water treatment plant	13,800	1,280

Regional variations

The approximate estimating costs are based on average Bratislava rates, where labour may be 20 to 30% higher and materials up to 5% higher than in other regions of the Slovak Republic.

Value added tax (VAT)

The standard rate of value added tax (VAT) is currently 23%, chargeable on general building work and materials.

Exchange rates and cost and price trends

The combined effect of exchange rates and inflation on prices within a country and on price comparisons between countries is discussed in section 2.

Exchange rates

The graph on the next page plots the movement of the Slovak koruna against sterling, the ECU/Euro, the US dollar and 100 Japanese yen since 1993. The values used for the graph are quarterly and the method of calculating these is described and general guidance on the interpretation of the graph provided in section 2. The average exchange rate in the first quarter of 1999 was Sk 60.2 to the pound sterling, Sk 35.9 to the US dollar and Sk 42.4 to the Euro.

Cost and price trends

The table on the next page presents the indices for consumer prices and building costs in the Slovak Republic since 1993. The consumer price index has been rebased to 1990=100. The building cost index is based to 1989=100. The annual change is the percentage change between the average index of consecutive years. Notes on inflation are included in section 2.

The Slovak koruna against sterling, the ECU/Euro, the US dollar and 100 Japanese yen

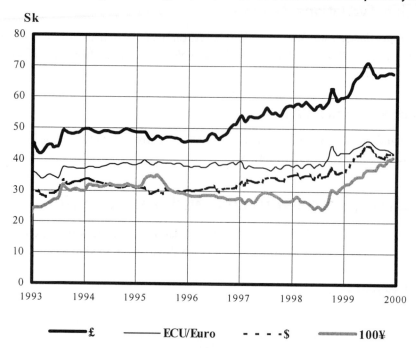

£   ———— ECU/Euro   - - - - $ ▬▬▬100¥

Consumer price and building cost indices

	Consumer prices		Building costs	
	annual	change	annual	change
Year	average	%	average	%
1993	218.5	23.1	199.7	24.1
1994	247.7	13.4	221.6	11.0
1995	272.2	9.9	248.3	12.0
1996	288.1	5.8	285.5	15.0
1997	305.7	6.1	295.2	3.4
1998	326.3	6.7	–	–

Sources: *International Monetary Fund, Yearbook 1998 and supplements 1999.*
Statistical Yearbook 1997.

Useful addresses

Government and public organization

Mnisterstvo výstavby a verejných prác Sliovenske republiky
Ministry of Building of the Slovak Republic

Špitálska 8, 81644 Bratislava. Tel: (421-7) 59751111

Trade and professional organizations

Asociácia súkromných stavebných podnikate'ov Slovenska
Private Construction Contractors
Association of Slovakia
Nobelova 18
83102 Bratislava
Tel: (421-7) 5031260
Fax: (421-7) 5031249

Zväz stavebnych podnikate'ov Slovenska
The Association of Construction
Entrepreneurs of Slovakia
Zahradnícka 46
82493 Bratislava
Tel: (421-7) 5261377
Fax: (421-7) 5261302

Slovak National Agency for Foreign Investment and Development
Sladkovicova 7
81106 Bratislava
Tel: (421-7) 54435175
Fax: (421-7) 54435022

Slovenská obchodná a priemyselná komora
Slovak Chamber of Commerce and
Industry
Gorkého 9
81603 Bratislava
Tel: (421-7) 333291

Other organization

Ustav ekonomiky a organizácie Stavebníctva
Institute of Economics and
Organization of Building Industry
Ruzová dolina 27
82469 Bratislava
Tel: (421-7) 5237111
Fax: (421-7) 5211603

Spain

All data relate to 1998 unless otherwise indicated.

Population

Population	39.3 million
Urban population (1997)	77%
Population under 15	15%
Population 65 and over	16%
Average annual growth rate (1985 to 1998)	0.2%

Geography

Land area	504,750 km^2
Agricultural area	60%
Capital city	Madrid

Economy

Monetary unit	Peseta (Pta)
Exchange rate (average first quarter 1999) to:	
the pound sterling	Pta 237
the US dollar	Pta 142
the Euro	Pta 166
the yen x 100	Pta 120
Average annual inflation (1995 to 1998)	2.4%
Inflation rate	1.8%
Gross Domestic Product (GDP)	Pta 80,060 billion
GDP PPP basis (1997)	$ (PPP) 642 billion
GDP per capita	Pta 2,050,000
GDP per capita PPP basis (1997)	$ (PPP) 16,340
Average annual real change in GDP (1995 to 1998)	3.3%
Private consumption as a proportion of GDP	62%
Public consumption as a proportion of GDP	16%
Investment as a proportion of GDP	21%

Construction

Gross value of construction output (1997)	Pta 9,303 billion
Gross value of construction output per capita	Pta 236,600
Gross value of construction output as a proportion of GDP	11.6%
Average annual real change in gross value of construction output (1985 to 1998)	2.3%
Annual cement consumption	28.6 million tonnes

PPP Purchasing power parity

The construction industry

Construction output

The value of gross output of the construction sector in Spain in 1998 was Pta 9,303 billion equivalent to 55.6 billion ECUs (nearly the same as the January 1999 Euro). This represents 11.6% of GDP. In addition, it is estimated that an extra Pta 1,380 (15%) is spent on construction by other sectors, in the black economy and in DIY. The breakdown of output in 1998 is shown in the table below.

Output of construction sector 1998 (current prices)

Type of work	Pta billion	ECUs billion	% of total
New work			
Residential building	3028	18.1	33
Non-residential building			
Office and commercial	554	3.3	6
Industrial	604	3.6	6
Other	202	1.2	2
Total	1,360	8.1	15
Civil engineering	2,156	12.9	23
Total new work	6,545	39.1	70
Repair and maintenance			
Residential	1,363	8.1	15
Non-residential	789	4.7	8
Civil engineering	609	3.6	7
Total repair and maintenance	2,760	16.5	30
Total	**9,303**	**55.6**	**100**

Note: the value of the Euro in January 1999 was veru similar to that of the ECU in 1998.
Source: based on Perez, J.M.i, 'Spain' Euroconstruct conference proceedings,
Prague, June 1998

The output of the construction sector in Spain increased by about 6% in 1998 largely because of a continuing rise in GDP. Residential construction has increased between 5 and 8% per annum over the last three years, whereas other sectors have 'taken off' only in 1998 or in the last part of 1997. Renovation work has been generally buoyant in all building work but civil engineering repair and maintenance became positive only in 1998. The boom in residential building has surprised observers in its magnitude. The economic environment is favourable with lower interest rates, falling unemployment and an increase in household formation and the rate of growth in housing output and in building permits is outstandingly high. From 1996 to 1998 the units for which permits were given rose from 319 thousand to about 460 thousand, an increase of 44%. Some of these will continue to boost output well into 1999. It is thought that the very high growth cannot be sustained, and a change to more stable levels is expected thereafter.

Growth in non-residential building rose to 4% in 1998 with increases spread fairly evenly across types of work, the highest being industrial building at just under 5% in 1998. Schools, universities and hospital building increased just oven 4% and office and commercial buildings by about 3.5% with favourable economic conditions, this solid but unspectacular performance may well continue.

The civil engineering sector is recovering from low levels and achieved growth in 1998 of about 6%. Recovery took place in all types of work, the highest increase coming from transport, especially roads and airports. Government tenders increased in 1998, which is encouraging for the future level of output.

Repair maintenance increased steadily across all types of work in 1998. The government is been to encourage renovation.

The geographical distribution of dwelling completions in 1996 compared with that of population is shown overleaf. The location of dwellings roughly follows that of the population distribution.

Characteristics and structure of the industry

The professions, particularly the architects, have a highly protected position in Spain. Every profession has its own *colegios* (colleges) to which all professionals must belong and the law defines their duties and practices. Only an architect registered with a college may put his signature to a project. Fee scales are recommended by the college but are no longer dictated by it. It is obligatory to employ an architect on all building projects; he is responsible for design and supervision, and he employs the other professionals. In addition, there are technical architects who are legally responsible for site organization, quality control and safety, but also undertake many of the functions of a quantity surveyor. Private sector architectural practices are nearly all very small.

There are 17 *Colegios Oficiales de Arquitectura* (COAs), one in each of the autonomous communities. The COAs are represented by the *Consejo Superior de los Colegios de Arquitectura* (CSCA) which is a very influential body with offices in Madrid. It represents the interests of COAs and hence architects at a

Population and dwelling completions 1996 regional distribution

Regions	Population (%)	Dwelling completions (%)
Andalucía	18.2	15.3
Aragón	3.0	2.5
Asturias	2.7	2.2
Baleares	1.9	1.9
Canarias	4.1	5.3
Cantabria	1.3	1.8
Castilla y León	6.3	6.2
Castilla-La Mancha	4.3	4.5
Cataluña	15.4	18.8
Comunidad Valenciana	10.1	10.6
Extremadura	2.7	1.7
Galicia	6.9	6.5
Comunidad de Madrid	12.7	13.0
Murcia	2.8	3.5
Navarra	1.3	1.5
País Vasco	5.3	3.8
La Rioja	0.7	0.6
Ceuta y Melilla	0.3	0.3
Total	**100.0**	**100.0**

Sources: *Espana en Cifras 1997, Instituto Nacional de Estadística.*
Estadística de Edificación y Vivienda, Ministerio de Fomento.

national and international level. The COAs are substantial organizations whose functions, in addition to those relating to the conduct, rights and interests of architects, include the control and approval (*visado*) of all project documents and the channelling of all payments to architects. This means that they can ensure that clients actually pay architects and that debtors cannot engage an architect. It also means that they are an integral part of the building control process and, by their certification that documents are prepared in compliance with the rules, ensure that architects conform to planning and building regulations. Technical Architects (*Arquitectos Técnicos* or *aparejadores*) have a similar organization being represented by the *Consejo General de los Colegios de Aparejadores y Arquitectos Técnicos de España*).

Engineers do not have an equivalent statutory role to that of architects. Some 40% to 50% of engineers are employed in the public sector, working in client organizations. The remainder are in the private sector. Some of the engineering firms are quite substantial and indeed most are stock companies, often with participation from banks, contractors or suppliers. Generally, engineers are involved in pure civil work but not in structural design of buildings.

All contractors operating in Spain must be legally registered and hold a licence. To qualify for public contracts, contractors must also be classified by the Treasury Ministry according to types of work, size of contract and maximum value of contract at any one time. Contractors from other European Community states need not be so qualified, but must present documentation to prove their status in their own countries, agree to submit to Spanish law, and furnish proof of financial and technical capability.

The contracting industry is dominated by a large number of strong and successful contractors. Ten Spanish companies are in *Building* magazine's list of 'Top 300 European Contractors' and there are five in the top fifty. Most of them have some export business. Eight Spanish contracting firms feature in *Engineering News Record's* 1997 list of 'Top 225 International Contractors'. They normally operate in both the building and civil engineering fields and as developers. The larger contractors are organizing themselves to undertake projects as managing contractors and their share of total construction employment is falling as a result. Sub-contracting is increasing.

The principal Spanish contractors and their percentage sales abroad are shown below.

Major Spanish contractors, 1997

Major contractors	Place in Building's 'Top 300 European Contractors'	Percentage of sales abroad
FCC	16	9.4
Dragados y Construcciones	19	–
Acciona	23	–
Ferrovial	30	10
ACS	35	15
Agroman	73	14
Construcciones Lain	100	5
Obrascon	123	2
Sacyr	160	-
Huarte & Hasa	161	8

Source: Building December 1998.

The complexity of requirements for the operations of construction firms and companies has led to the development of firms of *gestores administradores* which specialize in obtaining and filling in forms and obtaining permits.

Clients and finance

Public projects, including virtually all infrastructure, financed mainly from public funds, account for about one fifth of construction. The 17 Autonomous Communities have increased their share of this financing and are now responsible for over one third of the total.

Housing is largely financed by banks and savings banks although there is some small, direct local authority involvement. Most housing is constructed for sale under government-subsidized loan schemes for those earning less than two and a half times the minimum wage.

Nearly all commercial property is built for sale or for end-user clients. The normal pattern of speculative development is for developer companies to buy a site and sell property from plans, commencing construction only when the cash from sales is sufficient to cover financing costs. Construction is financed from purchasers' deposits and bank loans. Recently there has been a shift towards building commercial property for rent, in which case insurance companies and banks are usually the owners.

Selection of design consultants

The first professional to be employed by the client in the private sector is the architect. The architect appoints the other professionals, but the client is required by law also to appoint an independent technical architect to carry out the site supervision of quality and measurement and the safety plan. The technical architect is a sort of intermediary between the architect and the contractor and passes orders from the former to the latter. He also undertakes much of the work normally done by the quantity surveyor. The architect also has a substantial administrative role since he is in charge of obtaining approvals of documents from the *colegio* and of licences from the local authority. Along with the technical architect, he prepares the acceptance certificate.

Selection of the architect is almost always determined by past working relations or personal recommendation. Since there is little private owner-occupier development, clients are usually professionally competent private developers or in the public sector. Because of the regulatory powers of the COAs (architectural 'colleges') once an architect has been appointed it is very difficult to change the appointment. Any future extension or rehabilitation work has to be offered to the architect of the original building.

On public civil engineering projects, works are designed within the client body or by a consultant firm acting as a design bureau to whom the design package is let, usually by competitive tender. The contracting authority's own representative engineer is responsible for supervising the works and appoints a quality control company to carry out tests on materials and to monitor sites.

Contractual arrangements

The basic principles of construction contracts are laid down in the 'Civil Code' (*Real Decreto 24-VII-1809 Código Civil*) and the 'Commercial Code' (*Real Decreto 22-VIII-1885 Código de Comercio*). Because of the protection of these two codes contract documents are relatively simple. There are no standard forms of contract.

The 'Civil Code' establishes various fundamental principles of construction including the following:

- Prices cannot be varied except by client contract variation or by an explicit inflation clause (Civil Code Art. 1593).

- The main contractor is totally responsible to the client (Civil Code Art. 1596) and the subcontractors or suppliers cannot make any claim against the client except in exceptional circumstances (Civil Code Art. 1597).

- The client can cancel work at any time, as long as he or she indemnifies the contractor for all costs and damages (Civil Code Art. 1594).

- If the contractor is unable to finish the work for any reason outside his control, the contract is rescinded and the contractor is entitled to payment pro rata with work done for materials and labour but not for profit (Civil Code Art. 1595).

- A contractor is not permitted to retain possession of a building for which the client has not paid, but has preference over all other creditors (except the tax authorities and the building insurers) in claims against the client.

Three types of contract are quite common:

Fixed price lump sum contract: the contractor undertakes the job without any provision for inflation or variations. Materials are supplied at a fixed price. It is usually adopted for single family dwellings.

Unitary quantities contract: the unit price of the elements is fixed in the contract and the quantities are based on measured work. Optionally, an inflation adjustment formula may be added. It is published by the *Confederación Nacional de la Construcción* (National Construction Federation) and is widely adopted.

Management contract: the contractor deals with all aspects regarding work, equipment and materials, then passes the cost on to the client with a management margin, either as a percentage or a fixed amount. It is not in common use except for extensions.

Tenders may be 'open', in which case quantities are remeasured and valued on completion of the contract, or 'closed', where the bill of quantities is taken as correct, the contractor's 'lump sum' being affected only by the value of variations. A contractor will submit his tender only after discussions with the client and the technical personnel involved.

Contract disputes which cannot be settled by discussion between the client's representative and the contractor may be referred to an independent arbitrator. More serious disputes are dealt with in the courts of justice.

In common with other countries with a legal code based on the Napoleonic system, there is a separate body of law relating to public administrations and the provisions of the 'Civil Code' do not apply to public contracts.

Public contracts are regulated by:

- the 'Law of State Contracts' (LCE) set out in 'decree 923/1965', 'law 51/1973', 'law 50/1984' and 'decree 931/1986'
- the 'General Regulations for State Contracts' (RCE) set out in 'decree 2410/1975'
- the 'General Administrative Clauses' (PCAG) set out in 'decree 3854/1970'.

These provisions are contained in a book published by *Boletín Oficial del Estado* (BOE) (Official State Gazette). The three methods of tendering for public works are outlined below.

Subasta (auction): Tenderers are provided with documents which include the official estimate containing quantities, rates, extensions and totals. Each tenderer undertakes his own measurement and provides his own rates and prices in order to arrive at his tender figure, which is quoted at a discount from the official estimate. Normally the contract will be awarded to the lowest tenderer, though if a tender is more than 10% below the average discount offered by the other tenderers, it is judged to be 'imprudently low' and automatically disqualified.

Concurso (competition): In this case the contract is awarded to the most suitable contractor based on an assessment of his record, experience, personnel, plant, equipment, etc., and his tender figure. This method can be used where the contractor's work is to include design.

Contratación directa (negotiation): Under this method the government body concerned will calculate the contract price and will negotiate with a suitable contractor to carry out the work. It is commonly used for small works to speed up all pre-construction stage.

Both *subasta* and *concurso* may be 'open', in which case any qualified contractor may bid, or 'restricted', in which case it is confined to those already successful in a prequalification stage. Open *subasta* was mainly used by local authorities for rather small contracts. There has been a large increase in the 1980s in the number of competitions at the expense of *subastas*.

Also covered by regulations within the same decree are procedures to be followed in the awarding of contracts, setting out and approvals, penalties for delays in completion, credits to the contractor, variations, determination, handing over, final account, sub-contracts, dayworks and other conditions of contract. Interim payments are made using the official estimate as a basis, rates being adjusted by the appropriate percentage to take into account the difference between the tender figure and the official estimate.

Liability, insurance and inspection

The Spanish 'Civil Code' determines liability but it applies only to the main contractor and the architect and technical architect, who share responsibility between them. There is a ten year liability for serious defects – these include those which pose a threat of serious damage to any part of the building or render it unsuitable for its intended use. Liability does not require any proof of negligence. It is said that, in practice, tribunals usually allocate liability to the professionals unless the contractor has deliberately disobeyed an architect's instruction. Architects believe they carry an unfair share of the risk because they are insured and contractors are not.

Architects take responsibility for all design matters, and the technical architect bears a secondary responsibility for all quality control aspects of the contractor's work and materials on site. The architect is liable for the whole design even if part was carried out by a structural engineer. Developers can also be made responsible with the contractor and be found liable for defects to the owner.

In the case of civil engineering works, the civil engineer normally carries the responsibility assigned to the architect in the 'Civil Code', but this is a matter of contractual liability and not a 'Civil Code' obligation.

The time limitation for initiating claims is 15 years, that is, five years beyond the liability period. This also applies, unlike the UK and other countries, when there has been fraud or concealment. When a defect has been remedied, the 10-year decennial period begins to run again.

In *viviendas de protección oficial* (VPO), or subsidized housing built by private developers (for sale to low-income households with government-subsidized loans and grants), the developer has special responsibilities and must provide a five year guarantee, during which period any repairs resulting from construction defects must be carried out. A claim may then be made against the architect or contractor if appropriate.

Both architects and technical architects have public liability insurance arranged through their 'colleges'. This covers the decennial risk on an annual premium basis but, if the individual architect stops practising and does not pay his own premiums, no cover is available and plaintiffs have no way of obtaining redress. All architects pay a standard premium, related to the total limit of cover, that is independent of their claims record or experience and also of their level of fee income. Insurance for architects and technical architects is underwritten by mutual insurance companies and by private insurance companies.

There is no legal obligation to carry insurance cover, and some professionals who are not in independent practice or have low workloads do not bother to insure. Only those who have undersigned a project need the insurance for civil liability. In case of accident the Penal Code applies against the parties involved and there is no insurance for Penal issues.

Contractors do not carry any decennial insurance cover, but generally hold an all-risks policy which covers their public liability on site, fire and accident, and their risks during the guarantee period of twelve or six months.

Some insurance companies have also begun offering single-premium project decennial insurance policies along French lines, and large projects by international developers are now generally covered by these. This is complicating the professional roles, since now the insurance companies' risk control bureau takes over part of the quality control role of the client's independent technical architect. The liability is then unclear and some technical architects say that, contrary to the intention, this may lead to a reduction in the quality of building.

Development control and standards

The 17 autonomous communities have ultimate responsibility for planning and control of construction. Powers are exercised by local authorities of which there are some 8,000 grouped into 52 provinces.

Local authorities issue building permits which are required for all works, including even quite small extensions, and an architect needs to be involved. The operation of the system is not very clear and where there is no plan, the criteria for granting a building permit are not known in advance. The local authority is not concerned with technical standards and quality – these are the sole responsibility of the architect.

The legislation relating to building covers technical requirements plus both the functions of professionals and of institutions such as the technical approvals body. In practice, there has been very little regional legislation, and it can be taken that, in general, national legislation applies.

Since 1977 the legislation has been divided explicitly into two classes:

- the 'basic norms' (*normas básicas de la edificación* – NBEs) which are the only obligatory standards. NBEs can be passed only by royal decree and cover such things as structural steelwork, fire protection and seismic resistance.
- the 'technical norms' (*normas tecnológicas de la edificación* – NTEs) which are developed by the *Ministerio de Obras Públicas y Urbanismo* (MOPU) (the Ministry for Public Works and Planning) or the autonomous communities, and are advisory but not obligatory.

An index to the legislation on building, the *Indice de Disposiciones Relacionadas con la Edificación* is published by MOPU. There is also a compendium set of volumes of NTEs, but not of other regulations.

The Directorate-General of Architecture and Building Technology of MOPU produces or approves Codes of Practice (*Soluciones Homologadas de Edificación* – SHEs), whose use guarantees meeting the minimum requirements of the NBEs. The Catalan Institute of Building Technology (*Instituto de Tecnologica de la Construccio de Cataluña* – ITEC) in Barcelona plays the main part in their development.

Product standards in Spain, for all industries, are set by the Spanish Standards Institution (*Asociación Española de Normalización y Certificación* – AENOR), the equivalent of the British Standards Institute (BSI). A product approval mark – the 'N' mark – has been established by AENOR. As yet, very few construction products have passed through the certification and approval process for 'N' marks. There are, however, two other schemes for quality marks – one for reinforcement bars and one for prefabricated prestressed concrete beams. Both these are administered by the research and *agrément* body *Instituto Eduardo Torroja de la Construcción y del Cemento* (IETCC).

The Quality Control document specifies the minimum quality of the materials to be used for every specific project. Materials that do not have AENOR certification are required to be tested against UNE/ISO standards before being approved by the technical architect as suitable for the project.

There is, in general, no legal requirement to use approved products, and no import restriction on products which do not meet or are not approved by the Spanish standards. Neither do the existing insurance arrangements require the use of approved products. Since product suppliers do not carry any legal liability for building defects, the use or otherwise of approved products has no liability implications, except for the architect and technical architect. Many of the standards for steel and concrete plus those for design and testing procedures are, however, embodied in the obligatory NBE regulations.

Construction cost data

Cost of labour

The figures on the following page are typical of labour costs in the Barcelona area as at the first quarter 1999. The wage rate is the basis of an employee's income, while the cost of labour indicates the cost to a contractor of employing that employee. The difference between the two covers a variety of mandatory and voluntary contributions - a list of items which could be included is given in section 2.

	Wage rate (per hour) Pta	Cost of labour (per hour) Pta	Number of hours worked per year
Site operatives			
Mason/bricklayer	1,300	2,410	1,800
Carpenter	1,200	2,220	1,800
Plumber	1,200	2,220	1,800
Electrician	1,200	2,220	1,800
Structural steel erector	1,350	2,500	1,800
HVAC installer	1,400	2,600	1,800
Semi-skilled worker	1,050	1,950	1,800
Unskilled labourer	900	1,670	1,800
Equipment operator	900	1,670	1,800
Watchman/security	1,050	1,950	1,600
Site supervision			
General foreman	2,400	4,500	1,800
	(per year)	(per year)	
Clerk of works	3,000,000	4,050,000	1,800
Contractors' personnel	(per year)	(per year)	
Site manager	6,000,000	12,000,000	–
Resident engineer	6,000,000	12,000,000	–
Resident surveyor	5,000,000	10,000,000	–
Junior engineer	4,000,000	8,000,000	–
Junior surveyor	3,600,000	7,200,000	–
Planner	4,000,000	8,000,000	–

	Wage rate (per year) Pta	Cost of labour (per year) Pta	Number of hours worked per year
Consultants' personnel			
Senior architect	7,000,000	14,000,000	–
Senior engineer	7,000,000	14,000,000	–
Senior surveyor	6,500,000	18,000,000	–
Qualified architect	6,500,000	13,000,000	–
Qualified engineer	6,500,000	13,000,000	–
Qualified surveyor	6,500,000	13,000,000	–

Cost of materials

The figures that follow are the costs of main construction materials, delivered to site in the Barcelona area, as incurred by contractors in the first quarter of 1999. These assume that the materials would be in quantities as required for a medium sized construction project and that the location of the works would be neither constrained nor remote. All the costs in this section exclude value added tax (VAT) which is at 16%.

	Unit	Cost Pta
Cement and aggregates		
Ordinary portland cement in 50kg bags	tonne	14,200
Coarse aggregates for concrete	m^3	1,820
Fine aggregates for concrete	m^3	2,210
Ready mixed concrete (175kg/cm^2)	m^3	7,900
Ready mixed concrete (200kg/cm^2)	m^3	3,500
Steel		
Mild steel reinforcement	tonne	60,000
High tensile steel reinforcement	tonne	65,000
Structural steel sections	tonne	85,000
Bricks and blocks		
Common bricks (150 x 2800 x 100mm)	1,000	18,000
Good quality facing bricks (150 x 50 x 280mm)	1,000	35,000
Hollow concrete blocks (200 x 400 x 200mm)	each	135
Solid concrete blocks (150 x 300 x 100mm)	each	55
Precast concrete cladding units with exposed aggregate finish	m^2	10,000
Timber and insulation		
Softwood sections for carpentry	m^3	45,000
Softwood for joinery	m^3	55,000
Hardwood for joinery	m^3	90,000
Exterior quality plywood (18mm)	m^2	1,500

	Unit	Cost Pta
Plywood for interior joinery (18mm)	m²	1,050
Softwood strip flooring (16mm)	m²	5,500
Chipboard sheet flooring (16mm)	m²	2,300
100mm thick quilt insulation	m²	610
100mm thick rigid slab insulation	m²	1,650
Softwood internal door complete with frames and ironmongery	each	30,000
Glass and ceramics		
Float glass (mm)	m²	3,100
Sealed double glazing units (4/6/4)	m²	4,900
Good quality ceramic wall tiles (200 x 200mm)	m²	2,100
Plaster and paint		
Plaster in 50kg bags	tonne	10,400
Plasterboard (13mm thick)	m²	590
Emulsion paint in 5 litre tins	litre	465
Gloss oil paint in 5 litre tins	litre	1,450
Tiles and paviors		
Clay floor tiles (200x 200 x 80mm)	m²	2,250
Vinyl floor tiles (600x 600 x 30mm)	m²	1,600
Precast concrete paving slabs (400 x 400 x 40mm)	m²	2,300
Clay roof tiles	1,000	45,000
Precast concrete roof tiles	1,000	69,000
Drainage		
WC suite complete	each	19,000
Lavatory basin complete	each	17,000
100mm diameter clay drain pipes	m	1,180
150mm diameter cast iron drain pipes	m	4,900

Unit rates

The descriptions of work items following are generally shortened versions of standard descriptions listed in five languages (English, French, Italian, German and Spanish) in Appendix 3.

Where an item has a two digit reference number (e.g. 05 or 33), this relates to the full description against that number in Appendix 3. Where an item has an alphabetic suffix (e.g. 12A or 34B) this indicates that the standard description has been modified. Where a modification is major the complete modified description is included here and the standard description should be ignored.

Where a modification is minor (e.g. the insertion of a named hardwood) the shortened description has been modified here but, in general, the full description in Appendix 3 prevails.

The unit rates below are for main work items on a typical construction project in the Barcelona area in the first quarter of 1999. The rates include all necessary labour, materials and equipment, and allowances for contractor's overheads and profit (6%), preliminary and general items (16%) and contractor's profit and attendance on specialist trades (6%). All the rates in this section exclude VAT which is at 16%.

		Unit	Rate Pta
Excavation			
01	Mechanical excavation of foundation trenches	m^3	1,540
02	Hardcore filling making up levels	m^2	450
03	Earthwork support	m^2	2,300
Concrete work			
04	Plain insitu concrete in strip foundations in trenches	m^3	10,700
05	Reinforced insitu concrete in beds	m^3	11,200
06	Reinforced insitu concrete in walls	m^3	12,000
07	Reinforced insitu concrete in suspended floor or roof slabs	m^3	12,100
08	Reinforced insitu concrete in columns	m^3	12,700
09	Reinforced insitu concrete in isolated beams	m^3	12,700
10	Precast concrete slab	m^2	10,800
Formwork			
11	Softwood formwork to concrete walls	m^2	3,400
12	Softwood or metal formwork to concrete columns	m^2	2,900
13	Softwood or metal formwork to horizontal soffits of slabs	m^2	3,350
Reinforcement			
14	Reinforcement in concrete walls (16mm)	tonne	135,000
15	Reinforcement in suspended concrete slabs	tonne	130,000
16	Fabric reinforcement in concrete beds	m^2	405
Steelwork			
17	Fabricate, supply and erect steel framed structure	tonne	215,000
Brickwork and blockwork			
18	Precast lightweight aggregate hollow concrete block walls (100mm thick)	m^2	3,410
19	Solid (perforated) concrete bricks (priced at 490/m^2 delivered to site)	m^2	3,300
20	Solid (perforated) sand lime bricks (priced at 585/ m^2 delivered to site)	m^2	3,500
21	Facing bricks (priced at 950/ m^2 delivered to site)	m^2	7,000

		Unit	Rate Pta
Roofing			
22	Concrete interlocking roof tiles	m²	3,360
23	Plain clay roof tiles	m²	3,200
24	Fibre cement roof slates	m²	3,900
25	Sawn softwood roof boarding	m²	12,700
26	Particle board roof covering	m²	10,900
27	3 layers glass-fibre based bitumen felt roof covering include chippings	m²	5,200
28	Bitumen based mastic asphalt roof covering	m²	3,250
29	Glass-fibre mat roof insulation	m²	2,100
30	Rigid sheet loadbearing roof insulation 75mm thick	m²	2,450
31	Troughed galvanised steel roof cladding	m²	2,800
Woodwork and metalwork			
34A	Single glazed casement window in Tea Pine hardwood, size 650 x 900mm	each	45,000
35A	Two panel glazed door in Tea Pine hardwood, size 850 x 2000mm	each	75,000
36	Solid core half hour fire resisting hardwood internal flush doors, size 800 x 2000mm	each	77,000
37	Aluminium double glazed window, size 1200 x 1200mm	each	57,000
38	Aluminium double glazed door, size 850 x 2100mm	each	81,500
40	Framed structural steelwork in universal joist sections	tonne	255,000
41	Structural steelwork lattice roof trusses	tonne	275,000
Plumbing			
42	UPVC half round eaves gutter	m	2,350
43	UPVC rainwater pipes	m	2,050
44	Light gauge copper cold water tubing	m	1,200
45	High pressure plastic pipes for cold water supply	m	475
46	Low pressure plastic pipes for cold water distribution	m	450
47	UPVC soil and vent pipes	m	2,050
48	White vitreous china WC suite	each	25,500
49	White vitreous china lavatory basin	each	24,000
50	White glazed fireclay shower tray	each	29,000
51	Stainless steel single bowl sink and double drainer	each	21,000
Electrical work			
52	PVC insulated and copper sheathed cable	m	1,310
53	13 amp unswitched socket outlet	each	6,100
54	Flush mounted 20 amp, 1 way light switch	each	6,000
Finishings			
55	2 coats gypsum based plaster on brick walls	m²	1,150
56	White glazed tiles on plaster walls	m²	4,250
57	Red clay quarry tiles on concrete floor	m²	4,580
58	Cement and sand screed to concrete floors 50mm thick	m²	1,700

		Unit	Rate Pta
59	Thermoplastic floor tiles on screed	m^2	3,600
60	Mineral fibre tiles on concealed suspension system	m^2	3,150
Glazing			
61	Glazing to wood	m^2	3,300
Painting			
62	Emulsion on plaster walls	m^2	515
63	Oil paint on timber	m^2	1,680

Approximate estimating

The building costs per unit area given overleaf are averages incurred by building clients for typical buildings in the Barcelona area as at the first quarter 1999. They are based upon the total floor area of all storeys, measured between external walls and without deduction for internal walls.

Approximate estimating costs generally include mechanical and electrical installations but exclude furniture, loose or special equipment, and external works; they also exclude fees for professional services. The costs shown are for specifications and standards appropriate to Spain and this should be borne in mind when attempting comparisons with similarly described building types in other countries. A discussion of this issue is included in section 2. Comparative data for countries covered in this publication, including construction cost data, are presented in Part Three.

Approximate estimating costs must be treated with caution; they cannot provide more than a rough guide to the probable cost of building. All the rates in this section exclude VAT which is at 16%.

	Cost Pta per m^2	Cost Pta per ft^2
Industrial buildings		
Factories for letting (include lighting, power and heating)	35,000	3,250
Factories for owner occupation (light industrial use)	45,000	4,180
Factories for owner occupation (heavy industrial use)	55,000	5,110
Factory/office (high-tech) for letting (shell and core only)	60,000	5,570
Factory/office (high-tech) for letting (ground floor shell, first floor offices)	65,000	6,040
Factory/office (high-tech) for owner occupation (controlled environment, fully furnished)	70,000	6,500
High tech laboratory (air conditioned)	97,000	9,010

	Cost *Pta per m²*	Cost *Pta per ft²*
Warehouses, low bay (6 to 8m high) for letting (no heating)	35,000	3,250
Warehouses, low bay for owner occupation (including heating)	40,000	3,720
Warehouses, high bay for owner occupation (including heating)	44,000	4,090
Cold stores/refrigerated stores	57,500	5,340
Administrative and commercial buildings		
Civic offices, non air conditioned	85,000	7,900
Civic offices, fully air conditioned	100,000	9,290
Offices for letting, 5 to 10 storeys, non air conditioned	75,000	6,970
Offices for letting, 5 to 10 storeys, air conditioned	90,000	8,360
Offices for letting, high rise, air conditioned	120,000	11,100
Offices for owner occupation, 5 to 10 storeys, non air conditioned	130,000	13,900
Offices for owner occupation, 5 to 10 storeys, air conditioned	150,000	12,100
Offices for owner occupation, high rise, air conditioned	150,000	13,900
Prestige/headquarters office, 5 to 10 storeys, air conditioned	180,000	16,700
Prestige/headquarters office, high rise, air conditioned	180,000	16,700
Health and education buildings		
General hospitals (300 beds)	175,000	16,300
Teaching hospitals (200 beds)	180,000	16,700
Private hospitals (200 beds)	195,000	18,100
Health centres	170,000	15,800
Nursery schools	150,000	13,900
Primary/junior schools	150,000	13,900
Secondary/middle schools	130,000	12,100
University (arts) buildings	130,000	12,100
University (science) buildings	130,000	12,100
Management training centres	130,000	12,100
Recreation and arts buildings		
Theatres (over 500 seats) including seating and stage equipment	180,000	16,700
Theatres (less than 500 seats) including seating and stage equipment	190,000	17,700
Concert halls including seating and stage equipment	210,000	19,500
Sports halls including changing and social facilities	110,000	10,200
Swimming pools (international standard) including changing facilities	140,000	13,000
Swimming pools (schools standard) including changing facilities	130,000	12,100
National museums including full air conditioning and standby generator	225,000	20,900

	Cost Pta per m²	Cost Pta per ft²
Local museums including air conditioning	170,000	15,800
City centre/central libraries	120,000	11,100
Branch/local libraries	120,000	11,100
Residential buildings		
Social/economic single family housing (multiple units)	70,000	6,500
Private/mass market single family housing 2 storey detached/semidetached (multiple units)	80,000	7,430
Purpose designed single family housing 2 storey detached (single unit)	90,000	8,360
Social/economic apartment housing, low rise (no lifts)	65,000	6,040
Social/economic apartment housing, high rise (with lifts)	63,000	5,850
Private sector apartment building (standard specification)	80,000	7,430
Private sector apartment building (luxury)	110,000	10,200
Student/nurses halls of residence	120,000	11,100
Homes for the elderly (shared accommodation)	95,000	8,830
Homes for the elderly (self contained with shared communal facilities)	110,000	10,200
Hotel, 5 star, city centre	190,000	17,700
Hotel, 3 star, city/provincial	135,000	12,500
Motel	100,000	9,290

Regional variations

The approximate estimating costs are based on average Barcelona rates. Adjust these costs by the following factors for regional variations:

Bilbao	0%	Madrid	0%	Valencia	−10%
La Coruña	−15%	Sevilla	−15%		

Value added tax (VAT)

The standard rate of value added tax (VAT) is currently 16%, chargeable on general building work and materials.

Exchange rates and cost and price trends

The combined effect of exchange rates and inflation on prices within Spain and on price comparisons between countries is discussed in section 2.

Exchange rates

The graph below plots the movement of the Spanish peseta against sterling, the ECU/Euro, the US dollar and 100 Japanese yen since 1990. The values used for the graph are quarterly and the method of calculating these is described and general guidance on the interpretation of the graph provided in section 2. The average exchange rate in the first quarter of 1999 was Pta 237 to the pound sterling, Pta 142 to the US dollar and Pta 166 to the Euro.

The Spanish peseta against sterling, the Euro, the US dollar and 100 Japanese yen

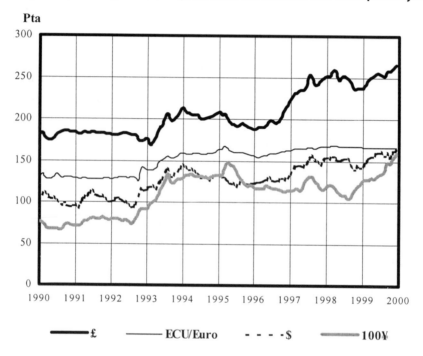

Cost and price trends

The table on the next page presents the indices for consumer price and residential building cost inflation in Spain since 1990. The indices have been rebased to 1990=100. The annual change is the percentage change between the average index of consecutive years. Notes on inflation are included in section 2.

Consumer price and residential building cost indices

	Consumer prices		Residential building costs	
Year	annual average	change %	annual average	change %
1990	100.0	6.7	100.0	9.6
1991	105.9	5.9	105.8	5.8
1992	112.2	5.9	109.5	3.4
1993	117.3	4.5	113.9	4.0
1994	122.9	4.8	119.0	4.5
1995	128.6	4.6	123.4	3.7
1996	133.2	3.6	127.0	3.0
1997	135.7	1.9	–	–
1998	138.2	1.8	–	–

Sources: International Monetary Fund, Yearbook 1998 and supplements 1999.
European Mortgage Federation, Hypostat 1986-1996.

Useful addresses

Government and public organizations

Consejo de Obras Publicas y Urbanismo
Committee for Public Works and
Town Planning
Avenida de Portugal 81, 1° floor
28071 Madrid
Tel: (34-91) 2531600

Dirección General de la Vivienda, la Architectura y el Urbanismo
General Directorate for Housing,
Architecture and Town Planning
Nuevo Ministerios
Madrid
Tel: (34-91) 5977000

Trade and professional associations

Confederación Nacional de la Construcción (CNC)
National Construction Federation
Diego de Leon 50
28006 Madrid
Tel: (34-91) 2619715

Asociación Española de Empresas Constructoras con Actividad (AECI)
National Syndicate of Public Works Companies
Serrano 174
28002 Madrid
Tel: (34-91) 5630504
Fax: (34-91) 5628544

Asociación de Empresas Constructoras de Ambito Nacional (SEOPAN)
Syndicate of National Companies
Serrano 174
28002 Madrid
Tel: (34-91) 5630504
Fax: (34-91) 5625844

Asociación de Arquitectos y Ingenieros Consultores
Association of Consulting Architects and Engineers
Orense 30
28020 Madrid
Tel: (34-91) 5551630

Consejo Superior de Los Colegios de Arquitectos
Representative body of the architecture 'colleges'
Paseo de la Castellana 12
28046 Madrid
Tel: (34-91) 4357723

Consejo General de Colegios Oficiales de Aparejadores y Arquitectos Técnicos
Representative body of the 'colleges' for technical architects
Paseo de la Castellana 155
28046 Madrid
Tel:(34-91) 5707317

Consejo General de Colegios Oficiales de Peritos y Ingenieros Tecnicos Industriales
Representative body of the 'colleges' for industrial engineers
Avenida Pablo Iglesias 2
28003 Madrid
Tel: (34-91) 5541806

The Professional Consultant at the centre of Europe

In addition to our many local projects for Swiss Clients, we have experience working together with international Clients and Planners building in and around Switzerland.

Services for building and civil engineering work:

- Cost Planning and Construction Economics
- Pre- and Postcontract Quantity Surveying
- Project Management
- Condition assessment and investment planning
- IT-consultancy and development for Cost Planning and Property Management
- Courses of further education

The quality factors:

- our qualified and motivated staff
- our extensive databases
- the efficient use of IT-Systems
- continuing professional development and the use of state of the art technology

The partners:

- Harry Diggelmann
  Structural Engineer/Bauingenieur ETH/SIA, Managing Director
- Martin Wright
  Chartered Quantity Surveyor ARICS, Bauökonom/Economiste de la construction AEC

Address:

Bauconsulting Center, Industriestrasse 11, CH-8808 Pfaeffikon/SZ, Switzerland
Telefon: +41 (0)55/415'48'68, Telefax: +41 (0)55/420'16'20

pbk@bauconsult.ch / www.swissonline.ch/pbk

Switzerland

All data relate to 1998 unless otherwise indicated.

Population

Population	7.3 million
Urban population (1997)	62%
Population under 15	17%
Population 65 and over)	15%
Average annual growth rate (1985 to 1998)	0.9%

Geography

Land area	41,290 km^2
Agricultural area	40%
Capital city	Berne

Economy

Monetary unit	Swiss Franc (SFr)
Exchange rate (average first quarter 1999) to:	
the pound sterling	SFr 2.27
the US dollar	SFr 1.37
the Euro	SFr 1.60
the yen x 100	SFr 1.16
Average annual inflation (1995 to 1998)	0.4%
Inflation rate	0.0%
Gross Domestic Product (GDP)	SFr 328 billion
GDP PPP basis (1997)	$ (PPP) 172.4 billion
GDP per capita	SFr 45,230
GDP per capital PPP basis (1997)	$ (PPP) 24,320
Average annual real change in GDP (1995 to 1998)	1.3%
Private consumption as a proportion of GDP	60%
Public consumption as a proportion of GDP	15%
Investment as a proportion of GDP	20%

Construction

Gross value of construction output	SFr 36.6 billion
Gross value of construction output per capita	SFr 5,010
Gross value of construction output as a proportion of GDP	11.1%
Average annual real change in gross value of construction output (1995 to 1998)	−1.7%
Annual cement consumption	3.4 million tonnes

PPP Purchasing power parity

The construction industry

Construction output

The value of gross output of the construction sector in Switzerland in 1998 is estimated at SFr. 36.6 billion equivalent to 22.4 billion ECUs (nearly the same as January 1999 Euros). This represents 11.1% of GDP. In addition it is estimated that an extra SFr 12.7 billion (35%) of construction work is undertaken by other sectors. No data are available on the amount of work in the black economy and by DIY. The breakdown of the output in 1998 is shown below.

Estimated output of construction sector 1998

Type of work	SFr billion	ECUs billion	% of total
New work			
Residential building	11.3	6.9	31
Non-residential building			
Office and commercial	2.1	1.3	6
Industrial	1.1	0.7	3
Other	2.0	1.2	5
Total	5.1	3.1	14
Civil engineering	5.6	3.4	15
Total new work	2,1.9	13.4	60
Repair and maintenance			
Residential	3.9	2.4	11
Non-residential	5.1	3.1	14
Civil engineering	5.9	3.6	16
Total repair and maintenance	14.9	9.1	41
Total	**36.6**	**22.4**	**100**

Note: the value of the Euro in January 1999 was very similar to that of the ECU in 1998.
Source: based ion Graf, H-G., 'Switzerland' in Euroconstruct conference proceedings Prague, June 1999.

The Swiss economy performed poorly over the years 1990 to 1996 with a small fall in GDP, in contrast to EU countries where growth was positive. Construction investment has also followed a downward trend with the only positive year being 1994 due to an increase in residential building. However, in 1997 and 1998 GDP turned up and construction output is thought to have followed with a small increase in 1998. Preliminary figures show the increase was mainly in renovation work with only civil engineering increasing in new work. In addition to a shortage of income and poor prospects, the low construction output was due to an over-supply of residential, commercial, office and industrial buildings. Construction prices fell from 1996.

The residential sector in 1998 was still suffering from a large number of empty dwellings. One reason is that the level of immigration has declined as enterprises set up premises in the EU. The large falls in new residential construction have been of flats which constitute over two thirds of total residential units. Construction of one or two family dwellings has remained very stable, helped by low interest rates. It seems unlikely that the new residential sector will change substantially in the immediate future. It is, however, possible that, as confidence in the economy improves and with interest rates low, more renovation will be undertaken.

Non-residential construction is also suffering from a supply of empty buildings, some because they are not suitable for modern IT. New development is thought to be hindered because Switzerland is outside the EU. Manufacturing in Switzerland has declined in favour of service industries and some industrial buildings are surplus.

There is a need for educational and health buildings but construction is unlikely to rise in the public sector due to budget problems. This similarly affects civil engineering work but some improvement may take place there. Private finance is increasing in importance.

It seems likely that in the near future any growth in output will come mainly from repair an maintenance rather than new work.

The geographical distribution of construction output in 1996 compared to population is shown on the following page. There is a close relationship between the location of construction output and population.

Characteristics and structure of the industry

Swiss cantons and municipalities enjoy a high degree of autonomy and this has implications for the construction industry. The cantons and municipalities impose their own taxes and arrive at their own spending policies and budgets, though moves are afoot to co-ordinate policies and harmonize taxation.

The construction industry in Switzerland comprises a large number of small firms, many of whom are specialist single trade contractors who operate only on a local basis, and large general contractors who operate throughout Switzerland and undertake a large proportion of the work. In 1991 there were 10,400 construction firms employing a total of 214,000 persons. Of these, six employed more than 1,000 persons. There were four Swiss contractors in the 1997

Population and construction output 1996 regional distribution

Region	Population (%)	Construction output (%)
Zurich	16.6	17.8
Bern	13.3	11.1
Luzern	4.8	4.8
Uri	0.5	0.8
Schwyz	1.7	1.9
Obwalden	0.4	0.7
Nidwalden	0.5	0.7
Glarus	0.6	0.6
Zug	1.3	1.9
Freiburg	3.2	3.5
Solothurn	3.4	3.1
Basel-Stadt	2.8	2.8
Basel-Landschaft	3.6	3.9
Schaffhausen	1.0	0.9
Appenzell A. Rh.	0.8	0.6
Appenzell I. Rh.	0.2	0.2
St. Gallen	6.3	6.1
Graubünden	2.6	5.1
Aargau	7.5	7.7
Thurgau	3.2	3.2
Tessin	4.3	4.1
Waadt	8.6	5.7
Wallis	3.8	4.9
Neuenburg	2.3	1.8
Genf	5.6	5.2
Jura	1.0	0.9
Total	**100.0**	**100**

Sources: Annuaire Statistique de la Suisse 1998, Office Fédéral de la Suisse.

Building magazine list of 'Top 300 European Contractors'. There were no Swiss firms in *Engineering News Record's* list of 'Top 225 International Contractors'. They are not shown as operating abroad. The main Swiss contractors are shown below.

Major Swiss contractors, 1997

Major Contractors	Place in Building's 'Top 300 European Contractors'
Zschokke Conrad	104
Batigroup	105
Walo Bertschinger	212
Frutiger	249

Source: Building; December 1998.

A large number of small specialist contractors have foreign origins. For example, many specialists in high quality finishings such as plastering, marble work, and terrazzo have entered Switzerland from Italy. Construction philosophy and techniques tend to be allied to those of West Germany. German trade press and publications are widely read and many German materials and components are available.

Clients and finance

The public sector as a client of the construction industry is not very important in Switzerland. In settlements of over 10,000 persons in 1991 only a small proportion (about 4%) of housing was constructed by the public sector. Of private sector housing in 1991, 54% was constructed by individuals, 8% by co-operatives and 38% by other private bodies.

Credit in Switzerland is relatively cheap, especially for long-term loans and mortgages. Loan conditions are generally more favourable than in many other European countries. Normally 50% of the total investment value (i.e. market value or construction costs including machinery and equipment) can be favourably financed.

Some public sector support for investment is available from some cantons which may procure credits, grant guarantees and direct loans, and arrange for contributions towards interest payments and other services. The Federal Government may also support investment plans for specific assisted regions.

Tendering and contract procedures

In Switzerland the architect has the responsibility of designer and project manager, planning and co-ordinating the activities of a number of (most commonly) single trade contractors. Larger contracts are often carried out by general contractors on the basis of bills of approximate quantities, with the onus of management transferred to a great extent to the contractor. A similar transfer of responsibility takes place in the case of 'design and build' contracts.

Competitive tendering based on drawings, specifications and bills of quantities is the norm. Bills of quantities are approximate and subject to remeasurement on completion of the work. Prequalification is not necessary though performance bonds are required from general contractors.

Development controls, standards and liability

Planning and development control is based on the principle that a right to build exists under the constitution, provided that no valid legal objection is raised. However, cantonal legislation covers the legal and procedural rules with respect to zoning, designation of development land and ownership. Such legislation may be supplemented at municipal level by further 'building ordinances' (*Bauordnungen*) mainly concerned with planning, but also governing building heights, spacings etc. and sometimes specific constructional requirements in respect of safety and health.

A construction permit procedure must be completed before a construction project can begin. Applications must be filed with the political community. The decisions on planning are very much a community matter in rural areas. It is not unusual for a framed structure to be erected on the site so that the public can see the effect of the dimensions of the proposed building.

Industrial buildings also require operations permits in the interest of work safety. These applications are filed with cantonal authorities.

Structural design and properties of building materials are covered by the standards of the Association of Engineers and Architects (SIA) which virtually serve as national regulations.

The building regulation system tends to be complex due to involvement, in varying degrees, of federal, cantonal and municipal governments. The power of federal government in this context is restricted by constitution to a limited number of areas of national importance, for example in civil defence where there is a requirement for shelters to be provided in all buildings. Other federal government regulations cover the safety of electrical systems and anti pollution measures.

Although smaller authorities exercise minimal supervision and inspection of building work, in larger cities, including Zurich and Basle, organized building inspectorates exist to check applications for permits and carry out on-site inspections. Resulting technical standards are generally high.

Much of the responsibility for technical standards in construction lies with the building engineer whose personal liability in civil and criminal law is regarded as an important safety factor, ensuring continuing high standards.

Construction cost data

Cost of labour

The figures on the following page are typical of labour costs in the Zurich area as at the first quarter 1999. The wage rate is the basis of an employee's income with an explanation in section 2.

	Wage rate (per month) SFr	Cost of labour (per hour) SFr	Number of hours worked per year
Site operatives			
Mason/bricklayer	5,100	64	1,933
Carpenter	5,100	71	1,928
Plumber	4,500	79	1,968
Electrician	4,600	70	1,980
Structural steel erector	5,000	85	1,968
HVAC installer	4,500	79	1,968
Semi-skilled worker	4,500	56	1,933
Unskilled labourer	4,500	56	1,933
Equipment operator	4,800	62	1,933
Site supervision			
General foreman	6,700	96	1,933
Trades foreman	6,000	81	1,933
Clerk of works	4,400	70	1,828
Site staff			
Site manager	6,500	125	1,828
Resident engineer	6,000	105	1,828
Junior engineer	4,600	90	1,828
Consultants' personnel			
Senior architect	7,500	150	1,828
Senior engineer	7,500	150	1,828
Qualified architect	5,700	105	1,828
Qualified engineer	5,700	105	1,828

Cost of materials

The figures that follow are the costs of main construction materials, delivered to site in the Zurich area, as incurred by contractors in the first quarter of 1999. These assume that the materials would be in quantities as required for a medium sized construction project and that the location of the works would be neither constrained nor remote. All the costs in this section exclude value added tax (VAT) which is 7.5%.

	Unit	Cost SFr
Cement and aggregates		
Ordinary portland cement in 50kg bags	bag	11.10
Coarse aggregates for concrete	m^3	54.00
Fine aggregates for concrete	m^3	54.00
Ready mixed concrete (20N/mm^2)	m^3	132.00

	Unit	Cost SFr
Steel		
Mild steel reinforcement	tonne	131.00
High tensile steel reinforcement	tonne	138.00
Structural steel sections	g	1.00
Bricks and blocks		
Common bricks (12 x 19 x 29 cm)	m^2	14.00
Good quality facing bricks (12 x 14 x 25 cm)	m^2	43.00
Hollow concrete blocks (12 x 25 x 50 cm)	m^2	29.90
Solid concrete blocks (12 x 13.5 x 25cm)	m^2	24.00
Precast concrete cladding units	m^2	160.00
Timber and insulation		
Softwood sections for carpentry	m^3	390.00
Softwood for joinery	m^3	720.00
Hardwood for joinery	m^3	2,200.00
Exterior quality plywood (22mm)	m^2	32.20
Plywood for interior joinery (22mm)	m^2	32.20
Softwood strip flooring (27mm)	m^2	42.00
Chipboard sheet flooring (25mm)	m^2	14.70
100mm thick quilt insulation	m^2	8.70
100mm thick rigid slab insulation	m^2	12.30
Softwood internal door complete with frames and ironmongery	each	410.00
Glass and ceramics		
Float glass (4mm)	m^2	63.00
Sealed double glazing units (1.3W/m²K)	m^2	70.00
Good quality ceramic wall tiles (15 x 15 cm)	m^2	25.00
Plaster in 50 kg bags	tonne	274.00
Plasterboard (12.5mm thick)	m^2	3.80
Emulsion paint in 5 litre tins	litre	4.00
Gloss oil paint in 5 litre tins	litre	13.80
Clay floor tiles (20 x 20 cm)	m^2	35.00
Vinyl floor tiles (61 x 61 x 61 cm)	m^2	36.00
Precast concrete paving slabs (50 x 50 x 4 cm)	m^2	24.00
Clay roof tiles	1,000	1,550.00
Precast concrete roof tiles	1,000	1,352.00
WC suite complete	no.	384.00
Lavatory basin complete	no.	438.00
100mm diameter clay drain pipes	m	40.00
150mm diameter cast iron drain pipes	m	17.20
Other materials		
Sheet metal decking (8mm thick)	m²	17.00
Drainpipe PVC (10mm)	m	5.75
Waterpipe steel (25mm)	m	5.05
Waterpipe steel (50mm)	m	10.50

	Unit	Cost SFr
Waterpipe plastic with plastic conduit (10mm)	m	4.50
Waterpipe plastic with plastic conduit (20mm)	m	6.30
Electrical wire, 4mm^2	m	0.50
Electrical conduit 20mm	m	1.30
Mercury vapour light, 400w	no.	850.00
Fluorescent light (2x36w)	no.	63.00

Unit rates

The descriptions of work items below are generally shortened versions of standard descriptions listed in five languages (English, French, Italian, German and Spanish) in Appendix 3.

Where an item has a two digit reference number (e.g. 05 or 33), this relates to the full description against that number in Appendix 3. Where an item has an alphabetic suffix (e.g. 12A or 34B) this indicates that the standard description has been modified. Where a modification is major the complete modified description is included here and the standard description should be ignored.

Where a modification is minor (e.g. the insertion of a named hardwood) the shortened description has been modified here but, in general, the full description in Appendix 3 prevails.

The unit rates below are for main work items on a typical construction project in the Zurich area in the first quarter of 1999. The rates include all necessary labour, materials and equipment, and, where appropriate, allowances for contractor's overheads and profit (7%), preliminary and general items (5%) and contractor's profit and attendance on specialist trades. All the rates in this section exclude VAT which is 7.5%.

		Unit	Rate SFr
Excavation			
01	Mechanical excavation of foundation trenches	m^3	16
02	Hardcore filling making up levels	m^2	10
03	Earthwork support	m^2	30
Concrete work			
04	Plain insitu concrete in strip foundations in trenches	m^3	180
05	Reinforced insitu concrete in beds	m^3	170
06	Reinforced insitu concrete in walls	m^3	190
07	Reinforced insitu concrete in suspended floor or roof slabs	m^3	180
08	Reinforced insitu concrete in columns	m^3	250

		Unit	Rate SFr
09	Reinforced insitu concrete in isolated beams	m³	240
10	Precast concrete slab	m²	140

Formwork
11	Softwood formwork to concrete walls	m²	28
12	Softwood or metal formwork to concrete columns	m²	60
13	Softwood or metal formwork to horizontal soffits of slabs	m²	29

Reinforcement
14	Reinforcement in concrete walls	tonne	1,250
15	Reinforcement in suspended concrete slabs	tonne	1,250
16	Fabric reinforcement in concrete beds	m²	5

Steelwork
| 17 | Fabricate, supply and erect steel framed structure | tonne | 2,130 |

Brickwork and blockwork
19	Solid (perforated) concrete blocks	m²	68
20	Solid (perforated) sand lime bricks	m²	72
21	Facing bricks	m²	120

Roofing
22	Concrete interlocking roof tiles, 430 x 380mm	m²	50
23	Plain clay roof tiles, 260 x 160mm	m²	60
24	Fibre cement roof slates, 600 x 300mm	m²	93
25	Sawn softwood roof boarding, presevative treated	m²	29
26	Particle board roof covering	m²	36
27	3 layers glass-fibre based bitumen felt roof covering	m²	51
28	Bitumen based mastic asphalt roof covering	m²	55
29	Glass-fibre mat roof insulation, 160mm thick	m²	38
30	Loadbearing Polyurethane roof insulation	m²	35
31	Troughed galvanised steel roof cladding	m²	30

Woodwork and metalwork
32	Preservative treated sawn softwood 50 x 100mm	m	9
33	Preservative treated sawn softwood 50 x 150mm	m	9
35	Two panel glazed door in hardwood, size 850 x 2000mm	each	1,300
36	Solid core half hour fire resisting hardwood internal flush doors, size 800 x 2000mm	each	670
37	Aluminium double glazed window, size 1000 x 1600mm	each	1,100
38	Aluminium double glazed door, size 850 x 2100mm	each	1,500
39	Hardwood skirtings	m	11
40	Framed structural steelwork in universal joist sections	tonne	2,130
41	Structural steelwork lattice roof trusses	tonne	2,130

		Unit	Rate SFr
Plumbing			
42	Copper half round eaves gutter	m	30
43	Copper rainwater pipes	m	30
45	High pressure UPVC pipes	m	14
46	Low pressure plastic pipes for cold water distribution	m	14
47	UPVC soil and vent pipes	m	50
48	White vitreous china WC suite	each	750
49	White vitreous china lavatory basin	each	600
50	White glazed fireclay shower tray	each	900
51	Stainless steel single bowl sink and double drainer	each	500
Electrical work			
52	PVC insulated copper sheathed cable	m	6
53	13 amp unswitched socket outlet	each	45
54	Flush mounted 20 amp, 1 way light switch	each	40
Finishings			
55	2 coats gypsum based plaster on brick walls	m^2	23
56	White glazed tiles on plaster walls	m^2	85
57	Red clay quarry tiles on concrete floor	m^2	110
58	Cement and sand screed to concrete floors	m^2	36
59	Thermoplastic floor tiles on screed	m^2	54
60	Mineral fibre tiles on concealed suspension system	m^2	80
Glazing			
61	Glazing to wood	m^2	280
Painting			
62	Emulsion on plaster walls	m^2	9
63	Oil paint on timber	m^2	15
Other unit rates			
	Fibre cement cladding to external walls	m^2	140
	External rendering to walls	m^2	110
	Plasterboard partitions with metal frame	m^2	100
	Gypsum block partitions , (100mm thick)	m^2	85

Approximate estimating

The building costs per unit area given overleaf are averages incurred by building clients for typical buildings in the Zurich area as at the first quarter 1999. They are based upon the total floor area of all storeys, measured to the outside face of external walls and without deduction for internal walls.

Approximate estimating costs generally include mechanical and electrical installations but exclude furniture, loose or special equipment, and external works; they also exclude fees for professional services. The costs shown are for specifications and standards appropriate to Switzerland and this should be borne in mind when attempting comparisons with similarly described building types in other countries. A discussion of this issue is included in section 2. Comparative data for countries covered in this publication, including construction cost data, are presented in Part Three.

Approximate estimating costs must be treated with caution; they cannot provide more than a rough guide to the probable cost of building. All the rates in this section exclude VAT which is 7.5%.

	Cost SFr/m^2	Cost SFr/ft^2
Industrial buildings		
Factories for owner occupation (light industrial use)	1,700	158
High-tech laboratory workshop centres (air conditioned)	1,150	107
Warehouses, for owner occupation (including heating)	1,700	158
Fire station		
Administrative and commercial buildings		
Offices for letting, non air conditioned	1,350	125
Offices for owner occupation, non air conditioned	2,100	195
Prestige/headquarters offices , air conditioned	2,750	255
Health and education buildings		
General hospitals	3,000	279
Nursery schools	2,100	195
Primary/junior schools	1,800	167
Secondary/middle schools	2,000	186
University (arts) buildings	2,150	200
Management training centres	2,150	200
Recreation and arts buildings		
Theatres and concert halls including seating and stage equipment	3,200	297
Sports halls including changing and social facilities	2,100	195
Public halls excl. seating and stage equipment	2,000	186
City centre/central libraries	1,800	167
Residential buildings		
Purpose designed single family housing 2 storey detached single unit	1,850	172
Private sector apartment building (standard specification)	1,450	135
Private sector apartment building (luxury)	2,000	186
Student/nurses halls of residence	1,600	149
Homes for the elderly (shared communal facilities)	2,000	186
Homes for the elderly (self contained)	1,750	163
Hotel, 5 star, city centre	2,450	228
Hotel, 3 star, city/provincial	2,000	186

Value added tax (VAT)

The standard rate of VAT is currently 7.5%, chargeable on general building work and materials.

Exchange rates and cost and price trends

The combined effect of exchange rates and inflation on prices within a country and on price comparisons between countries is discussed in section 2.

Exchange rates

The graph below plots the movement of the Swiss franc against sterling, the ECU/Euro, the US dollar and 100 Japanese yen since 1990. The values used for the graph are quarterly and the method of calculating these is described and general guidance on the interpretation of the graph provided in section 2. The average exchange rate in the first quarter of 1999 was SFr 2.27 to the pound sterling, SFr 1.37 to the US dollar and SFr 1.60 to the Euro.

The Swiss franc against sterling, the ECU/Euro, the US dollar and 100 Japanese yen

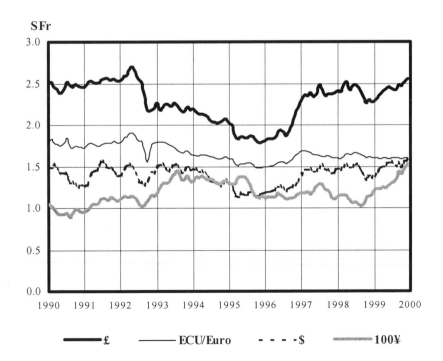

Cost and price trends

The table below presents the indices for consumer prices and building costs in Switzerland since 1990. The indices have been rebased to 1990=100. The annual change is the percentage change between the average index of consecutive years. Notes on inflation are included in section 2.

Consumer price and building cost indices

	Consumer prices		Building costs	
	annual	change	annual	change
Year	average	%	average	%
1990	100.0	5.4	100.0	5.0
1991	105.8	5.8	108.1	8.1
1992	110.1	4.1	98.8	−8.6
1993	113.8	3.4	95.8	−3.0
1994	114.7	0.8	95.8	0.0
1995	117.7	2.6	96.0	0.2
1996	119.2	1.3	93.3	−2.8
1997	119.2	0.0	90.7	−2.9
1998	119.2	0.0	84.1	−1.3

Sources: International Monetary Fund, Yearbook 1998 and supplements 1999. Annuai Statistiquere.

It will be seen that inflation fell to nil in 1997 and 1998 while building costs fell from 1996 to 1998.

Useful addresses

Trade and professional associations

Schweizerischer Baumeisterverband
Swiss Contractors' Association
Weinbergstrasse 49
CH-8035 Zürich
Tel: (41-1) 2588111
Fax: (41-1) 2588335

Schweizerischer Ingenieur und Architekten Verein
Swiss Society of Engineers and Architects
Selnaustrasse 16
800239 Zürich
Tel: (41-1) 2831515
Fax: (41-1) 2016335

Bund Schweizer Architekten
Federation of Swiss Architects
Pfluggässlein 3
4001 Basel
Tel: (41-61) 2621010
Fax: (41-61) 2621009

Verband Schweizerischer
Generalunternehmer (VSGU)
Swiss General Contractors Association
Schweizergasse 20
8001 Zürich
Tel: (41-1) 2103333
Fax: (41-1) 2103334

Schweizerischer Technischer Verband
(STA)
Swiss Technical Association
Weinbergstrasse 41
8006 Zürich
Tel: (41-1) 2683711

ENKA

world constructor in the world city Istanbul

ENKA was founded in Istanbul 42 years ago, and ENKA has been placing new black dots on the world map by building new settlements and cities.

constructing the future,

combining genuine technical capability, fast and qualified labour, and superior equipment, ENKA is striving to fulfil this mission. Moreover, ENKA is currently investing in education, culture, arts and sports, and thus constructing the future.

achieving highest quality with maximum speed,

ENKA's employees have been performing in various environments and climates ranging from –40°C to 55°C. Regardless of the surroundings and the difficulties, the aim has always been fulfilled and ENKA's corporate philosophy has been a guiding light through all the years.

keeping the promises,

the most important aspect of business is to keep promises. For the last 42 years and through over 400 ventures, ENKA has always fulfilled every clause of the contract. Moreover. ENKA has always finished the job on time and sometimes even earlier.

satisfying client needs with a personal touch,

ENKA's workforce is highly conscious of responsibilities as if they were not constructing for the client, but for themselves. Every ENKA employee combines experience and spirit with long hours of hard work to achieve the corporate mission.

contracting every kind of work in every amount,

ENKA is a leading general contractor in business, engaging in every possible kind of engineering and construction activity, from the initial feasibility studies to the turn-key operation, including maintenance and training. With the contributions of more than 20 subsidiaries, each specialising in different fields, ENKA aims at providing services fitting the needs of each client.

ahead in the world,

with an annual turnover of 400 million US Dollars, over 1200 technical and administrative personnel, 10,000 workers and a total equipment value of 90 million US Dollars, ENKA ranks among the leading construction companies, sometimes taking a place in the top 10 international contractors of the world.

ENKA

INSAAT VE SANAYI A.S.
Head Office
Balmumcu, Enka Binalari, 80780
Besiktas_ ISTANBUL / TURKEY
Tel: 90 (212) 274 25 40
Fax: 90 (212) 272 88 69
Telex: 26490 enas tr

Turkey

All data relate to 1998 unless otherwise indicated

Population

Population	64.6 million
Urban population (1997)	70%
Population under 15	31%
Population 65 and over	6.0%
Average annual growth rate (1984 to 1998)	1.9%

Geography

Land area	780,580 km^2
Agricultural area	52%
Capital city	Ankara

Economy

Monetary unit	Turkish Lira (TL)
Exchange rate (average first quarter 1999) to:	
the pound sterling	TL 513,000
the US dollar	TL 303,000
the Euro	TL 361,000
the yen x 100	TL 256,000
Average annual inflation (1995 to 1998)	83.6%
Inflation rate	84.6%
Gross Domestic Product (GDP)	TL 57,320,000 billion
GDP PPP basis (1997)	$ (PPP) 388 billion
GDP per capita	TL 888 million
GDP per capita PPP basic (1997)	$ (PPP) 6,090
Average annual real change in GDP (1995 to 1998)	6.2%
Private consumption as a proportion of GDP	69%
Public consumption as a proportion of GDP	12%
Investment as a proportion of GDP	26%

Construction

Gross value of construction output	TL 5,730,000 billion
Gross value of construction output per capita	TL 59.4 million
Gross value of construction output as a proportion of GDP	10%
Annual cement consumption	36.0 million tonnes

PPP Purchasing power parity

The construction industry

Construction output

The gross value of output of new construction in Turkey in 1998 was estimated at about TL 5,732 trillion, or about 17.8 billion ECUs (in terms of Euros of January 1999 this would be equivalent to 15.4 billion) or about $24.9 billion at the mid 1998 exchange rate. This represents around 10% of GDP, a fall from 13% in 1997. However this may well be an understatement of the real value of output, because of undercounting both of GDP and construction and because the exchange rate is unrealistic and inflation rampant. On a purchasing power parity basis the value of construction output in 1997 was around double that calculated on an exchange rate basis, at $ 50.5 billion. The breakdown by type of work in 1997 was estimated as follows:

Estimated output of new construction, 1997

Type of work	TI tiillion	ECUs billion	% of total
New work			
Residential building	1,295	7.5	34
Non-residential building			
Office and commercial	371	2.2	10
Industrial	242	1.4	6
Other	133	0.8	4
Total	746	4.3	20
Civil engineering*	1,746	10.2	46
Total	**3,788**	**22.0**	**100**

Note: * includes repair and maintenance
Source: based on Euroconstruct conference paper, Berlin, December 1998.

One of the notable features of the Turkish construction industry is the substantial output abroad. From the 1970s to 1997, the total turnover of foreign construction projects by Turkish firms reached about £40 billion. Most of the projects were undertaken in the Middle East, Russia, Turkic Republics and North Africa. After the breakdown of the USSR, work done by Turkish contractors in CIS countries has increased substantially and now around one third of contracts by Turkish companies are for work undertaken in these countries.

The evolution of the trend in construction in 1996 and 1997 is shown below:

Construction output % real change, 1995–96 and 1996–97

Type of work	1996–96	1996–97
Commercial building	25.5	7.0
Industrial building	1.2	29.9
Medical, social culture building	56.8	−12.1
Other building	−5.9	1.8
Total	**17.4**	**14.1**

Source: State Institute of Statistics, 1998.

Turkey has experienced rapid urbanization and the trend is that more dwellings will be needed in the future, to balance the growing demand for housing in cities. Investment is expected to grow in the industrial and commercial building sub-sectors and for major infrastructure work, especially for motorways and tourism industry.

Characteristics and structure of the industry

Turkish contractors, in general, are not large by international standards in spite of their international operations.

In 1997 there were nine Turkish contractors listed in *Engineering News Record's* list of the 'Top 225 International Contractors' for 1997.

Main contractors in Turkey, 1997

Contractor	Place in ENR *list*
Sezai Turkes Feyzi Akkaia Const. Co.	72
ENKA Construction & Industry Co. Inc.	74
Tekfen Construction & Installation Co.	110
Yapi Merkezi Construction & Industry Inc.	111
TEKSER Const. Industry and Trading Inc.	136
GAMA Endustri Tesisleri Imalat Ve Montaj	145
Soyak International Construction Co. Ltd.	146
Hazinedaroglu Construction Group	160
Baytur Construction & Contracting Co.	162

Source: Engineering News Record 17.8.1998.

Turkish firms tend to be family-run in a centralized, informal way by the founder. This provides flexibility in overseas activities, with over 300 firms operating abroad in 1987. Firms tend to specialize in certain types of work using directly employed labour and traditional trade subcontractors. The strongest selling point of Turkish firms abroad has been their large, flexible labour force.

There are about 55,500 registered contractors in the domestic market, of which over 44,000 are registered private contractors and over 11,000 are registered state owned contractors.

The Ministry of Public Works and Resettlement groups contractors into six categories:

- Group A (capable of undertaking large and special technical projects which require special and important quantities of machinery and equipment, know-how and large organization)
- Group B (capable of undertaking medium sized technical projects which do not require any special and large quantity of machinery and equipment, know-how and organization as required in Group A)
- Group C (capable of undertaking construction and repair works which are excluded from Group B).
- Group D (capable of undertaking the repair and restoration works of the old and historical buildings)
- Group G (capable of undertaking electrical and mechanical installation works of the projects mentioned in Group A)
- Group H (capable of undertaking electrical and mechanical installation works of the projects mentioned in Groups B and C and the smaller ones of the Group G).

There is also a sub-group nominated as 'Unlimited'. There are 1,000 such registered contractors as Group A, with Unlimited Certificate in Turkey.

As of the end of 1996, according to the Chamber of Engineers and Architects, there were 167,225 engineers and 25,084 architects registered in these chambers. In reality, their number is more than 200,000. Engineering companies produce all kinds of designs, including execution drawings.

Clients and finance

Of the total construction investment in 1997 22.7% was by the public sector and 77.3% by the private sector. The public sector is an important client of the construction industry for work other than housing and especially for civil engineering work. It also undertakes significant investment in manufacturing industry, agriculture, mining, education, health and tourism. The major public sector client is the Ministry of Public Works and Resettlement; others include the Ministry of Energy and Natural Resources, the Ministry of Communications and Transportation, the Ministry of National Defence and the Ministry of Health. The public sector pays the contractors as per the budget allocated to such projects.

'Design and build' contracts are being used increasingly. A variant is the 'build-operate-transfer' (BOT) project, in which the contractor operates the building or works for a specified period of time and retains the profits. At the end of the period the building or other work is transferred to the Government, which therefore receives it free in exchange for the original concession to build and operate. There is also a 'mutual construction contract' whereby the landowner enters into partnership with a contractor who will develop the land. This is used mainly in private residential and commercial developments. Some energy projects are contracted on a Build, Own and Operate (BOO) basis. For these, the supply of natural gas and the purchase of produced energy are guaranteed.

Contractor selection

Apart from the arrangements mentioned above, competitive bids are the norm in the public sector while negotiation is more usual in the private sector. For public sector projects, bidding takes place on specification, drawings, schedules of quantities (on major projects) and a priced schedule of rates (for smaller works). The contractor quotes on the basis of those rates.

Some big projects are contracted to the lowest bidder. The public sector now selects a bid near to the average after assessing financial strength, reputation, experience and reliability.

The normal contract is let either on a unit rate basis or 'lump sum'. There is cost escalation reimbursement based on cost indices. Sometimes the private sector gives contracts in which the price is quoted in US$ but these contracts have no escalation clauses. Bid bonds and performance bonds are normal.

Development control

The 'Town Development Act' of 1985 gave most of the planning and development control powers to those municipalities with a population inside their boundaries of over 10,000 . The main settlement control is in the hands of the Government. Organized Industrial Zones exist now in many privileged towns. Regional plans prepared by local authorities have to be approved by the Ministry. The municipal authority acts as approval authority for all levels of plans for construction works inside the cities and gives the necessary construction permit. For all other areas the Ministry of Public Works is the approval authority.

Construction cost data

Cost of labour

The figures below are typical of labour costs in the Istanbul area as at the first quarter 1999. The wage rate is the basis of an employee's income, while the cost of labour indicates the cost to a contractor of employing that employee. The difference between the two covers a variety of mandatory and voluntary contributions - a list of items which could be included is given in section 2.

	Wage rate (per hour) 1000TL	Cost of labour (per hour) 1000TL	Number of hours worked per year
Site operatives			
Mason/bricklayer	320	670	2,574
Carpenter	320	670	2,574
Plumber	320	670	2,574
Electrician	340	710	2,574
Structural steel erector	320	670	2,574
HVAC installer	340	710	2,574
Semi-skilled worker	226	474	2,574
Unskilled labourer	210	440	2,574
Equipment operator	365	765	2,574
Watchman/security	210	440	2,574
Site supervision			
General foreman	486	1,020	2,574
Trades foreman	458	960	2,574
Clerk of works	250	524	2,574
Contractors' personnel *	(per month)	(per month)	
Site manager	900,000	1,065,000	2,574
Resident engineer	750,000	890,000	2,574
Resident surveyor	550,000	657,000	2,574
Junior engineer	475,000	570,000	2,574
Junior surveyor	350,000	408,000	2,574
Planner	500,000	600,000	2,574

* monthly average salary/labour cost.

Cost of materials

The figures that follow are the costs of main construction materials, delivered to site in the Istanbul area, as incurred by contractors in the first quarter of 1999. These assume that the materials would be in quantities as required for a medium sized construction project and that the location of the works would be neither constrained nor remote. All the costs in this section exclude value added tax (VAT) which is at 15%.

	Unit	Cost 1000 TL
Cement and aggregates		
Ordinary portland cement in 50kg bags	tonne	11,000
Coarse aggregates for concrete	m³	3,010
Fine aggregates for concrete	m³	2,800
Ready mixed concrete (mix B14)	m³	9,800
Ready mixed concrete (mix B25)	m³	11,500
Steel		
Mild steel reinforcement	tonne	61,500
High tensile steel reinforcement	tonne	66,000
Structural steel sections	tonne	83,500
Bricks and blocks		
Common bricks (190 x 190 x 135 mm)	1,000	30,000
Good quality facing bricks (190 x 90 x 85mm)	1,000	35,000
Hollow concrete blocks (200 x 190 x 390mm)	each	68
Solid concrete blocks (200 x 190 x 390 mm)	Each	110
Timber and insulation		
Softwood sections for carpentry	m³	63,000
Softwood for joinery	m³	75,200
Hardwood for joinery	m³	134,000
Exterior quality plywood (21mm)	m²	4,500
Plywood for interior joinery (21mm)	m²	4,800
Softwood strip flooring (22mm)	m²	2,100
Chipboard sheet flooring (25mm)	m²	480
100mm thick rigid slab insulation	m²	3,600
Softwood internal door with frames (80 x 200 cm)	No.	35,000
Glass and ceramics		
Float glass (4mm thick)	m²	1,100
Sealed double glazing units (4 x 4 mm)	M²	5,000
Good quality ceramic wall tiles (300 x 300mm)	m²	1,800
Plaster and paint		
Plaster in 50kg bags	tonne	120,000
Plasterboard (15mm thick)	m²	600
Emulsion paint in 5 litre tins	Litre	700
Gloss oil paint in 5 litre tins	litre	900

	Unit	*Cost* *1000 TL*
Tiles and paviors		
Clay floor tiles (300 x 300 x 9 mm)	m²	2,000
Vinyl floor tiles (300 x 300 x 2 mm)	m²	600
Precast concrete paving slabs (200 x 200 x 60 mm)	m²	1,150
Precast concrete roof tiles (250 x 250 x 50 mm)	1,000	85,000
Drainage		
Lavatory basin complete	each	25,000
100mm diameter clay drain pipes	m	700
150mm diameter cast iron drain pipes	m	4,900

Unit rates

The descriptions of work items below are generally shortened versions of standard descriptions listed in five languages (English, French, Italian, German and Spanish) in Appendix 3.

Where an item has a two digit reference number (e.g. 05 or 33), this relates to the full description against that number in Appendix 3. Where an item has an alphabetic suffix (e.g. 12A or 34B) this indicates that the standard description has been modified. Where a modification is major the complete modified description is included here and the standard description should be ignored.

Where a modification is minor (e.g. the insertion of a named hardwood) the shortened description has been modified here but, in general, the full description in Appendix 3 prevails.

The unit rates below are for main work items on a typical construction project in the Istambul area in the first quarter of 1999. The rates include labour, materials and equipment and an allowance (25%) to cover contractor's overheads and profit and preliminary and general items. No allowance to cover contractor's profit and attendance on specialist trades has been made. All the rates in this section exclude VAT which is at 15%.

		Unit	*Rate* *1000 TL*
Excavation			
01	Mechanical excavation of foundation trenches	m³	1,050
02	Hardcore filling making up levels	m²	750
03	Earthwork support	m²	700

		Unit	*Rate*
			1000 TL
Concrete work			
04	Plain insitu concrete in strip foundations in trenches 20N/m^2	m^3	15,100
05	Reinforced insitu concrete in beds 20N/m^2	m^3	15,300
06	Reinforced insitu concrete in walls	m^3	15,400
07	Reinforced insitu concrete in suspended floor or roof slabs 20N/m^2	m^3	15,600
08	Reinforced insitu concrete in columns 20N/m^2	m^3	15,600
09	Reinforced insitu concrete in isolated beams 20N/m^2	m^3	15,600
Formwork			
11	Softwood formwork to concrete walls	m^2	2,500
12	Softwood or metal formwork to concrete columns	m^2	2,800
13	Softwood or metal formwork to horizontal soffits of slabs	m^2	3,850
Reinforcement			
14	Reinforcement in concrete walls (16mm)	tonne	123,000
15	Reinforcement in suspended concrete slabs (25mm)	tonne	123,000
16	Fabric reinforcement in concrete beds	m^2	430
Steelwork			
17	Fabricate, supply and erect steel framed structure	tonne	360,000
Brickwork and blockwork			
18	Precast lightweight aggregate hollow concrete block walls (100m thick)	m^2	1,500
19	Solid (perforated) concrete bricks	m^2	2,200
21	Facing bricks	m^2	3,000
Roofing			
22	Concrete interlocking roof tiles 430 x 380mm	m^2	5,000
25	Sawn softwood roof boarding	m^2	3,900
26	Particle board roof coverings	m^2	4,200
27	3 layers glass-fibre based bitumen felt roof covering include chippings	m^2	2,900
29	Glass-fibre mat roof insulation 160mm thick	m^2	3,000
30	Loadbearing glass-fibre roof insulation	M^2	3,200
31	Troughed galvanised steel roof cladding	m^2	3,800
Woodwork and metalwork			
32	Preservative treated sawn softwood	M	600
33	Presevative treated sawn softwood, pitched roof members	m	900
34	Single glazed casement window in hardwood, size 650 x 900mm	each	25,000
35	Two panel glazed door in hardwood, size 850 x 2000mm	each	75,000

		Unit	Rate
			1000 TL
36	Solid core half hour fire resisting hardwood internal flush doors, size 800 x 2000mm	each	65,000
37	Aluminium double glazed window, size 1200 x 1200mm	each	85,000
38	Aluminium double glazed door, size 850 x 2100mm	each	110,000
39	Hardwood skirtings 20 x 100mm	m	650
40	Framed structural steelwork in universal joist sections	Tonne	400,000
41	Structural steelwork lattice roof trusses	tonne	440,000

Plumbing

42	UPVC half round eaves gutter (110mm)	m	1,480
43	UPVC rainwater pipes (100mm)	m	930
44	Light gauge copper cold water tubing (15mm)	m	870
45	High pressure plastic pipes for cold water supply (15mm)	m	660
46	Low pressure plastic pipes for cold water distribution (20mm)	m	910
47	UPVC soil and vent pipes	m	1,930
48	White vitreous china WC suite	each	17,000
49	White vitreous china lavatory basin	each	21,000
50	White glazed fireclay shower tray	each	43,800
51	Stainless steel single bowl sink and double drainer	each	19,500

Electrical work

52	PVC insulated and PVC sheathed copper cable core and earth	m	465
53	16 amp unswitched socket outlet	each	2,490
54	Flush mounted 18 amp, 1 way light switch	each	2,970

Finishings

55	2 coats gypsum based plaster on brick walls	m^2	850
56	White glazed tiles on plaster walls	m^2	2,200
58	Cement and sand screed to concrete floors 50mm thick	m^2	1,200
59	Thermoplastic floor tiles on screed	m^2	2,250
60	Mineral fibre tiles on concealed suspension system	m^2	3,200

Glazing

61	Glazing to wood 4mm	m^2	2,000

Painting

62	Emulsion on plaster walls	m^2	500
63	Oil paint on timber	m^2	1,350

Exchange rates and cost and price trends

The combined effect of exchange rates and inflation on prices within a country and on price comparisons between countries is discussed in section 2.

Exchange rates

The graph below plots the movement of the Turkish lira against sterling, the ECU/Euro, the US dollar and 100 Japanese yen since 1990. The values used for the graph are quarterly and the method of calculating these is described and general guidance on the interpretation of the graph provided in section 2. The average exchange rate in the first quarter of 1999 was TL 513,000 to the pound sterling, TL 303,000 to the US dollar and TL 361,000 to the Euro.

The Turkish lira against sterling, the ECU/Euro, the US dollar and 100 Japanese yen

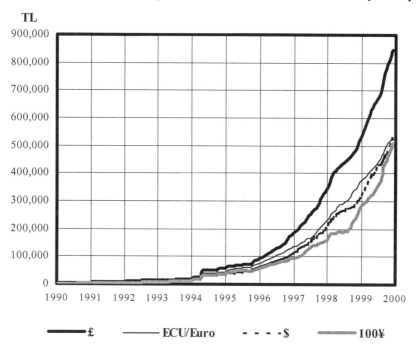

Cost and price trends

The table following presents the indices for consumer prices, building material costs and building costs in Turkey since 1991, using the base of 1991=100. The annual change is the percentage change between the fourth quarter index of consecutive years. Notes on inflation are included in section 2.

Consumer price, building cost and material cost indices

Year	Consumer prices *		Building costs		Building material costs	
	annual average	change %	annual average	change %	annual average	change %
1991	100	n.a.	100	n.a.	100	n.a.
1992	170	70	170	70	164	64
1993	282	66	288	69	257	57
1994	583	107	588	104	681	165
1995	1,096	88	1,007	71	980	44
1996	1,976	80	1,781	77	1,711	75
1997	3,671	86	3,365	89	3,154	84
1998	6,778	85	5,888	75	5,254	67

Note: * rebased to 1991.
Sources: International Monetary Fund, Yearbook 1998 and supplements 1999, State Institute of Statistics and Ministry of Public Works.

The building construction cost index is based on factor costs of materials, labour and machinery. No taxes are included and the prices are net of discounts. The index covers nearly all types of building accounting for 90% of building activity in Turkey. Weights are determined by an examination of a sample of bills of quantities from projects in 24 provinces of Turkey.

The building materials cost index represents the materials component of the above index.

Inflation is very high in Turkey often approaching 100% per annum. Building costs have increased in line with consumer prices.

Useful addresses

Government and public organizations

Ministry of Public Works and Settlement
Vekaletler Cad. N. 1
Bakanhklar – Ankara
Tel: (90-312) 4179260
Fax: (90-312) 4180406

General Directorate of Foreign Investment
Ankara
Tel: (90-312) 2125876
Fax: (90-312) 2128916

State Institute of Statistics
Ankara
Tel;: (90-312) 4176440
Fax: (90-312) 4253387

Professional and trade associations

**Union of Chambers of Commerce
and Industry of Turkey**
Atatürk Bulvari 149
Bakanliklar, Ankara
Tel: (90-312) 4177700
Fax: (90-312) 4183268

**Turkish Union of Engineers and
Architects' Chambers**
Konur Sokak 4 Kizilay
Ankara
Tel: (90-312) 4181275
Fax: 90-3124174824

**Turkish Contractors Association and
Union of International Contractors**
Ahmet Mithat Efendi Sok. 21/3
Çankaya 06550 Ankara
Tel: (90-312) 4391712
Fax: (90-312) 4400253

**Istanbul Chamber of Commerce
(ITO)**
Ragip Gümüspala Cad. N. 84
Eminönü – Istanbul
Tel: (90-212) 5114150
Fax: (90-212) 5262197

United Kingdom

All data relate to 1998 unless otherwise indicated

Population

Population	58.9 million
Urban population (1997)	89%
Population under 15	19%
Population 65 and over	16%
Average annual growth rate (1985 to 1998)	0.3%

Geography

Land area	244,820 km^2
Agricultural area	71%
Capital city	London

Economy

Monetary unit	Pound Sterling (£)
Exchange rate (average first quarter 1999) to:	
the US dollar	£ 0.60
the Euro	£ 0.70
the yen x 100	£ 0.51
Average annual inflation (1995 to 1998)	2.8%
Inflation rate	2.6%
Gross Domestic Product (GDP)	£844 billion
GDP PPP basis (1997)	$ (PPP) 1,242 billion
GDP per capita	£14,310
GDP per capita PPP basic (1997)	$ (PPP) 21,120
Average annual real change in GDP (1995 to 1998)	2.8%
Private consumption as a proportion of GDP	63%
Public consumption as a proportion of GDP	20%
Investment as a proportion of GDP	17%

Construction

Gross value of construction output (1997)	£63.1 billion
Gross value of construction output per capita	£1,070
Gross value of construction output as a proportion of GDP	7.5%
Average annual real change in gross value of construction output (1995 to 1998)	2.3%
Annual cement consumption	13.0 million tonnes

PPP Purchasing power parity basis.

The construction industry

Construction output

The value of the gross output of the construction sector in the United Kingdom in 1998 was £63.1 billion equivalent to 91.6 ECUs (nearly the same as January 1999 Euros). This represents 7.5% of GDP, a very low proportion compared to the rest of Europe. In addition, it is estimated that a further £12.6 billion (20%) of construction was undertaken by other sectors, the black economy and DIY. The breakdown by type of work is shown below:

Output of construction sector , 1998 (current prices)

Type of work	£ billion	ECUs billion	% of total
New work			
Residential building	8.4	12.2	13
Non-residential building +			
Office and commercial	9.6	13.9	15
Industrial	4.1	5.9	6
Other	4.5	6.6	7
Total	18.2	26.4	29
Civil engineering*	6.3	9.2	10
Total new work	33.0	47.9	52
Repair and maintenance			
Residential	16.4	23.8	26
Non-residential	10.1	14.6	16
Civil engineering	3.7	5.3	6
Total repair and maintenance	30.1	43.7	48
Total	**63.1**	**91.6**	**100**

Note: the value of the ECU is similar to the value of the Euro in January 1999.
Source: Cannon, J., 'United Kingdom' in Euroconstruct conference proceedings, Prague, June 1998.

The level of construction output increased only slightly in 1998 following a pause in the development of the whole economy. GDP increased by just over 2% compared with 3.5% in 1997 and construction by 1.6% compared with 3.0% in 1997.

Residential building actually fell in 1998 by nearly 2% but it is unlikely to fall much further unless the present low rates of interest are raised. The effect of rising prices of existing houses on the private new housing market is uncertain. It could be beneficial to new house sales by the year 2000. Public sector housing in the UK accounts for only about 15% of dwellings. Most of these were built for Registered Social Landlords (RSLs) mainly housing associations. New public sector housing is unlikely to increase. In 1998 only about 22% of dwelling completions were flats.

In non-residential buildings there was an increase of about 6% in output of which the largest share was taken by commercial buildings, which account for 30% of this type of work. They increased by about 18% after a similar rise in 1997. Office buildings have increased by about 4% in 1997 and 6% in 1998 while industrial buildings had a lower rate of increase of 4% in 1998. One of the reasons for the high level of commercial construction is the number of projects for the millennium, which presumably will all be completed by the end of 1999. Industrial buildings are suffering from the difficulties faced by manufacturing industry, which are exacerbated by the high value of the pound.

Public sector non-residential building has not been active, except that in 1998 school and university construction increased. Government is pressing hard to increase the number of Private Finance Initiative (PFI) projects but although a number are starting, many of them for hospitals, their full impact on output will not appear until 1999 or 2000. Once such projects are underway, they should correct the hiatus which has dogged this sector while the way of operating PFI projects was worked out.

Civil engineering work too is increasingly funded from private sources and many of the former public types of work are now private, for example, electricity, water, gas and railways. They are however regulated by government in some way. Overall civil engineering work fell in 1998 and no increase is expected in the near future. In particular transport infrastructure work, including roads declined by about 11% in 1998. As this accounts for over half the civil engineering work it made a fall overall almost inevitable. Only the energy and water sectors increased by about 13% together.

Repair and maintenance expenditure is very much influenced by the health of the economy and in residential buildings by consumer incomes. Apart from an expected pause in 1999 both are buoyant. Moreover unemployment is low and so also are interest rates.

Government policy is to encourage improvement and regeneration of public sector housing which has been neglected in the past. Private sector housing repair and maintenance rose in 1998 and because of the underlying favourable conditions prospects are good for modest growth. Residential repair and maintenance accounts for well over half the total in the UK. Other sectors are expected also to have modest growth. In general the swings in repair and maintenance are less than in new work.

Renovation output in housing grew steadily, especially in the private sector, due to high economic growth and the bonuses from the de-mutualization of building societies. The Government is planning to promote rehabilitation programmes which are partially financed by private parties. Renovation in the civil engineering sector increased, due to the railways and water and sewerage sectors, while the trend has been negative for roads.

The table below shows the percentage distribution of population and contractors output in 1996 in the main UK regions. It will be seen that the South East accounts for over a third of contractors' output – marginally more than the share of population.

Regional population and contractors' output, 1997

Region	Population %	Output %
North	5.5	4.8
Yorks and Humberside	8.9	8.2
East Midlands	7.2	7.6
East Anglia	3.7	3.8
South East	31.5	34.8
South West	8.5	8.0
West Midlands	9.3	8.6
North West	11.3	10.1
Wales	5.1	4.6
Scotland	9.0	9.4
Total	**100.0**	**100.0**

Sources: DETR, Housing and Construction Statistics 1987-1997.
CSO, Annual Abstract of Statistics 1996.

The work abroad by British contractors present in the top 225 international contractors in Engineering News Record in 1997 was valued at £ 7.7 billion or more than 13% of total domestic work as follows:

British construction work abroad, 1997

Area	£ billion	% of total
Europe	2.3	29.3
Africa	0.1	1.9
Middle East	0.4	5.8
Asia	1.9	24.2
USA	2.5	31.9
Canada	0.2	3.2
Latin America	0.3	3.7
Total	**7.7**	**100.0**

Source: Engineering News Record 12-8-98.

Characteristics and structure of the industry

The bulk of building work in the UK is undertaken by general contractors who traditionally employed their own labour force but now increasingly use labour-only subcontractors. Specialist subcontractors may be nominated by the client's consultant team or employed by the general contractor.

Traditionally, building work in the UK has been administered by professional consultants appointed by the building client. The consultants are responsible for the design and specification of the work, the contractual arrangements and the supervision of the contract. However, over the last 20 years, design-and-build contractors offering a single point of responsibility have also established a sizeable share of the market as have other, less traditional, arrangements (see 'Contractual arrangements').

The construction industry in the UK consists of a large number of small firms and a few very large firms. In 1997 there were estimated to be 160,148 construction firms of which 38 employed over 1,200 persons and 205 employed over 300. However, figures do not truly indicate the importance of large firms because of the prevalence of labour-only sub-contractors.

In *Building* magazine's 'Top 300 European Contractors' list for 1998 there are 63 UK companies listed overall, and 22 in the top 100 – more than any other country but France. The *Engineering News Record's* 1997 'Top 225 International Contractors' lists seven UK contractors in the top 100 in 1997. In both the *Building* and *Engineering News Record* lists, some contractors are very specialized, for example in engineering or housing development, rather than building and civil engineering.

The principal UK contractors and their sales abroad are shown in the table below.

Major UK contractors, 1997

Major contractors	Place in Building's 'Top 300 European Contractors' 1997	% of sales abroad
Amec	12	28
Tarmac	13	23
Balfour Beatty	15	25
Bovis	20	-
Laing	27	13
Mowlem	28	28
Taylor Woodrow	34	38
Wimpey	37	18
Barratt	49	6
Keir Goup	52	11

Sources: Building, December 1998.

For consistency, contractors are listed in the order quoted in Building magazine's list which is based on total turnover. This does not, however, really represent their construction nor their contracting size. One feature of the major British contractors is the extent of their diversification beyond general contracting into property development, speculative housebuilding, material production and other businesses such as mining or airports. This was thought to have enabled the larger contractors to weather the fluctuations in construction demand and, in many cases, gives a better return on capital employed. However, in recent years, some of the non-construction businesses have been the cause of financial failure amongst medium size contractors and the larger firms have found it necessary to divest themselves of such subsidiary businesses.

Construction design work in the UK is undertaken mainly by architects and engineers. In 1994 there were about 28,000 architects in full-time employment but many are underemployed. More than 80% are members of the Royal Institute of British Architects (RIBA). Most practices are small but some practices employ over 100 staff. Architects must register with the Architects Registration Council of the United Kingdom (ARCUK) to be permitted to use the designation.

Practising civil engineers are normally members of the Institution of Civil Engineers (ICE) whose membership is about 80,000 worldwide. Most structural engineers are members of the Institution of Structural Engineers, with a membership of about 23,000, and most building services engineers are members of the Chartered Institution of Building Services Engineers, having a membership of about 15,000. All these institutions have substantially increased their membership in the last few years. The title of Chartered Engineer is registered and protected, either by the professional institution or by the Engineering Council.

Most construction design work in the UK is undertaken by private firms of professional architects or engineers. The amount of in-house work has shrunk considerably in the last ten years and the remaining contractors' design departments are relatively small, being mostly concerned with building rather than civil engineering work. In 1994 there were 750 independently registered consulting engineering firms, many of whom were members of the Association of Consulting Engineers (ACE).

The surveying profession is very important in the UK. The Royal Institution of Chartered Surveyors (RICS) is an umbrella organization for quantity surveyors and building surveyors as well as a number of other surveying disciplines more concerned with property than the construction industry. In total, RICS membership in the UK is about 76,000 and there are a further 19,000 in training. The quantity surveyor plays a key role in the UK construction industry. Originally his role was to prepare a bill of quantities and measure work on site.

The profession has, however, developed a range of consultancy services for clients and has a full professional status equivalent to that of designers. Nowadays quantity surveyors advise at every stage of the property life-cycle; from raw land, through measurement, planning, funding, design and construction, management, refurbishment and redevelopment. Quantity surveyors work mainly in private practice, but also in the public sector and commercial organizations. Most quantity surveying practices are small but there are a number of very large firms employing several hundred staff.

Clients and finance

Since the mid-1970s there has been a marked decline in public sector investment. Although the decline in public work has been for all types of public construction it has been most dramatic in the public housing sector and infrastructure – the latter largely as a result of the privatization of public utilities. Nearly all publicly funded housing is undertaken by housing associations. They are replacing local authorities in new housing provision and in 1997 accounted for 99% of public sector starts.

A large part of the purchase of private housing, both existing and new, is financed by mortgages, with building societies most important but competing hard with the banks to retain their position.

Other non-residential buildings in the private sector may be financed in a number of ways and may be built and owned by owner occupiers or be built by developers/investors and then let. It is estimated, for example, that owner occupiers account for up to 80% of new construction of industrial buildings. However, the amount of other private buildings built and owned by owner occupiers is much less. Available statistics suggest that the majority of non-industrial building and non-house building is financed by the banking, pension and insurance sectors or by property developers' own funds.

Selection of design consultants

In the past, much of the design work of the public sector was done in-house by professional teams; as a result of the privatization process, these teams have increasingly been privatized or disbanded, thus providing more work for private design firms.

In the last few years there have been major changes in the method of selection of design consultants in the public sector. Firstly, public clients are now required to select on the basis of a fee competition, although it is not mandatory to accept the lowest tender if greater value for money is achieved by the acceptance of another. And secondly, the EU Directive relating to the selection of professional consultants came into effect on 1st July 1994 and appointments have to be advertised. The EU directive can affect privatized utilities too. It is however possible to have a prequalification process to identify a list of suitable consultants and then draw from that list on the basis of fee competition. In fact most public sector clients interview potential consultants and select on the basis of capability and experience as well as on price.

Nevertheless there is considerable diversity in the approach of the public sector. A large number of substantial and regular public sector clients have in-house project managers and they have their own views as to the way in which they will manage a project. In many cases they appoint the architect first and adopt the traditional process except that they themselves are the lead consultants. Others will appoint other professionals first, most commonly a project manager but perhaps a specialized engineer or a quantity surveyor if cost control is especially important.

In the private sector there is more flexibility in the method of selection of consultants with personal attributes of the main player being very important and experience and reputation of the firm of great significance. Fee competition is less usual, although it has been used by some and is being considered by others. Work is secured more and more by competition. It is usual for fees to be negotiated. Regular clients rarely adopt the fee scales recommended by the institutions. In the private sector too there is great variety in the order in which the consultants are appointed and their responsibilities.

In the last few years the whole ethos of the organization of the construction process has changed from one of a well trodden path to one of choice and flexibility.

Contractual arrangements

As in the case of the selection of design consultants, the contractual arrangements in the UK are undergoing substantial change. The 'traditional' system of a main contractor appointed by the client is still used for the majority of projects, but alternative contractual arrangements are increasing in number and the popularity of each one fluctuates according to: the relative bargaining power of the client and other parties to the process, the size, type and complexity of work being undertaken, and fashion. Recently two reports have influenced thinking both in the private and the public sector on the way to organize projects. The first is the Latham Report, *Constructing the Team* and the second is known as the Egan report *Rethinking Construction*. Partly as a result of these the variety of arrangements is increasing. For large projects there are two main types. The first is where the contractor alone or in some form of partnership is selling a building or other facility, often together with some other service, such as maintenance of the building or even its management. These include:

- project management where a project manager, organizes the whole process from inception to completion to meet the clients needs.

- design and build where the contractor is responsible for design and construction of the project. Sometimes novated design and build is used where the preliminary design is undertaken by the architect working to the client's instructions but the responsibility for design and execution are then transferred to the contractor working direct to the client.

- management contracting where the client enters into separate contracts with a designer and a management contractor and the latter then enters into subcontracts with trade contractors.

- construction management where the client enters into separate contracts with the designer, construction manager and trade contractors.

- prime contracting – a system recently developed by the Ministry of Defence and others which has one point of contact to the industry for the client. The system aims to incorporate optimal full-life costs, to deliver exactly what the client requires and, through a target cost-incentive scheme, to be fair to all parties.

- build, operate, transfer (BOT) – a system in which the contractor builds (and possibly designs) the project and then runs it and is remunerated by a regular income based on expected costs over time. The project may later be handed back to the client.

The second type of arrangement is where the contractor, alone or in some form of partnership is selling a service which includes ownership and financing of whatever assets are required to provide that service. These include:

- build, own, operate transfer (BOOT). In this arrangement the client is being provide with a service which happens to require a building or other works to enable the service to be provided. The supplier is a service provider and may or may not be a construction contractor. BOOT carries a high level of risk for the service providers.

- private finance initiative (PFI). This is the system, still being developed but already in use, by which contractors, usually in consortia are asked to bid in competition to finance, build and operate a facility traditionally provided by the public sector, transferring to government ownership at the end of a period of years.

Data on types of contract documentation used suggest that for large projects, and excluding speculative housing construction which is all design and build, that in 1997 about 20% of these projects were design and build but they accounted for about 40% of value. Guidelines to government departments encourage the use of design and build and prime contracting, both of which give the contractor a lead role. They also say that traditional procurement should be used only if there are very good reasons in terms of value for money. At the same time it is pressing on with the development of PFI.

The Construction Industry Board (CIB) has issued codes of practice for the selection of contractors and sub-contractors for all procurement routes. The National Joint Consultative Committee, which comprises a group of client and consultant bodies, publishes codes for selective tendering. In the case of lump sum contracts prices are based on firm bills of quantities though sometimes approximate bills of quantities are used. For smaller projects specifications and drawings are more commonly used in place of bills of quantities.

The predominant form of contract used in the UK is the 'Joint Contracts Tribunal (JCT) 1998 Standard Form of Contract with quantities' (known as 'JCT98'). This contract assumes the use of measured bills of quantities that are normally prepared by the quantity surveyor. The JCT also produces the '1998 Intermediate Form of Building Contract' (known as 'IFC98') for works of simpler content. The tender documentation under both these forms of contract might comprise:

- drawings
- specification
- bills of quantities, schedules of works or schedules of rates.

The JCT building contracts are produced in private and local authority editions. Government buildings and civil engineering contracts are often placed using the 'General Conditions of Government Contracts for Building and Civil Engineering Works', otherwise known as 'Form GC/Works/1' (1998).

For civil engineering, local authority and private clients generally use the 'Institution of Civil Engineers Conditions of Contract, sixth edition', but an amended version of 'GC/Works/1' is also sometimes used. The seventh edition of the ICE form, introducing a number of minor changes was published in October 1999.

There are separate contract forms prepared by these bodies for the other types of construction arrangements (e.g. design and build). The JCT introduced a standard form for management contracting in 1987 but standard forms of contract for construction management do not exist and construction lawyers usually use *ad hoc* versions.

The Latham report entitled *Constructing the Team* recommended the increased use of a form of contract known as the 'New Engineering and Construction Contract' (NEC).

Contracts of up to 18 months duration or less are generally let on a fixed price basis. Fluctuations in labour, materials and plant costs on longer term contracts can be adjusted, where the contract permits. Such increased costs on private contracts may be paid on the basis of invoices for materials and plant, and time sheets for labour. Local authority government contracts and some private contracts have generally involve the use of an adjustment formula based on monthly published indices.

Both contractors' quantity surveyors and quantity surveyors representing the client need to have an agreed method for the measurement of building works. The first such agreed method was produced in 1922 and a new completely revised edition of the 'Standard Method of Measurement of Building Works, Seventh Edition', commonly known as 'SMM7' was issued in 1987. It was prepared by a Development Unit set up by the Royal Institution of Chartered Surveyors and the Building Employers Confederation. It embodies the essentials of good practice and is generally followed for all UK building work. Civil engineers have a comparable document – the Civil Engineering Standard Method of Measurement sponsored by the Institution of Civil Engineers and the Federation of Civil Engineering Contractors.

There are a number of different definitions of area for commercial offices and some other types of building. Gross External Area (GEA) is an RICS term defined in the Code of Measuring Practice published by the RICS and the ISVA (Incorporated Society of Valuers and Auctioneers) and describes the office floor space as per the 'Town and Country Planning Act' (1971). It is the area on each floor measured from the outside face of the external walls, i.e. the complete footprint of the building. However, when calculating this area the following should be excluded: open balconies and fire escapes; atria and areas with a height of less than 1.50m, e.g. under roof slopes; open covered ways or minor canopies; open vehicle parking areas; terraces, party walls beyond the centre line. Structural elements and spaces such as partitions, columns, lift wells, plant rooms and the like should be included.

Gross Internal Area (GIA) is an RICS term measured on the same basis as GEA but between the inside faces of the external walls to all enclosed spaces fulfilling the functional requirements of the building including all circulation areas, voids, staircases (to be measured flat on plan) and other non office areas such as plant rooms, toilets and enclosed car parks.

Net Lettable Area (NLA) refers to the gross internal building area less the building core area and any other common areas. This equates with the letting agents lettable floor area.

Net Internal Area (NIA) is the usable space within a building measured to the internal face of external walls and excludes ancillary and auxilliary spaces such as toilets, lifts, plant rooms, stairwells, corridors and other circulatory areas, internal structural walls and columns, the space occupied by air-conditioning/heating plant, areas with a height of less than 1.5m and parking areas.

The Maximum Usable Area (MUA) is the benchmark for comparing empty buildings and their space efficiency and is measured as the net internal area excluding only the minimum primary circulation, i.e. the base minimum circulation which would satisfy the local fire authority. Typically most cases would be covered by assuming routes 1.50m wide, joining vertical routes and fire exits with no point further than 12m from a primary fire route.

Treated Area is measured on the same basis as GIA but excludes those areas which are not directly heated, i.e. plant and lift motor rooms, car parks, unheated storage rooms, etc.

Liability and insurance

Since the mid-1970s there has been a marked increase in litigation on professional liability. Liability may arise in contract or in tort. In the first case, proof is required of a breach of contract. However, contracts do not normally define clearly the limits of work or duties of the professional and are in any case often informal, thus giving plenty of opportunity for litigation. The number of claims under tort, though increasing in the 1970s, have more recently declined as a result of various legal judgements.

Largely because of the reliance on case law made by the courts, the great problem of the English system is its uncertainty and complexity. There is no certainty on whether liability exists, on the amount involved, the period of liability or the time lag before any liability is determined.

It is mandatory for professional practices to carry professional indemnity insurance. The cost of premiums is now very high.

The proposals of the European Commission for a standard liability throughout the Community may substantially alter the UK position and create greater certainty.

Development control and standards

The system for planning and control of development was introduced by the 'Town and Country Planning Act 1947'. Although this has been amended, notably by the 'Town and Country Planning Act 1971' and then in the Town and Country Planning Act 1990, the principles remain the same. The main responsibility for planning lies with local authorities. A development plan for each area must be produced and every development (which is very widely defined) must receive permission from the relevant authority.

In general, applications for development must be made to the relevant district council . The total volume of applications to district councils in England is over 500,000 a year. Most planning applications are simple and go through a process which varies from district to district but is usually straightforward. However, quite a lot of weight lies on the judgement of the local planning officer and ultimately of the elected council members and therefore there is no certainty that an application will be granted. Nevertheless, there is usually considerable room for negotiation on changes to the original application.

According to statistics produced by the Department of the Environment, Transport and the Regions (DETR) *Development Control* Statistics, for England 1997/98, in 1997/98, 88% of all planning applications were granted. 62% of planning applications in 1997/98 were dealt with within eight weeks and 79% within 13 weeks. About 9% of applications refused came to appeal in 1996/97.

All new construction in the UK has to comply with Building Regulations, which are couched in terms of a series of technical requirements. These requirements are backed up by 'Approved Documents' which set out ways in which the requirements can be met. These documents give guidance but are not mandatory. In many areas the 'Approved Document' will refer a designer to an appropriate British Standard for complex issues. At present, approval for construction work is generally given by local authorities except that in the case of private sector housing the National House Builders Council (NHBC) acts as an alternative 'Approved Inspector' and is reported to have about 40% of that market. Since 1997 it has no longer been a statutory requirement for approval for non-housing work to be obtained from a local authority. In 1997 three corporate bodies were granted approval to act as 'Approved Inspectors'.

Construction cost data

Cost of labour

The figures on the next page are typical of labour costs in the London area as at the first quarter 1999. The wage rate is the basis of an employee's income, while the cost of labour indicates the cost to a contractor of employing that employee. The difference between the two covers a variety of mandatory and voluntary contributions - a list of items which could be included is given in section 2.

	Wage rate (per hour) £	Cost of labour (per hour) £	Number of hours worked per year
Site operatives			
Mason/bricklayer	5.50	7.37	1,802
Carpenter	5.50	7.37	1,802
Plumber	7.54	10.60	1,733
Electrician	8.88	11.10	1,733
Structural steel erector	7.03	9.76	1,756
HVAC installer	8.46	11.40	1,802
Semi-skilled worker	4.19	5.63	1,802
Unskilled labourer	4.23	5.65	1,802
Equipment operator	4.56	6.05	1,802
Watchman/security	4.23	5.65	1,802
Site supervision			
General foreman	7.00	9.34	1,802
Trades foreman	5.83	7.78	1,802
Clerk of works*	25,700 (per year)	32,800 (per year)	–
Contractors' personnel **			
Site manager	37,200	46,600	–
Resident engineer	26,900	33,900	–
Resident surveyor	26,500	33,800	–
Junior engineer	20,700	23,800	–
Junior surveyor	15,750	18,200	–
Planner	28,500	36,000	–
Consultants' personnel **			
Senior architect	34,700	43,700	–
Senior engineer	32,200	40,900	–
Senior surveyor	35,000	44,100	–
Qualified architect	27,000	34,300	–
Qualified engineer	23,000	29,800	–
Qualified surveyor	31,800	39,800	–

* London average annual salary/labour cost.   *** national average annual salary/labour cost.*

Cost of materials

The figures that on the next page are the costs of main construction materials, delivered to site in the London area, as incurred by contractors in the first quarter of 1999. These assume that the materials would be in quantities as required for a medium sized construction project and that the location of the works would be neither constrained nor remote. All the costs in this section exclude value added tax (VAT) which is at 17.5%.

	Unit	Cost £
Cement and aggregates		
Ordinary portland cement in 50kg bags	tonne	90.00
Shingle aggregates for concrete (40mm)	m^3	11.50
Fine aggregates for concrete (sharp sand)	m^3	11.00
Ready mixed concrete (10N/mm^2)	m^3	56.30
Ready mixed concrete (25N/mm^2)	m^3	58.90
Steel		
Mild steel reinforcement (12mm)	tonne	240.00
High tensile steel reinforcement (12mm)	tonne	237.00
Structural steel sections	tonne	430.00
Bricks and blocks		
Common bricks (215 x 102.5 x 65mm)	1,000	167.00
Good quality facing bricks (215 x 102.5 x 65mm)	1,000	340.00
Hollow concrete blocks (450 x 225 x 140mm)	1,000	1,070.00
Solid concrete blocks (450 x 225 x 140mm)	1,000	780.00
Precast concrete cladding units with exposed aggregate finish	m^2	170.00
Timber and insulation		
Softwood sections for carpentry	m^3	220.00
Softwood for joinery	m^3	290.00
Hardwood for joinery (Iroko)	m^3	925.00
Exterior quality plywood (18mm)	m^2	11.20
Plywood for interior joinery (6mm)	m^2	2.40
Softwood strip flooring (22mm)	m^2	7.68
Chipboard sheet flooring (18 mm t&g)	m^2	2.65
100mm thick quilt insulation	m^2	3.13
100mm thick rigid slab insulation	m^2	6.29
Softwood internal door complete with frames and ironmongery (826 x 2040 x 40mm)	each	80.10
Glass and ceramics		
Float glass (6mm)	m^2	21.50
Sealed double glazing units	m^2	49.00
Good quality ceramic wall tiles (198 x 64.5 x 6 mm)	m^2	21.50
Plasterboard (9.5mm thick)	m^2	1.62
Emulsion paint in 5 litre tins	litre	1.78
Gloss oil paint in 5 litre tins	litre	2.42
Tiles and paviors		
Clay floor tiles (150 x 150 x 12.5 mm)	m^2	11.80
Vinyl floor tiles(300 x 300 x 2.0mm)	m^2	3.62
Precast concrete paving slabs (200 x 100 x 60mm)	m^2	7.00
Clay roof tiles (plain 265 x 165 mm)	1,000	201.00
Precast concrete roof tiles (419 x 330mm)	1,000	512.00

	Unit	Cost £
Drainage		
WC suite complete	each	170.00
Lavatory basin complete	each	90.00
100mm diameter clay drain pipes	m	4.64
150mm diameter cast iron drain pipes	m	30.50

Unit rates

The descriptions of work items following are generally shortened versions of standard descriptions listed in five languages (English, French, Italian, German and Spanish) in Appendix 3.

Where an item has a two digit reference number (e.g. 05 or 33), this relates to the full description against that number in Appendix 3. Where an item has an alphabetic suffix (e.g. 12A or 34B) this indicates that the standard description has been modified. Where a modification is major the complete modified description is included here and the standard description should be ignored. Where a modification is minor (e.g. the insertion of a named hardwood) the shortened description has been modified here but, in general, the full description in Appendix 3 prevails.

The unit rates following are for main work items on a typical construction project in the London area in the first quarter of 1999. The rates include all necessary labour, materials and equipment, and allowances for preliminary and general items (10%). No allowance has been made for either contractor's general overheads and profit or profit and attendance on specialist trades, reflecting the competitive market conditions that prevail. All the rates in this section exclude VAT which is at 17.5%.

		Unit	Rate £
Excavation			
01	Mechanical excavation of foundation trenches	m^3	17.70
02	Hardcore filling making up levels	m^2	2.87
03	Earthwork support	m^2	1.14
Concrete work			
04	Plain insitu concrete in strip foundations in trenches	m^3	64.40
05	Reinforced insitu concrete in beds	m^3	66.40
06	Reinforced insitu concrete in walls	m^3	75.30
07	Reinforced insitu concrete in suspended floor or roof slabs	m^3	78.90
08	Reinforced insitu concrete in columns	m^3	87.50
09	Reinforced insitu concrete in isolated beams	m^3	82.20
10	Precast concrete slab	m^2	30.10

		Unit	Rate £
Formwork			
11	Softwood formwork to concrete walls (four uses)	m^2	46.20
12	Softwood or metal formwork to concrete columns (four uses)	m^2	30.40
13	Softwood or metal formwork to horizontal soffits of slabs	m^2	24.40
Reinforcement			
14	Reinforcement in concrete walls (16mm)	tonne	503.00
15	Reinforcement in suspended concrete slabs	tonne	465.00
16	Fabric reinforcement in concrete beds	m^2	2.07
Steelwork			
17	Fabricate, supply and erect steel framed structure	tonne	1,280.00
Brickwork and blockwork			
18	Precast lightweight aggregate hollow concrete block walls (100mm thick)	m^2	13.70
19	Solid (perforated) concrete bricks (£100/1000)	m^2	23.50
20	Solid (perforated) sand lime bricks (£130/1000)	m^2	23.40
21	Facing bricks (£275/1000)	m^2	35.60
Roofing			
22	Concrete interlocking roof tiles	m^2	18.40
23	Plain clay roof tiles	m^2	32.30
24	Fibre cement roof slates	m^2	25.40
25	Sawn softwood roof boarding	m^2	11.30
26	Roof board coverings	m^2	13.50
27	3 layers glass-fibre based bitumen felt roof covering include chippings	m^2	16.80
28	Bitumen based mastic asphalt roof covering	m^2	12.40
29	Glass-fibre mat roof insulation, 150mm thick	m^2	4.10
30	Rigid sheet loadbearing roof insulation 75mm thick	m^2	12.20
31	Troughed galvanised steel roof cladding	m^2	14.10
Woodwork and metalwork			
32	Preservative treated sawn softwood 50 x 100mm	m	3.04
33	Preservative treated sawn softwood 50 x 150mm	m	3.82
34	Single glazed casement window in Meranti hardwood, size 650 x 900mm	each	121.00
35	Two panel glazed door in West African Mahogany hardwood, size 850 x 2000mm	each	498.00
36	Solid core half hour fire resisting hardwood internal flush doors, size 800 x 2000mm	each	217.00
37	Aluminium double glazed window, West African Mahogany, size 1200 x 1200mm	each	444.00
38	Aluminium double glazed door, West African Mahogany size 850 x 2100mm	each	1,670.00
39	Hardwood skirtings in West African Mahogany	m	7.32

		Unit	Rate £
40	Framed structural steelwork in universal joist sections	tonne	1,270.00
41	Structural steelwork lattice roof trusses	tonne	1,430.00
Plumbing			
42	UPVC half round eaves gutter	m	6.31
43	UPVC rainwater pipes	m	8.64
44	Light gauge copper cold water tubing	m	3.52
45	High pressure plastic pipes for cold water supply	m	4.02
46	Low pressure plastic pipes for cold water distribution	m	3.82
47	UPVC soil and vent pipes	m	9.43
48	White vitreous china WC suite	each	176.00
49	White vitreous china lavatory basin	each	88.70
50	White glazed fireclay shower tray	each	139.00
51	Stainless steel single bowl sink and double drainer	each	138.00
Electrical work			
52	PVC insulated and copper sheathed cable, 600/1000 volt grade	m	3.82
53	13 amp unswitched socket outlet	each	32.00
54	Flush mounted 20 amp, 1 way light switch	each	18.30
Finishings			
55	2 coats gypsum based plaster on brick walls	m^2	6.62
56	White glazed tiles on plaster walls	m^2	23.90
57	Red clay quarry tiles on concrete floor	m^2	22.60
58	Cement and sand screed to concrete floors 50mm thick	m^2	6.28
59	Thermoplastic floor tiles on screed	m^2	5.66
60	Mineral fibre tiles on concealed suspension system	m^2	18.60
Glazing			
61	Glazing to wood	m^2	22.30
Painting			
62	Emulsion on plaster walls	m^2	1.71
63	Oil paint on timber	m^2	4.25

Approximate estimating

The building costs per unit area given below are averages incurred by building clients for typical buildings in the London area as at the first quarter 1999. They are based upon the total floor area of all storeys, measured between external walls and without deduction for internal walls.

Approximate estimating costs generally include mechanical and electrical installations but exclude furniture, loose or special equipment, and external works; they also exclude fees for professional services. The costs shown are for specifications and standards appropriate to the United Kingdom and this should be borne in mind when attempting comparisons with similarly described building types in other countries. A discussion of this issue is included in section 2. Comparative data for countries covered in this publication, including construction cost data, are presented in Part Three.

Approximate estimating costs must be treated with caution; they cannot provide more than a rough guide to the probable cost of building. All the rates in this section exclude VAT which is at 17.5%.

	Cost £ per m²	Cost £ per ft²
Industrial buildings		
Factories for letting (include lighting, power and heating)	295	27
Factories for owner occupation (light industrial use)	385	36
Factories for owner occupation (heavy industrial use)	635	59
Factory/office (high-tech) for letting (shell and core only)	395	37
Factory/office (high-tech) for owner occupation (controlled environment, fully furnished)	630	59
High tech laboratory (air conditioned)	1,705	159
Warehouses, low bay (6 to 8m high) for letting (no heating)	210	20
Warehouses, low bay for owner occupation (including heating)	310	29
Warehouses, high bay for owner occupation (including heating)	645	60
Administrative and commercial buildings		
Civic offices, non air conditioned	790	73
Civic offices, fully air conditioned	960	89
Offices for letting, 5 to 10 storeys, non air conditioned	750	70
Offices for letting, 5 to 10 storeys, air conditioned	1,065	99
Offices for letting, high rise, air conditioned	1,500	139
Offices for owner occupation, 5 to 10 storeys, non air conditioned	970	90
Offices for owner occupation, 5 to 10 storeys, air conditioned	1,355	126
Offices for owner occupation, high rise, air conditioned	1,790	166
Prestige/headquarters office, 5 to 10 storeys, air conditioned	1,570	146
Prestige/headquarters office, high rise, air conditioned	2,030	189
Health and education buildings		
General hospitals	845	79
Teaching hospitals	715	66
Private hospitals	880	82
Health centres	760	71
Nursery schools	780	73

	Cost £ per m²	Cost £ per ft²
Primary/junior schools	715	66
Secondary/middle schools	690	64
University (arts) buildings	710	66
University (science) buildings	800	74
Management training centres	825	77
Recreation and arts buildings		
Theatres (over 500 seats) including seating and stage equipment	1,450	135
Theatres (less than 500 seats) including seating and stage equipment	965	90
Concert halls including seating and stage equipment	1,710	159
Sports halls including changing and social facilities	565	53
Swimming pools (international standard) including changing facilities	1,355	126
Swimming pools (schools standard) including changing facilities	715	66
National museums including full air conditioning and standby generator	2,090	194
Local museums including air conditioning	845	79
City centre/central libraries	860	80
Branch/local libraries	690	64
Residential buildings		
Social/economic single family housing (multiple units)	310	29
Private/mass market single family housing 2 storey detached/ semidetached (multiple units)	395	37
Purpose designed single family housing 2 storey detached (single unit)	605	56
Social/economic apartment housing, low rise (no lifts)	480	45
Social/economic apartment housing, high rise (with lifts)	570	53
Private sector apartment building (standard specification)	480	45
Private sector apartment building (luxury)	770	72
Student/nurses halls of residence	565	53
Homes for the elderly (shared accommodation)	615	57
Homes for the elderly (self contained with shared communal facilities)	630	59
Hotel, 5 star, city centre	1,410	131
Hotel, 3 star, city/provincial	1,090	101
Motel	690	64

Regional variations

The approximate estimating costs are based on average London rates. Adjust these costs by the following factors for regional variations:

South East	-6%	Yorkshire and Humberside	-11%
South West	-9%	North West	-7%
East Midlands	-12%	Northern	-11%
West Midlands	-12%	Scotland	-10%
East Anglia	-10%	Wales	-12%

Value added tax (VAT)

The standard rate of value added tax (VAT) is currently 17.5%, chargeable on general building work and materials.

Exchange rates and cost and price trends

The combined effect of exchange rates and inflation on prices within a country and on price comparisons between countries is discussed in section 2.

Exchange rates

The graph on the next page plots the movement of the pound sterling against the ECU/Euro, the US dollar and 100 Japanese yen since 1990. The values used for the graph are quarterly and the method of calculating these is described and general guidance on the interpretation of the graph provided in section 2. The average exchange rate in the first quarter of 1999 was £0.70 to the Euro and £0.60 to the US dollar.

Cost and price trends

The table overleaf presents the indices for consumer price, building cost and building tender price inflation in the United Kingdom since 1990. The indices have been rebased to 1990=100. The annual change is the percentage change between the average index of consecutive years. Notes on inflation are included in section 2.

The pound sterling against the ECU/Euro the US dollar and 100 Japanese yen

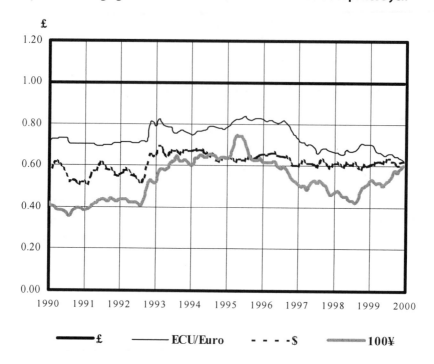

£

£   ECU/Euro   - - - -$ 100¥

Consumer price, building cost and building tender price indices

Year	Consumer prices annual average	change %	Building costs annual average	change %	Building tender prices annual average	change %
1990	100.0	9.5	100.0	7.3	100.0	−9.1
1991	105.9	5.9	105.3	5.3	84.8	−15.2
1992	109.8	3.7	108.3	2.8	78.6	−7.3
1993	111.6	1.6	110.4	1.9	76.1	−3.3
1994	114.3	2.4	113.4	2.7	81.6	7.2
1995	118.2	3.5	119.0	5.0	85.8	5.2
1996	121.1	2.4	122.0	2.5	86.4	0.8
1997	124.9	3.1	124.9	2.4	91.6	6.0
1998	128.2	2.6	130.3	4.3	101.3	10.6

Sources: International Monetary Fund Yearbook 1998 and supplements 1999.
           Davis Langdon & Everest.

Useful addresses

Government and public organizations

Department of the Environment, Transport and the Regions (DETR)
Eland House
Bressenden Place
London SW1E 5DU
Tel: (44-207) 890 3300
Fax: (44-207) 890 3000

British Standards Institution
389 Chiswick Road
London W4 4AJ
Tel: (44-208) 996 9000

Office for National Statistics
Great George Street
London SW1P 3AQ
Tel: (44-207) 270 6363

Trade and Professional Associations

The Association of Consulting Engineers (ACE)
Alliance House
12 Caxton Street
London SW1H 0QL
Tel: (44-207) 222 6557
Fax: (44-207) 222 0750

Chartered Institute of Building Services Engineering (CIBSE)
Delta House
222 Balham High Road
London SW12 9BS
Tel: (44-208) 675 5211

Building Employers Confederation
82 New Cavendish Street
London W1M 8AD
Tel: (44-207) 580 5588

Engineering Council
Canberra House
10–16 Maltravers Street
London WC2 3ER
Tel: (44-207) 240 7891

Institution of Civil Engineers (ICE)
1–7 Great George Street
London SW1P 3AA
Tel: (44-207) 222 7722
Fax: (44-207) 222 7500

**Royal Institute of British Architects
(RIBA)**
66 Portland Place
London W1N 4AD
Tel: (44-207) 580 5533
Fax: (44-207) 436 1197

**Royal Institution of Chartered
Surveyors (RICS)**
12 Great George Street
London SW1P 3AD
Tel: (44-207) 222 7000
Fax: (44-207) 2229430

The Housing Corporation
149 Tottenham Court Road
London W1P 0BN
Tel: (44-207) 393 2000
Fax: (44-207) 393 2011

Chartered Institute of Building (CIOB)
Englemere
Kings Ride
Ascot SL5 8BJ
Berks
Tel: (44-1344) 23355

Other organizations

British Board of Agrément (BBA)
PO Box 195
Bucknalls Lane
Garston, Watford WD2 7JR
Herts
Tel: (44-1923) 665300

**Construction Industry Research and
Information Association (CIRIA)**
6 Storey's Gate
London SW1P 3AU
Tel: (44-207) 222 8891
Fax: (44-207) 222 1708

The Building Centre
26 Store Street
London WC1E 7BT
Tel: (44-207) 637 1022
Fax: (44-207) 580 9641

**Building Research Establishment
(BRE)**
Bucknalls Lane, Garston
Watford WD2 7JR
Herts
Tel: (44-1923) 664000
Fax: (44-1923) 664010

United States of America

All data relate to 1998 unless otherwise indicated.

Population

Population	270.3 million
Urban population (1997)	77%
Population under 15	22%
Population 65 and over	13%
Average annual growth rate (1985 to 1998)	1.0%

Geography

Land area	9,629,901 km^2
Agricultural area	44%
Capital city	Washington DC

Economy

Monetary unit	US Dollar ($)
Exchange rate (average first quarter 1999) to:	
the pound sterling	$ 1.67
the Euro	$ 1.18
the yen x 100	$ 0.85
Average annual inflation (1995 to 1998)	1.6%
Inflation rate	1%
Gross Domestic Product (GDP)	$9,756 billion
GDP PPP basis (1997)	$ (PPP) 8,080 billion
GDP per capita	$36,090
GDP per capita PPP basis (1997)	$ (PPP) 30,170
Average annual real change in GDP (1995 to 1998)	3.8%
Private consumption as a proportion of GDP	73%
Public consumption as a proportion of GDP	19%
Investment as a proportion of GDP	15%

Construction

Gross value of construction output	$558 billion
Gross value of construction output per capita	$2,063
Gross value of construction output as a proportion of GDP	5.7%
Average annual real change in gross value of construction output (1995 to 1998)	2.7%
Annual cement consumption	103.0 million tonnes

PPP Purchasing power parity

The construction industry

Construction output

The total value of gross output of new work of the US construction industry in 1998 was $ 558 billion equivalent to 485 billion ECUs. This total new construction output represents 5.7% of GDP which, even allowing for repair and maintenance, is low compared with most European countries. The table below shows the breakdown of output by type of work for 1998.

Output of new construction, 1998 (current prices)

Type of work	$ billion	ECUs billion	% of total
Residential private			
New housing units	167	145	30
Improvements	67	58	12
Total private residential	234	203	42
Non-residential private			
Offices and banks	22	19	4
Industrial buildings and warehouses	24	21	4
Stores	16	15	3
Other	28	24	5
Improvements	50	43	9
Total private non-residential	141	123	25
Structures, non buildings private	39	34	7
Public building			
Housing	6	5	1
Industrial	1	1	0
Other	61	53	11
Total public building	68	59	12
Public non-building			
Highways	39	34	7
Sewers and water supply	17	15	3
Other	22	19	4
Total public non-building	77	67	14
Total	**558.0**	**485**	**100**

Source: DLC estimates based on Engineering News Record.

Construction output grew by almost 1% in 1997 but fell by about 3% in 1998. The main falls have been in the private sector. Housing and industrial output fell in the public sector but other public and civil engineering work rose. The evolution of the trend in construction in 1997 and 1998 is shown below:

Construction output % real change, 1997 and 1998

Type of work	1997	1998
Residential private		
New housing units	−2.2	−5.1
Improvements	2.4	−7.4
Total private residential	-0.9	−5.8
Non-residential private		
Offices	3.5	−1.7
Industrial buildings	−1.4	−5.4
Other commercial	4.3	6.2
Other	19.1	15.0
Total private non-residential	3.3	4.6
Structures, non buildings private	−8.9	−7.3
Public buildings		
Housing	−6.3	−1.0
Industrial	−2.7	−1.4
Other	9.6	10.0
Total public buildings	7.9	8.9
Public non-buildings		
Highways	2.1	2.0
Sewers and water supply	4.8	6.0
Other	0.5	2.2
Total public non-buildings	1.3	2.1
Total	**0.8**	**−2.7**

Source: DLC estimates based on US Department of Commerce
- Technology Administration.

The geographical distribution of construction work compared to that of population in 1996 is shown overleaf.

The construction contracts are broadly in line with population but nearly all the states with a high percentage of population have a lower percentage of construction output. This applies to California, Illinois and New York. It does not apply to Texas.

Population and construction output, 1996 regional distribution

States	Population (%)	Construction contracts (%)
Alabama	1.6	1.5
Alaska	0.2	0.3
Arizona	1.7	3.0
Arkansas	0.9	0.8
California	12.0	9.6
Colorado	1.4	2.4
Connecticut	1.2	1.0
Delaware	0.3	0.2
District of Columbia	0.2	0.4
Florida	5.4	6.8
Georgia	2.8	3.7
Hawaii	0.4	0.5
Idaho	0.4	0.6
Illinois	4.5	3.8
Indiana	2.2	2.7
Iowa	1.1	0.8
Kansas	1.0	1.1
Kentucky	1.5	1.4
Louisiana	1.6	1.5
Maine	0.5	0.3
Maryland	1.9	2.0
Massachusetts	2.3	2.3
Michigan	3.7	3.3
Minnesota	1.8	1.6
Mississippi	1.0	1.1
Missouri	2.0	1.8
Montana	0.3	0.3
Nebraska	0.6	0.6
Nevada	0.6	2.1
New Hampshire	0.4	0.4
New Jersey	3.0	2.2
New Mexico	0.6	0.7
New York	6.8	4.2
North Carolina	2.8	3.8
North Dakota	0.2	0.2
Ohio	4.2	4.2
Oklahoma	1.2	1.0
Oregon	1.2	1.6
Pennsylvania	4.5	2.8
Rhode Island	0.4	0.2
South Carolina	1.4	1.7
South Dakota	0.3	0.2
Tennessee	2.0	2.5
Texas	7.2	7.6
Utah	0.8	1.1
Vermont	0.2	0.2
Virginia	2.5	2.8
Washington	2.1	2.6
West Virginia	0.7	0.4
Wisconsin	1.9	1.8
Wyoming	0.2	0.2
Total	**100.0**	**100.0**

Source: US Census Bureau - Department of Commerce, 1998 Statistical Abstract of the United States.

The work abroad by US contractors present in the top 225 international contractors in Engineering News Record in 1997 was valued at $ 22.5 billion or 4% of total domestic work as follows:

US construction work abroad, 1997

Area	$ billion	% of total
Europe	6.4	26.1
Africa	1.5	6.1
Middle East	3.4	13.8
Asia	7.5	30.6
Canada	1.2	4.9
Latin America	4.5	18.5
Total	**24.5**	**100.0**

Source: Engineering News Record 12-8-98.

Characteristics and structure of the industry

The construction industry has over a million firms employing about 3.5 million persons and has 15 million working partners or self-employed proprietors. Some states require contractors to be licensed, though this is rarely strictly administered. It is, in same cases, a way of gaining some revenue from the licence fee. About 40% of main contractors and trade contractors use union registered employees and negotiate wages with the unions. 'Open-shop contracting' has grown significantly over the last 20 years, especially in housebuilding. This growth has moderated the behaviour and wage demands of unions. One adverse effect of the decline in union influence has been a decline in training, as the unions traditionally run good education programmes.

There are a large number of specialist trade contractors and they play an important role. They usually have to provide working drawings, organize and manage the work on site with little direction from the main contractor, and they often supply major items of plant and equipment. Labour-only subcontracting is rarely used.

In 1997 there were sixty-five US contractors listed in *Engineering News Record's* list of the 'Top 225 International Contractors' and eighteen in the top 100, more than any other country. The table on the next page shows the principal US contractors according to the *Engineering News Record's* list.

Major US Contractors, 1997

Major contractors	Place in ENR's 'Top 225 International Contractors'
Bechtel Group Inc, California	2
Fluor-Daniel Inc, California	3
Foster Wheeler, NJ	13
Brown and Root Inc, Texas	14
M W Kellogg Co, Texas	25
ABB Lummus Crest Inc., NJ	28
McDermott Inc., Louisiana	34
Raytheon Engrs. & Constructors Int'l, Massachusetts	41
Black & Veatch, Kansas	42
Chicago Bridge & Iron Co., Illinois	50

Source: Engineering News Record 17.08.98.

The title of architect is protected in the USA and the regulations for registration vary from state to state. However, the National Council of Architectural Registration Boards grants a certificate to a qualified architect which is usually recognized in all states. There are about 60,000 architects, mostly working in very small practices. The architect in the USA tends not to get involved in site operation and much detailed design is done by contractors. There are very few quantity surveyors in the USA as it is the architect who is principally concerned with the cost of projects. However, there are construction cost consultants who may originally have been architects or engineers but are increasingly being augmented by quantity surveyors. Contractors are often prepared to give cost advice to the architect.

There are many more building engineers than architects in the USA, the total being about 340,000. Engineers have to be registered, which generally requires a recognized engineering degree and four years work experience. There are a number of substantial multi-disciplinary practices in the USA.

Clients and finance

More than three fifths of new construction is commissioned by the private sector and for housing the figure is around 98%. Fifty-five per cent of all housing units were owner occupied in 1950, but this rose to 64% by 1990. Private rented property accounts for the bulk of the remainder. Publicly provided housing, known as 'project housing', is relatively unimportant. Most mortgages are now of variable interest.

Selection of design consultants

In 1972 Congress established as federal law a policy to select architects and engineers on the basis of the highest qualification for each project and at a fair and reasonable price. For large public projects, invitations are published for interested architects and engineers (usually only in the state where the project is located) to indicate their interest and to submit detailed, specific information on their qualifications. A panel of private sector architects and engineers, who are not paid, recommends five firms. These five candidates make presentations, attend interviews, and so on, and three are shortlisted. The most favoured of these enters into negotiations with the client. If these break down, the second candidate enters negotiations, and so on. The process is costly to firms .

Several states have followed the federal example and the American Institute of Architects (AIA) recommends the procedure for private clients. It is often followed by the large corporations, although sometimes in a form which gives earlier prominence to estimated construction prices and fees.

Contractual arrangements

The most usual methods of selecting a general contractor are by competitive bidding, by negotiation or by a combination of the two. There are two types of competitive bidding, 'open' and 'closed'. Open is the predominant type, where all contractors use the same proposal form. In the closed type, the competing contractors are required to submit their qualifications along with their bids and are encouraged to suggest cost saving proposals. There are numerous forms of negotiated contracts, but most are of the 'cost-plus-fee' type. Negotiated contracts are normally limited to privately financed work, since competitive bidding is a legal requirement for most public projects.

'Fixed price' contracts are the most common. Tenders for buildings are customarily prepared on a 'lump sum' basis, whereas engineering projects are generally bid as a series of unit prices. It is standard practice for contractors to prepare their own quantities which do not form part of the contract. With few exceptions, bids are accompanied by a bid bond guaranteeing that the contractor will enter into a contract if declared successful.

Standard contract conditions have been developed by various bodies, including the American Institute of Architects, the National Society of Professional Engineers, the Associated General Contractors of America and various federal, state and municipal governments. Where a contract provides for arbitration, most stipulate that it shall be conducted under the auspices of the Construction Industry Arbitration Association.

There has been an increasing use for large projects of 'management fee' or 'construction management' arrangements, but often still retaining a guaranteed maximum price. However, this tendency is now less apparent. 'Design and build' projects are also becoming more popular, although there are often other provisions for the contractor to offer advice at the design stage.

Specialist trade contractors are usually invited to bid, often from a list selected or approved by the architect. Eight to ten bidders are the norm. Nomination is virtually unknown.

The lien laws in the USA provide a large degree of protection for the contractors and subcontractors working on a project. Under their provisions a contractor can place a lien on the real property if he has not received payment for goods and services provided. This lien is registered on the title deed of the property and if not resolved can be a major impediment for subsequent sale or mortgage financing on the property. The owner is therefore obliged to ensure all payments are properly effected to each supplier of goods or services. In the event that the employer has made a payment to the general contractor, but the general contractor has not paid his sub-contractors, then the sub-contractors are entitled to place a lien on the property. In this case the employer may have to pay for the works twice to radiate (remove) the lien, unless he has a labour and materials payment bond in force, in which case he can recover the double payment from the bond company. Standard bond forms are available and in common use throughout the USA. Employers and their agents need to monitor payments carefully on projects to avoid lien actions.

Development control and standards

The planning process in the USA is known as planning control and zoning control. It is very fragmented and every town has its own system. There may be 50 separate zoning authorities in one state. There is normally a Zoning Commission Board, a Zoning Board of Appeal and often also a Planning Commission or Board in each town. The ease with which development zones of a town can be changed varies according to the attitude of the town or the state.

There is no single national building code for the whole of the USA. Approximately 19,000 municipalities are involved and many have their separate codes. Nevertheless, various national codes have been prepared. The most widely used is the 'International Conference of Building Officials (ICBO) Uniform Building Code'. Others are the codes of Building Officials and Code Administrators International (BOCA) and Southern Building Code Congress International (SBCC). There are also specialist codes for fire safety, etc. The codes are basically performance codes rather than specifications for the form of construction. Several organizations are working on harmonization of codes, notably the National Institute of Building Sciences (NIBS), a non-governmental institution set up with representation from all parts of the building community.

The specific arrangements for obtaining planning permission and the statutory period for approval varies from state to state. Once the plans have been passed and construction has commenced, field inspection takes place. This is generally regarded as very important and the number of visits is often specified in the codes.

Standards are continually referred to in the building codes. They may be mandatory or discretionary. There are some 150 organizations which develop standards of which perhaps a dozen or so are important. These include the American Society for Testing and Materials (ASTM), the American National Standards Institute (ANSI) and the American Insurance Association (AIA).

Liability and insurance

The contractor is liable for damages caused by his own acts or omissions. He must therefore obtain comprehensive liability insurance to protect himself and his subcontractors.

The liability of designers and contractors varies with the contract used and from state to state. In the USA the architect or engineer has a contractual obligation to check the shop drawings of specialist trade contractors and this affects the liability. Normally, professional liability extends three to four years, but in some circumstances it can extend up to ten years.

Professional indemnity insurance covers the liability of parties involved in design, except that trade contractors may not be covered or, if they are, may be insufficiently so. Professional indemnity insurance is, in any case, very expensive in the USA.

Construction cost data

Cost of labour

The figures overleaf are typical of national average labour costs at the first quarter 1999. The wage rate is the basis of an employee's income, while the cost of labour indicates the cost to a contractor of employing that employee. The difference between the two covers a variety of mandatory and voluntary contributions - a list of items which could be included is given in section 2.

	Wage rate (per hour) $	Cost of labour (per hour) $
Site operatives		
Mason/bricklayer	27.60	42.90
Carpenter	27.30	43.00
Plumber	32.60	49.40
Electrician	31.90	47.70
Structural steel erector	30.60	56.10
HVAC installer	31.80	49.30
Semi-skilled worker	21.50	33.80
Unskilled labourer	20.90	32.80
Equipment operator	28.40	43.20
Watchman/security		14.20
Site supervision		
General foreman	28.60	45.10
Trades foreman	30.10	47.40
Clerk of works *	725.00	1,150.00
Contractors' personnel *	(weekly)	(weekly
Site manager	1,250.00	1,970.00
Resident engineer	815.00	1,290.00
Resident surveyor	815.00	1,290.00
Junior engineer	625.00	985.00
Junior surveyor	625.00	985.00
Planner	725.00	1,150.00

* average weekly wage rate/labour cost.

Cost of materials

The figures that follow are the US national average costs of main construction materials delivered to site, as incurred by contractors in the first quarter of 1999. These assume that the materials would be in quantities as required for a medium sized construction project and that the location of the works would be neither constrained nor remote.

	Unit	Cost $
Cement and aggregates		
Ordinary portland cement in 50kg bags	bag	7.00
Coarse aggregates for concrete	tonne	12.20
Fine aggregates for concrete	tonne	13
Ready mixed concrete (mix 17MPa)	m^3	75
Ready mixed concrete (mix 21MPa)	m^3	80

	Unit	Cost $
Steel		
Mild steel reinforcement	tonne	535.00
High tensile steel reinforcement	tonne	595.00
Bricks and blocks		
Common bricks (8"x 2.67" x 4")	1,000	250.00
Good quality facing bricks (8" x 2.67" x 4")	1,000	330.00
Hollow concrete blocks (8" x 8" x 16")	1,000	900.00
Solid concrete blocks (4" x 8" x 16")	1,000	950.00
Precast concrete cladding units with exposed aggregate finish	m^2	102.00
Timber and insulation		
Exterior quality plywood (13mm)	m^2	7.10
Plywood for interior joinery (6mm)	m^2	4.84
Softwood strip flooring (25 x 102mm)	m^2	24.80
89mm thick quilt insulation	m^2	1.83
100mm thick rigid slab insulation	m^2	6.03
Softwood internal door complete with frames and ironmongery	each	190.00
Glass and ceramics		
Float glass (5mm)	m^2	36.10
Sealed double glazing units (16mm)	m^2	75.40
Good quality ceramic wall tiles	m^2	23.80
Plaster and paint		
Plaster in 36kg bags	bag	13.00
Plasterboard (10mm thick)	m^2	1.72
Emulsion paint in 5 litre tins	gallon	10.80
Gloss oil paint in 5 litre tins	gallon	21.50
Tiles and paviors		
Clay floor tiles (102x 102 x 13mm)	m^2	39.50
Vinyl floor tiles (305x 305 x 3mm)	m^2	20.90
Clay roof tiles	m^2	52.70
Precast concrete roof tiles	m^2	11.10
Drainage		
WC suite complete	each	380.00
Lavatory basin complete	each	320.00
100mm diameter clay drain pipes	m	6.33
150mm diameter cast iron drain pipes	m	41.00

Unit rates

The descriptions of work items below are generally shortened versions of standard descriptions listed in five languages (English, French, Italian, German and Spanish) in Appendix 3.

Where an item has a two digit reference number (e.g. 05 or 33), this relates to the full description against that number in Appendix 3. Where an item has an alphabetic suffix (e.g. 12A or 34B) this indicates that the standard description has been modified. Where a modification is major the complete modified description is included here and the standard description should be ignored.

Where a modification is minor (e.g. the insertion of a named hardwood) the shortened description has been modified here but, in general, the full description in Appendix 3 prevails.

The unit rates below are for main work items on a typical construction project as the US national average in the first quarter of 1999. The rates include all necessary labour, materials and equipment, and where appropriate, allowances for contractor's overheads and profit, preliminary and general items and contractor's profit and attendance on specialist rates.

		Unit	Rate $
Excavation			
01	Mechanical excavation of foundation trenches	m^3	6.12
02	Hardcore filling making up levels	m^3	31.40
03	Earthwork support	m^2	14.70
Concrete work			
04	Plain insitu concrete in strip foundations in trenches	m^3	220.00
05	Reinforced insitu concrete in beds	m^3	195.00
06	Reinforced insitu concrete in walls	m^3	319.00
07	Reinforced insitu concrete in suspended floor or roof slabs	m^2	20.60
08	Reinforced insitu concrete in columns	m^3	1,050.00
09	Reinforced insitu concrete in isolated beams	m^3	942.00
10	Precast concrete slab	m^2	71.60
Formwork			
11	Softwood formwork to concrete walls	m^2	27.50
12	Softwood or metal formwork to concrete columns	m^2	85.30
13	Softwood or metal formwork to horizontal soffits of slabs	m^2	39.60
Reinforcement			
14	Reinforcement in concrete walls (16mm)	tonne	1,125.00
15	Reinforcement in suspended concrete slabs	tonne	1,200.00
16	Fabric reinforcement in concrete beds	m^2	5.54

		Unit	Rate $
Steelwork			
17	Fabricate, supply and erect steel framed structure	tonne	1,750.00
Brickwork and blockwork			
18	Precast lightweight aggregate hollow concrete block walls	m²	49.20
19	Solid (perforated) concrete bricks	m²	67.80
20	Solid (perforated) sand lime bricks	m²	93.70
21	Facing bricks	m²	108.20
Roofing			
22	Concrete interlocking roof tiles	m²	38.80
23	Plain clay roof tiles	m²	79.70
27	3 layers glass-fibre based bitumen felt roof covering include chippings	m²	19.60
28	Bitumen based mastic asphalt roof covering	m²	1.19
29	Glass-fibre mat roof insulation	m²	8.18
30	Rigid sheet loadbearing roof insulation 75mm thick	m²	16.90
31	Troughed galvanised steel roof cladding	m²	23.03
Woodwork and metalwork			
34	Single glazed casement window in hardwood, size 650 x 900mm	each	208.00
36	Solid core half hour fire resisting hardwood internal flush doors, size 800 x 2000mm	each	216.00
37	Aluminium double glazed window, size 1200 x 1200mm	each	360.00
38	Aluminium double glazed door, size 850 x 2100mm	each	905.00
40	Framed structural steelwork in universal joist sections	tonne	1,750.00
41	Structural steelwork lattice roof trusses	tonne	2,050.00
Plumbing			
42	UPVC half round eaves gutter	m	12.70
43	UPVC rainwater pipes, 76mm DWC PVC	m	33.70
44	Light gauge copper cold water tubing 13mm	m	20.50
45	High pressure plastic pipes for cold water supply, Sch. 80 CPVC, 76mm	m	86.90
46	Low pressure plastic pipes for cold water distribution, Sch. 40 CPVC, 38mm	m	51.80
47	UPVC soil and vent pipes, 102mm	m	65.50
48	White vitreous china WC suite	each	615.00
49	White vitreous china lavatory basin	each	320.00
50	White glazed fireclay shower tray	each	730.00
51	Stainless steel single bowl sink and double drainer	each	430.00
Electrical work			
53	13 amp unswitched socket outlet	each	16.60
54	Flush mounted 20 amp, 1 way light switch	each	21.00

	Unit	Rate $
Finishings		
55 2 coats gypsum based plaster on brick walls	m²	21.50
56 White glazed tiles on plaster walls	m²	58.10
57 Red clay quarry tiles on concrete floor	m²	94.70
60 Mineral fibre tiles on concealed suspension system	m²	10.30
Glazing		
61 Glazing to wood	m²	101.00
Painting		
62 Emulsion on plaster walls	m²	4.95
63 Oil paint on timber	m²	9.47

Approximate estimating

The building costs per unit area given below are US national averages incurred by building clients for typical buildings as at the first quarter 1999. They are based upon the total floor area of all storeys, measured between external walls and without deduction for internal walls.

Approximate estimating costs generally include mechanical and electrical installations but exclude furniture, loose or special equipment, and external works; they also exclude fees for professional services. The costs shown are for specifications and standards appropriate to the USA and this should be borne in mind when attempting comparisons with similarly described building types in other countries. A discussion of this issue is included in section 2. Comparative data for countries covered in this publication, including construction cost data, are presented in Part Three.

Approximate estimating costs must be treated with caution; they cannot provide more than a rough guide to the probable cost of building.

	Cost $ per m²	Cost $ per ft²
Industrial buildings		
Factories for letting (include lighting, power and heating)	348	32
Factories for owner occupation (light industrial use)	513	48
Factories for owner occupation (heavy industrial use)	793	74
Factory/office (high-tech) for letting (shell and core only)	801	74
Factory/office (high-tech) for letting (ground floor shell, first floor offices)	1,030	95

	Cost $ per m^2$	Cost $ per ft^2$
Factory/office (high-tech) for owner occupation (controlled environment, fully furnished)	1,260	117
High tech laboratory (air conditioned)	1,340	124
Warehouses, low bay (6 to 8m high) for letting (no heating)	271	25
Warehouses, low bay for owner occupation (including heating)	378	35
Warehouses, high bay for owner occupation (including heating)	563	54
Cold stores/refrigerated stores	654	61
Administrative and commercial buildings		
Civic offices, non air conditioned	727	68
Civic offices, fully air conditioned	919	85
Offices for letting, 5 to 10 storeys, non air conditioned	662	62
Offices for letting, 5 to 10 storeys, air conditioned	802	75
Offices for letting, high rise, air conditioned	801	74
Offices for owner occupation, 5 to 10 storeys, non air conditioned	802	75
Offices for owner occupation, high rise, air conditioned	1.030	96
Prestige/headquarters office, 5 to 10 storeys, air conditioned	1,090	101
Prestige/headquarters office, high rise, air conditioned	1,260	117
Health and education buildings		
General hospitals	1,350	125
Teaching hospitals	1,560	145
Private hospitals	2,310	215
Health centres	975	91
Nursery schools	664	62
Primary/junior schools	819	76
Secondary/middle schools	834	78
University (arts) buildings	1,120	104
University (science) buildings	1,630	151
Management training centres	846	79
Recreation and arts buildings		
Theatres (over 500 seats) including seating and stage equipment	1,230	114
Theatres (less than 500 seats) including seating and stage equipment	835	78
Concert halls including seating and stage equipment	1,230	114
Swimming pools (international standard) including changing facilities	1,820	169
Swimming pools (schools standard) including changing facilities	1,300	121
City centre/central libraries	1,300	121
Branch/local libraries	1.020	95
Residential buildings		
Social/economic single family housing (multiple units)	533	50
Private/mass market single family housing 2 storey detached/semidetached (multiple units)	575	53
Purpose designed single family housing 2 storey detached (single unit)	828	77
Social/economic apartment housing, low rise (no lifts)	533	50
Social/economic apartment housing, high rise (with lifts)	717	67

	Cost $ per m²	Cost $ per ft²
Private sector apartment building (standard specification)	715	66
Private sector apartment building (luxury)	875	81
Student/nurses halls of residence	1,180	110
Homes for the elderly (shared accommodation)	712	66
Homes for the elderly (self contained with shared communal facilities)	951	88
Motel	702	65

Regional variations

The approximate estimating costs are based on average US national rates. Adjust these costs by the following factors for regional variations:

Los Angeles, CA	11%	Dallas, TX	−13%	Fort Worth, TX	−16%
Hartford, CT	4%	Providence, RI	5%	Columbia, SC	−24%
Miami, FL	−13%	Philadelphia, PA	12%	Seattle, WA	5%
New York, NY	34%				

Exchange rates and cost and price trends

The combined effect of exchange rates and inflation on prices within the United States and on price comparisons between countries is discussed in section 2.

Exchange rates

The graph on the next page plots the movement of the US dollar against sterling, the ECU/Euro, and 100 Japanese yen since 1990. The values used for the graph are quarterly and the method of calculating these is described and general guidance on the interpretation of the graph provided in section 2. The average exchange rate in the first quarter of 1999 was $1.67 to the pound sterling and $1.18 to the Euro.

The US dollar against the £ sterling, the ECU/ Euro and 100 Japanese yen

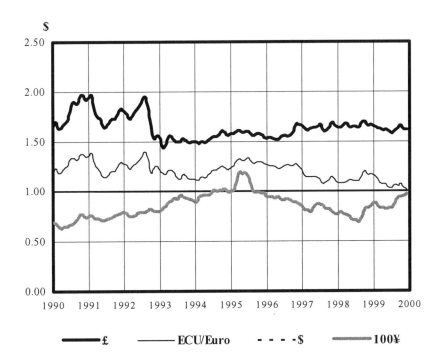

Cost and price trends

The table opposite presents the indices for consumer prices and material and labour costs in the USA since 1990. The indices have been rebased to 1990=100. The annual change is the percentage change between the average index of consecutive years. Notes on inflation are included in section 2.

The material inputs index is derived from data from the producer price index programme. The labour cost index is obtained from the employment cost indices. The construction indices together have moved approximately in line with consumer prices.

Consumer price, construction materials and construction employment cost indices

Year	Consumer prices annual average	Consumer prices change %	Material costs annual average	Material costs change %	Labour costs annual average	Labour costs change %
1990	100.0	5.4	100.0	1.3	100	2.5
1991	104.2	4.2	102.3	2.3	102.9	2.9
1992	107.4	3.1	103.6	1.3	105.0	2.0
1993	110.6	3.0	107.5	3.8	107.4	2.3
1994	113.4	2.5	110.6	2.9	110.5	2.9
1995	116.6	2.8	114.6	3.6	112.8	2.1
1996	119.0	2.1	116.7	1.8	116.3	3.1
1997	120.9	1.6	119.1	2.1	120.2	3.4
1998	122.1	1.0	118.5	−0.5	124.3	3.4

Sources: International Monetary Fund, Yearbook 1998 and supplements 1999.
Bureau of Labour Statisiics.

Useful addresses

Government and public organizations

General Service Administration (GSA) Central Office
18th & F Streets, NW
Washington DC 20314-1000
Tel: (1-202) 5123000
For Regional Government
Construction Ministries:Regional GSA
Office in each state

Department of Housing and Urban Development Headquarters (HUD)
451 7th Street, NW
Washington DC 20004
Tel: (1-202) 4635460

Department of Transportation
400 7th Street, SW
Washington DC 20024
Tel: (1-202) 5435004

American National Standards Institute
11 West 42nd Street, 13th floor
New York N.Y. 10036
Tel: (1-212) 624254900

Army Corps of Engineers
20 Massachusetts Avenue, NW
Washington DC 20314-1000
Tel: (1-202) 761660

Small Business Administration
1441 L Street NW
Washington DC 20006
Tel: (1-202) 7272540

National Statistics Organization
Department of Commerce
14th Street NW
Washington DC 20418
Tel: (1-202) 4823727

National Institute of Standards and Technology
100 Bureau Drive
Gaithersburg MD 20899-0001
Tel: (1-301) 9753058
Fax:(1-301) 9754032

National Academy of Science
Building Research Advisory Board
2101 Constitution Avenue NW
Washington DC 20418
Tel: (1-202) 3342000

National Bureau of Standards
US Department of Commerce
Centre for Building Technology
Gaithersburg MD 20899
Tel: (1-301) 9213377

Trade and professional associations

Associations of General Contractors (AGC) (Union)
10101 Bacon Drive
Beltsville MO 20006
Tel: (1-202) 3932040

Associations of Building Contractors of America (ABC) (non-union)
1300 N. 17th Street
Rosslyn VA 22209
Tel: (1-703) 8122000

American Institute of Architects (AIA)
1735 New York Avenue, NW
Washington DC 20006
Tel: (1-202) 6267300

For Contractors Registration Board:
Respective county/city Chamber of Commerce

American Association of Cost Engineers (AACE)
209 Prairie Avenue, Ste 100
Morgantown WV 26501
Tel: (1-304) 2968444

National Society of Professional Engineers (NSPE)
1420 King Street
Alexandria, VA 22314
Tel: (1-703) 6842800

Part Three: Comparative Data

4 Introductory notes

Part Three brings together data from a variety of sources but mainly Part Two, and presents them in the form of tables to allow rapid comparison among the countries included in the book. This also helps place countries, their main statistical indicators and their construction costs in an international context.

There are ten tables derived from Part Two arranged in three sections:

5   Key Macroeconomic indicators

- Population
- Geography
- Finance and economics

6   Construction output indicators

- Construction output
- Construction output per capita

7   Construction cost data

- Skilled and unskilled labour costs
- Material costs – cement and aggregates for concrete
- Material costs – reinforcing steel and softwood
- Approximate estimating – warehouses and offices
- Approximate estimating – housing

The first five tables are based on the statistics on the first page of each country section, the remainder are drawn from construction cost data in each country section. Each table is prefaced by explanatory notes. There are inherent dangers in attempting to compare international data, particularly where two sets of data are used (e.g.: construction output and population) and, even more so, when exchange rates are used. While these tables can provide useful initial comparisons between countries they should, nevertheless, be used with caution.

5 Key macroeconomic indicators

Population

The table below summarises population statistics from the first page of each country section. The table highlights not only the differences in total population between the countries included in this book but also the variations in population growth rates, the proportions of the population above and below working age and the proportions of the population living in urban areas.   Most European countries have low population growth rates, the exceptions here are Germany and Turkey.   In the case of Germany the dominant factor is the incorporation of the German Democratic Republic into the Federal Republic.   Developed countries in Europe tend to have rather similar proportions of their population under 15 and over 65 years of age. This is less true in Eastern and Central Europe and less developed countries.   The proportion of urban population is generally high throughout Europe; only Portugal has less than 50% of urban population.

Country	Population				
	Total (million)	Urban % 1997	Under 15 %	65 and over %	Growth % pa 1985 - 1998
Austria	8.1	64	17	15	0.6
Belgium	10.2	97	17	17	0.2
Cyprus	0.75	70	25	10	0.2
Czech Republic	10.3	66	18	13	0[1]
Finland	5.2	64	19	14	0.4
France	58.6	75	19	16	0.5
Germany	82.1	87	16	16	2.3
Greece	10.6	60	17	16	0.5[2]
Ireland	3.6	58	17	16	0.2
Italy	57.5	67	15	17	0.1
Japan	125.9	78	15	16	0.3
Netherlands	15.7	89	18	14	0.6
Poland	38.7	64	21	12	0.3
Portugal	9.9	37	17	15	0.3
Slovak Republic	5.4	60	21	11	0.2[3]
Spain	39.3	77	15	16	0.2
Switzerland	7.3	62	17	15	0.9
Turkey	63.6	70	31	6	1.9[4]
United Kingdom	58.9	89	19	16	0.3
United States	270.3	77	22	13	1.0

[1] 1995 – 1998   [2] 1984 – 1997v   [3] 1995 – 1997   [4] 1984 - 1998

pa = per annum

Geography

The table below summarises geographical statistics from the first page of each country section. As with the previous table, this table highlights the differences between countries. The figures for population density are probably more useful indicators of land use than the total area of a country. Belgium, Japan and the Netherlands all have population densities of 300 persons or more per km^2. Finland, Ireland and the USA have population densities of around 50 persons or less per km^2.

Country	Area km²	Agricultural %	Population per km²
Austria	83,850	42	97
Belgium	30,510	45	334
Cyprus	5,895	17	127
Czech Republic	78,703	54	131
Finland	337,030	8	15
France	547,030	59	107
Germany	356,910	49	230
Greece	131,940	68	80
Ireland	70,280	81	51
Italy	301,230	56	191
Japan	377,835	14	333
Netherlands	37,330	59	421
Poland	312,683	61	124
Portugal	92,391	44	107
Slovak Republic	48,845	51	111
Spain	504,750	60	78
Switzerland	41,290	40	177
Turkey	780,580	52	82
United Kingdom	244,820	71	241
United States	9,629,901	44	28

*1997

Finance and economics

This table summarises financial and economic data from the first page of each country section. Gross Domestic Product (GDP) and GDP per capita for each country are given in £ sterling, US$ and ECU. The average real rate of growth in GDP and the average inflation rate are given for the years 1995 to 1998. The United States, Japan, Switzerland, Austria, Germany, Finland, Belgium and the Netherlands all have relatively high GDP per capita (US$25,000 or more in 1998). All the countries showed a positive average growth in GDP over the three year period. In Turkey, the Czech Republic, Poland and Greece inflation was substantially higher than the growth in GDP per capita, in the United Kingdom and Japan GDP growth per capita equalled inflation while in Ireland , Finland and the United States GDP growth per capita was 2% or more higher than inflation.

Country	1998						1995 to 1998	
	GDP			GDP per capita			GDP change	Average
	£	US$	ECU	£	US$	ECU	(real)	Inflation
	Billions	billions	billions				% pa	% pa
Austria	135	225	192	16,600	27,700	23,700	2.6	1.3
Belgium	156	260	223	15,300	25,600	21,900	2.4	1.6
Cyprus	6	9	8	7,470	12,500	10,700	4.9	3.0
Czech Republic	35	58	50	3,370	5,630	4,820	0.7	9.3
Finland	80	134	115	15,600	26,000	22,200	4.8	1.1
France	911	1,520	1,300	15,600	26,000	22,200	2.4	1.3
Germany	1,360	2,260	1,940	16,500	27,600	23,600	2.1	1.4
Greece *	70	117	100	6,590	11,000	9,410	2.7	7.6
Ireland	49	81	70	13,400	22,400	19,200	9.1	1.8
Italy	749	1,250	1,070	13,000	21,700	18,600	1.7	2.4
Japan	2,700	4,510	3,850	21,400	35,800	30,600	0.8	0.8
Netherlands	239	398	341	15,200	25,400	21,700	3.3	2.1
Poland	92	154	132	2,380	3,980	3,400	4.8	15.9
Portugal	68	113	97	6,830	11,400	9,750	2.4	2.7
Slovak Republic	12	20	17	2,220	3,700	3,170	5.8	6.2
Spain	337	564	482	8,640	14,400	12,300	3.3	2.4
Switzerland	144	241	206	19,900	33,300	28,500	1.3	0.4
Turkey	112	187	160	1,730	2,890	2,470	6.2	83.6
United Kingdom	844	1,410	1,210	14,300	23,900	20,400	2.8	2.8
United States	5,840	9,760	6,830	21,600	36,100	25,300	3.8	1.6

*GDP change 1994 to 1997   pa = per annum

6 Construction output indicators

Construction output

The table below summarises construction output statistics from the first page of each country section. In this summary table, figures in national currency for 1998 are listed but, in addition, in order to facilitate (crude) comparisons, £ sterling, US dollar and ECU equivalents are presented for each figure. The currency conversions have been carried out using the average exchange rates for the first quarter of 1999. As noted earlier, construction statistics, including those for construction output, are notoriously unreliable and, in addition, national definitions of construction output vary widely. It would, therefore, be unwise to draw too many conclusions from the table.

Country	National unit of currency	Gross construction output (1998) National Currency	£	US$	Billions ECU
Austria	ASch	357	18.20	30.40	26.00
Belgium	BFr	1,089	18.90	31.60	27.00
Cyprus	C£	0.6	0.72	1.21	1.03
Czech Republic	Kcs	167	3.33	5.56	4.76
Finland	Fmk	77.2	9.10	15.20	13.00
France	FFr	666.1	71.20	119.00	102.00
Germany	DM	389	139.00	233.00	199.00
Greece *	Dr	3930	8.40	14.00	12.00
Ireland	IR£	9.4	8.39	14.00	12.00
Italy	L	200,200	72.50	121.00	104.00
Japan	¥	75,570	386.00	645.00	552.00
Netherlands	G	78.6	25.00	41.80	35.80
Poland	Zl	57.5	9.88	16.50	14.10
Portugal	Esc	2,714	9.50	15.90	13.60
Slovak Republic	Sk	63.7	1.06	1.77	1.51
Spain	Pta	9,303	39.20	65.50	56.00
Switzerland	SFr	36.6	16.10	26.90	23.00
Turkey	TL	5,730,000	11.20	18.70	16.00
United Kingdom	£	63.1	63.10	105.00	90.10
United States	$	558	334.00	558.00	391.00

*1997

Construction output per capita

This table is based on the previous one, but has each figure for construction output divided by the population in that country. Bearing in mind the uncertainty of both construction and population data and the limitations of exchange rates, the table reveals some useful indicators of construction activity. Twelve of the 30 countries covered in the book have construction output per capita figures for 1998 less than 1,000 US$, and eight of these are former Eastern Block countries; the remaining countries which have per capita figures greater than 1,000 US$, include all but two countries from the European Union (Greece and Portugal). Eleven countries have a per capita output above 2,000 US$.

Country	National unit of currency	Gross construction output per capita (1998)			
		National currency	£	US$	ECU
Austria	ASch	44,100	2,250	3,750	3,210
Belgium	BFr	107,000	1,850	3,100	2,650
Cyprus	C£	800	960	1,610	1,370
Czech Republic	Kcs	16,200	323	540	462
Finland	Fmk	15,100	1,780	2,970	2,540
France	FFr	11,400	1,220	2,030	1,740
Germany	DM	4,740	1,690	2,840	2,420
Greece*	Dr	371,000	792	1,320	1,130
Ireland	IR£	2,610	2,330	3,890	3,330
Italy	L	3,480,000	1,260	2,100	1,810
Japan	¥	600,000	3,070	5,120	4,380
Netherlands	G	5,010	1,590	2,660	2,280
Poland	Zl	1,490	256	428	366
Portugal	Esc	274,000	959	1,600	1,370
Slovak Republic	Sk	11,800	196	328	280
Spain	Pta	237,000	997	1,670	1,420
Switzerland	SFr	5,010	2,210	3,680	3,150
Turkey	TL	88,700,000	173	289	247
United Kingdom	£	1,070	1,070	1,780	1,530
United States	$	2,060	1,240	2,060	1,440

* 1997

7 Construction cost data

Skilled and unskilled labour costs

This table summarises average hourly labour costs for each country for which they are available. The figures in national currency are taken from each country's construction cost data and have been converted into pound sterling, US dollar and ECU equivalents using first quarter 1999 exchange rates.

It is probable that the definitions of skilled and unskilled, and what is included in labour costs varies between countries, thus these figures should not be taken as strictly comparable.

Country	National unit of currency	Skilled labour costs* National currency	£	(per hour) US$	Euro	Unskilled labour costs* National currency	£	(per hour) US$	Euro
Austria	ASch	409	20.80	34.80	29.80	258	13.20	22.00	18.80
Belgium	BFr	1,230	21.40	35.70	30.50	1,010	17.60	29.30	25.10
Cyprus	C£	6.54	7.88	13.20	11.30	4.90	5.90	9.86	8.43
Czech Republic	Kcs	93.70	1.87	3.12	2.67	62.10	1.24	2.07	1.77
Finland	Fmk	101	12.00	20.00	17.10	n.a.	n.a.	n.a.	n.a.
France	FFr	156	16.70	27.90	23.80	86.10	9.21	15.40	13.20
Germany	DM	58.40	20.90	35.00	29.90	50.50	18.10	30.20	25.90
Ireland	IR£	14	12.50	20.90	17.90	11.00	9.80	16.40	14.00
Italy	L	36,700	13.30	22.20	19.00	33,400	12.10	20.20	17.30
Netherlands	G	59.10	18.80	31.40	26.90	52.90	16.80	28.10	24.10
Slovak Republic	Sk	199	3.30	5.51	4.71	117	1.94	3.23	2.77
Spain	Pta	2,360	9.95	16.60	14.20	1670	7.04	11.80	10.10
Switzerland	SFr	74.70	32.90	54.90	47.00	55.50	24.40	40.80	34.90
Turkey	TL	683,000	1.33	2.23	1.90	440,000	0.86	1.43	1.23
United Kingdom	£	9.60	9.60	16.00	13.70	5.65	5.65	9.44	8.07
United States	$	48.10	28.80	48.10	33.60	32.80	19.60	32.80	23.00

* average.

Materials costs – cement and aggregates for concrete

The table below summarises for countries for which they are available, costs per tonne for cement and costs per m³ for aggregates for concrete. The figures in national currency are taken from each country's construction cost data and have been converted into pound sterling, US dollar and ECU equivalents using first quarter 1999 exchange rates. Converted figures have been rounded to the nearest whole number. Despite the fact that there are internationally recognized standards of quality for cement and that it is one of the few internationally traded construction materials, the variation in cost between countries is remarkably large. This may well be a result of controlled prices or of import protection. The variation in aggregate costs is less surprising; quality and availability can be expected to vary widely within, let alone between, countries and a large proportion of aggregate costs can be in their transportation.

Country	National unit of currency	Cement National Currency	£	US$	tonne Euro	Aggregates for concrete National currency	£	US$	m³ Euro
Austria	ASch	1,500	76.50	128.00	109.00	170	8.67	14.50	12.40
Belgium	BFr	n.a.	n.a.	n.a.	n.a.	304	5.28	8.82	7.55
Cyprus	C£	33	39.80	66.40	56.80	5	6.60	11.00	9.42
Czech Republic	Kcs	2,240	44.70	74.60	63.80	183	3.65	6.09	5.21
Finland	Fmk	n.a.	n.a.	n.a.	n.a.	18	2.09	3.48	2.98
France	FFr	896	95.80	160.00	137.00	228	24.30	40.70	34.80
Germany	DM	182	65.20	109.00	93.10	152	54.30	90.70	77.60
Ireland	IR£	99	88.30	147.00	126.00	19	16.60	27.60	23.60
Italy	L	140,000	50.70	84.70	72.40	30,500	11.00	18.40	15.80
Japan	¥	16,800	85.90	143.00	123.00	4,530	23.10	38.60	33.00
Netherlands	G	271	86.20	144.00	123.00	63	20.10	33.60	28.70
Poland	Zl	155	26.60	44.50	38.00	40	6.87	11.50	9.82
Portugal	Esc	15,300	53.60	89.50	76.60	2,050	7.16	11.90	10.20
Slovak Republic	Sk	1,600	26.60	44.40	37.90	350	5.81	9.70	8.30
Spain	Pta	14,200	59.90	100.00	85.50	2,010	8.48	14.20	12.10
Switzerland	SFr	222	97.80	163.00	140.00	54	23.80	39.70	34.00
Turkey	TL	11,000,000	21.50	35.80	30.60	2,910,000	5.67	9.46	8.09
United Kingdom	£	90	90.00	150.00	129.00	11	11.30	18.80	16.10
United States	$	140	83.80	140.00	98.00	19	11.20	18.70	13.10

Materials costs – reinforcing steel and softwood

The table below summarises costs per tonne for reinforcing steel and costs per m³ for softwood for carpentry where available. The figures in national currency are taken from each country's construction cost data and have been converted into pound sterling, US dollar and ECU equivalents using first quarter 1999 exchange rates.

Similar to cement, the range of costs for reinforcing steel is surprisingly large, from a low of 191 US$ per tonne in Cyprus to a high of 1,450 US$ per tonne in Japan, the latter being high tensile steel. The lowest costs for timber are in the Slovak Republic and Poland.

Country	National unit of currency	Mild steel reinforcement			tonne	Softwood sections for carpentry			m³
		National Currency	£	US$	Euro	National Currency	£	US$	Euro
Austria	ASch	4,000	204	341	291	1,900	97	162	138
Belgium	BFr	15,000	261	435	372	n.a.	n.a.	n.a.	n.a.
Cyprus	C£	95	114	191	164	240	289	483	413
Czech Republic	Kcs	n.a.	n.a.	n.a.	n.a.	4,000	80	133	114
Finland	Fmk	n.a.	n.a.	n.a.	n.a.	1,590	188	314	269
France	FFr	3,610	386	644	551	2,060	220	368	315
Germany	DM	1,140	409	683	584	293	105	175	150
Ireland	IR£	294	263	439	375	237	212	353	302
Italy	L	515,000	186	311	266	350,000	127	212	181
Japan *	¥	170,000	869	1,450	1,240	53,000	271	452	387
Netherlands	G	871	277	463	396	808	257	430	368
Poland **	Zl	n.a.	n.a.	n.a.	n.a.	650	112	187	160
Portugal	Esc	45,000	157	263	225	n.a.	n.a.	n.a.	n.a.
Slovak Republic	Sk	13,400	223	372	319	5,800	96	161	138
Spain	Pta	35,000	148	246	211	45,000	190	317	271
Switzerland	SFr	1,700	749	1,250	1,070	390	172	287	245
Turkey	TL	n.a.	n.a.	n.a.	n.a.	63,000,000	123	205	176
United Kingdom	£	210	210	351	300	220	220	367	314
United States	$	271	162	271	190	n.a.	n.a.	n.a.	n.a.

* high tensile steel reinforcement
** softwood for joinery

Approximate estimating – warehouses and offices

This table summarises for countries where they are available, approximate estimating costs per square metre for two non-residential building types: warehouses and offices. The aim has been to take typical costs, where available, for one of the most basic building types, the speculative low-bay warehouse, and one of the more common – but more sophisticated –ones, the speculative air-conditioned, medium rise office. In countries where building types do not exactly match these descriptions, the nearest equivalent has been taken. Where a range of costs has been given, the mid point is shown. It must be borne in mind that even where costs are given under the same description in one or more countries, this is not to say that they are identical, or even physically similar. Approximate estimating costs for a particular country are for the normal standards prevailing in that country. Quality and technical standards vary widely and there are differences between countries in what is, and is not, included. The table, therefore, should be used with care.

Country	National unit of currency	Warehouses National currency	£	US$	m² Euro	Medium rise offices National currency	£	US$	m² Euro
Austria	ASch	4,000	204	341	291	10,500	535	894	765
Belgium	BFr	15,000	261	435	372	37,000	643	1,070	919
Cyprus	C£	95	114	191	164	330	398	664	568
France	FFr	3,610	386	644	551	7,900	845	1,410	1,210
Germany	DM	1,140	409	683	584	2,770	992	1,660	1,420
Ireland	IR£	294	263	439	375	856	764	1,280	1,090
Italy	L	620,000	225	375	321	2,300,000	833	1,390	1,190
Japan	¥	170,000	869	1,450	1,240	347,000	1,770	2,960	2,530
Netherlands	G	871	277	463	396	2,770	881	1,470	1,260
Portugal	Esc	45,000	157	263	225	110,000	385	643	550
Slovak Republic *	Sk	13,400	223	372	319	19,500	323	540	462
Spain	Pta	35,000	148	246	211	90,000	379	634	542
Switzerland	SFr	1,700	749	1,250	1,070	2,100	925	1,540	1,320
United Kingdom	£	210	210	351	300	1,070	1,070	1,780	1,520
United States	$	271	162	271	190	802	480	802	561

* cost per m³

Approximate estimating – housing

This table summarises approximate estimating costs per square metre for housing. Two types of housing have been taken: mass market, single family houses and medium quality, medium rise apartments. In countries where housing types did not match exactly these descriptions, the nearest equivalent has been taken. Where a range of costs has been given, the mid point is shown. Approximate estimating costs for a particular country are for the normal standards prevailing in that country. It should be noted that costs per square metre for housing are particularly sensitive to levels of specification, total area of dwelling unit and the rules of measurement adopted; these all vary significantly among countries and this table should, therefore, be used with care.

Country	National unit of currency	Single family housing National currency	£	US$	m² Euro	Apartment/flats National currency	£	US$	m² Euro
Austria	ASch	15,000	765	1,280	1,090	12,000	612	1,020	874
Belgium	BFr	28,000	487	813	695	32,000	556	929	795
Cyprus	C£	290	349	583	499	290	349	583	499
Finland	Fmk	3,750	442	738	631	4,250	502	838	717
France	FFr	5,150	551	920	787	5,150	551	920	787
Germany	DM	1,230	442	738	631	1,290	463	774	662
Ireland	IR£	562	502	838	717	803	717	1,200	1,020
Italy	L	1,500,000	543	907	776	1,700,000	616	1,030	879
Japan	¥	n.a.	n.a.	n.a.	n.a.	301,000	1,540	2,560	2,190
Netherlands	G	1,230	392	654	560	1,440	457	763	653
Portugal	Esc	70,000	245	409	350	90,000	315	526	450
Slovak Republic	Sk	16,100	267	445	381	13,500	224	374	320
Spain	Pta	80,000	337	563	482	80,000	337	563	482
Switzerland	SFr	n.a.	n.a.	n.a.	n.a.	1,450	639	1,070	913
United Kingdom	£	395	395	660	564	480	480	802	686
United States	$	575	344	575	403	715	428	715	501

Appendix one: Abbreviations

Abbreviations

LENGTH

km .... kilometre
m .... metre
mm .... millimetre
yd ....yard
ft.. foot
in..inch

VOLUME

l .... litre

WEIGHT (MASS)

t.. tonne
kg..kilogram

NUMBERS

th .... thousand
mn..million
bn .... billion

FORCE

N .... newton

OTHER

CIA..Central Intelligence Agency
CIS ....Commonwealth of Independent States
DIY ....Do-it-yourself
DLC ....Davis Langdon Consultancy
ECU ....European Currency Unit
EMS...European Monetary System
EMU ....European Monetary Union
ERM...Exchange Rate Mechanism
EU... European Union
FSU...Former Soviet Union
GDP..Gross domestic product
n.a ....not available
PPP.....................................purchasing power party
UK... United Kingdom
USA/US.............United States of America/United States
VAT ....Value added tax

Appendix two: Conversion factors

Conversion Factors

LENGTH

Metric		Imperial equivalent
1 kilometre	1000 metres	0.6214 miles
		1093.6 yards
1 metre	100 centimetres	1.0936 yards
	1000 millimetres	3.2808 feet
		39.370 inches
1 centimetre	10 millimetres	0.3937 inches
1 millimetre		0.0394 inches

Imperial		Metric equivalent
1 mile	1,760 yards	1.6093 kilometres
	5,280 feet	1609.3 metres
1 yard	3 feet	0.9144 metres
	36 inches	914.40 millimetres
1 foot	12 inches	0.3048 metres
		304.80 millimetres
1 inch		25.400 millimetres

AREA

Metric		Imperial equivalent
1 square kilometre	100 hectares	0.3861 square miles
	10^6 square metres	247.11 acres
1 hectare	10,000 square metres	2.4711 acres
		11960 square yards
1 square metre	10,000 square centimetres	1.1960 square yards
		10.764 square feet
1 square centimetre	100 square millimetres	0.1550 square inches
1 square millimetre		0.0016 square inches

Imperial		Metric equivalent
1 square mile	640 acres	2.5900 square kilometres
		259.00 hectares
1 acre	4840 square yards	0.4047 hectares
		4046.9 square metres
1 square yard	9 square feet	0.8361 square metres
1 square foot		0.0929 square metres
1 square inch		6.4516 square centimetres
		645.16 square millimetres

VOLUME

Metric		**Imperial equivalent**
1 cubic metre or	10 hectolitres	1.3080 cubic yards
1 kilolitre	1000 cubic decimetres	35.315 cubic feet
	1000 litres	
1 hectolitre	100 litres	3.5315 cubic feet
		21.997 gallons
1 cubic decimetre or	1000 cubic centimetres	61.023 cubic inches
1 litre	1000 millilitres	0.2200 gallons
		1.7598 pints
		0.2642 US gallons
		2.1134 US pints
1 cubic centimetre or	1000 cubic millimetres	0.0610 cubic inches
1 millilitre		

Imperial		**Metric equivalent**
1 cubic yard	27 cubic feet	0.7646 cubic metres
1 cubic foot	1,728 cubic inches	28.317 litres
	6.2288 gallons	
	7.4805 US gallons	
1 cubic inch		16.387 cubic centimetres
1 gallon	8 pints	4.5461 litres
1 pint		0.5683 litres

US		
1 barrel	42 gallons	158.99 litres
1 gallon	8 pints	3.7854 litres
1 pint		0.4732 litres

WEIGHT (MASS)

Metric		**Imperial equivalent**
1 tonne	1,000 kilograms	0.9842 tons
		1.1023 US tons
		2204.6 pounds
1 kilogram	1,000 grams	2.2046 pounds
		35.274 ounces
1 gram		0.0353 ounces

Imperial		Metric equivalent
1 ton	20 hundredweights	1.0160 tonnes
	2,240 pounds	1016.0 kilograms
1 hundredweight	112 pounds	50.802 kilograms
1 pound	16 ounces	0.4536 kilograms
		453.59 grams
1 ounce		28.350 grams

US		
1 ton	20 hundredweights	0.9072 tonnes
	2,000 pounds	907.18 kilograms
1 hundredweight	100 pounds	45.359 kilograms

FORCE

Metric		Imperial equivalent
1 kilonewton	1000 newtons	0.1004 tons force
		0.1124 US tons force
1 newton		0.2248 pounds force

Imperial		Metric equivalent
1 ton force	2240 pounds force	9.9640 kilonewtons
1 pound force		4.4482 newtons

US		
1 ton force	2000 pounds force	8.8964 kilonewtons

PRESSURE

Metric	Imperial equivalent
1 newton per square millimetre	145.04 pounds force per square inch
1 kilonewton per square metre	20.885 pounds force per square foot

Imperial	Metric equivalent
1 pound force per square inch	6.8948 kilonewtons per square metre
	0.0069 newtons per square millimetre
1 ton force per square inch	107.25 kilonewtons per square metre
	0.1073 newtons per square millimetre

US	
1 ton force per square foot	95.761 kilonewtons per square metre
	0.9576 newtons per square millimetre

Appendix three:

Multilingual descriptions of construction items

Excavation
(Assume excavation in firm soil)

1. Mechanical excavation of foundation trenches. Starting from ground level (including removal of excavation material from site). Over 0.30m wide, not exceeding 2.00m deep.

2. Hardcore filling in making up levels. Hard brick, broken stone (or sand where appropriate). Crushed to pass a 100mm ring 150mm deep.

3. Earthwork support. Sides of trench excavation. Distance between opposing faces not exceeding 2.00m. Maximum depth 2.00m.

Concrete work
(Formwork and reinforcement measured separately)

4. Plain insitu concrete in strip foundations in trenches 20N/mm². Ordinary Portland Cement, 20mm coarse aggregate. Size 500mm wide x 300mm thick.

5. Reinforced insitu concrete in beds 20N/mm². Ordinary Portland Cement, 20mm coarse aggregate. 200mm thick.

6. Reinforced insitu concrete in walls 20N/mm². Ordinary Portland Cement, 20mm coarse aggregate. 200mm thick.

Terrassement
(Terrassement en terrain consistant)

1. Fouille en tranchée, à l'engin mécanique, pour fondations commençant au niveau du sol existant (Y compris chargement des déblais et enlèvement à la décharge). Dimensions des tranchées: 0,30m de largeur minimum et 2,00m de profondeur maximum.

2. Remblai de mise à niveau réalisé en tout venant, granulométrie de 100mm maximum. Epaisseur du remblai: 0,15m.

3. Blindage de tranchée. La distance entre les parois n'excédant pas 2,00m. Profondeur maximum de 2,00m.

Béton
(Le coffrage et les aciers sont comptés séparément)

4. Béton non armé de remplissage de tranchées de fondation 20N/mm². Ciment de Portland ordinaire, agrégats de 20mm maximum. Dimensions: 0,5m de largeur x 0,30m de profondeur.

5. Béton armé pour planchers en dalle pleine 20N/mm². Ciment de Portland ordinaire, agrégats de 20mm dimension maximum. Dimensions: 0,20m d'épaisseur.

6. Béton armé pour mur 20N/mm². Ciment de Portland ordinaire, agrégats de 20mm dimension maximum. Dimensions: 0,20m d'épaisseur.

Scavi
(In terreno consistente)

1. Scavo di fondazioni in trincea con mezzi meccanici, di oltre 0,30m di larghezza, e non superiore a 2,00m di profondità dal livello terra. E'compreso il trasporto del materiale di risulta fuori dal cantiere.

2. Riempimento in sottofondo con vecchi scarti di muratura, pietre spaccate o sabbia, frantumato per passare un anello di diametro 100mm, in uno strato di 150mm spessore.

3. Sbadacchiature ai lati di trincee, distanza tra le faccie opposte non superiore a 2,00m, fino a una profondità massima 2,00m.

Opere in calcestruzzo
(Casseforme e ferri di armatura misurati a parte)

4. Calcestruzzo semplice in trincea di fondazione 20N/mm²: cemento tipo Portland normale, inerti di 20mm. Dimensioni 500mm di larghezza x 300mm spessore.

5. Calcestruzzo armato gettato in opera in solette 20N/mm²: cemento tipo Portland normale, inerti di 20mm. Dimensione 200mm spessore.

6. Calcestruzzo armato gettato in opera in muri 20N/mm²: cemento tipo Portland normale, inerti di 20mm. Dimensione 200mm spessore.

Aushub

(Anzunehmen ist, daß der Aushub in festem Baugrund erfolgt)

1. Maschineller Aushub der Fundamentgräben, ausgehend vom bestehenden Planum (mit Abtransport des Aushubmaterials vom Grundstück). Breite über 0.30m, Tiefe bis maximal 2.00m.

2. Grobkörnige Unterpackung zur Erhöhung der Baufläche. Harteziegel, Steinschlag (wenn angemessen auch Sand), zerschlagen, um durch einen Ring, Durchmesser 100mm, zu kommen. Tiefe 150mm.

3. Grabenspriessungen. Grabenbreite maximal 2.00m, Grabentiefe maximal 2.00m.

Betonarbeiten

(Schalungen und Bewehrung seperat gemessen)

4. Unbewehrter Ortbeton für Streifenfundamente in Gräben, 20N/mm². Normaler Portlandzement, Korngröße 20mm. Breite 500mm, Dicke 300mm.

5. Ort-Stahlbeton für Bodenplatten, 20N/mm². Normaler Portlandzement, Korngröße 20mm. Dicke 200mm.

6. Ort-Stahlbeton für Wände, 20N/mm². Normaler Portlandzement, Korngröße 20mm. Dicke 200mm.

Excavacion

(Excavaciones en terreno compacto)

1. Excavación mecánica de zanjas de cimentación. Iniciandose a nivel de tierra e incluyendo el transporte de tierras. De más de 30cm de ancho, y una profundidad máxima de 2m.

2. Sub-base granular de cantera (o arena en su caso). Granulometría de 100 a 150mm.

3. Entibación de zanjas de cimentación de menos de 2m de ancho y 2m de profundidad.

Trabajos en hormigón

(Encofrados y armaduras, medidas aparte)

4. Hormigón 20N/mm² de cemento portland y áridos de 20mm en zanjas de cimentación de 500 x 300mm.

5. Hormigón para armar de 20N/mm² de cemento portland y áridos de 20mm en pavimento de 200mm de espesor.

6. Hormigón para armar de cemento portland y áridos de 20mm en muros de 200mm de espesor.

Excavation

(Assume excavation in firm soil)

1. Mechanical excavation of foundation trenches. Starting from ground level (including removal of excavation material from site). Over 0.30m wide, not exceeding 2.00m deep.

2. Hardcore filling in making up levels. Hard brick, broken stone (or sand where appropriate). Crushed to pass a 100mm ring 150mm deep.

3. Earthwork support. Sides of trench excavation. Distance between opposing faces not exceeding 2.00m. Maximum depth 2.00m.

Concrete work

(Formwork and reinforcement measured separately)

4. Plain insitu concrete in strip foundations in trenches 20N/mm². Ordinary Portland Cement, 20mm coarse aggregate. Size 500mm wide x 300mm thick.

5. Reinforced insitu concrete in beds 20N/mm². Ordinary Portland Cement, 20mm coarse aggregate. 200mm thick.

6. Reinforced insitu concrete in walls 20N/mm². Ordinary Portland Cement, 20mm coarse aggregate. 200mm thick.

7. Reinforced insitu concrete in suspended floor or roof slabs 20N/mm^2. Ordinary Portland Cement, 20mm coarse aggregate. 150mm thick.

7. Béton armé pour dalle de toiture 20N/mm^2. Ciment de Portland ordinaire, agrégats de 20mm dimension maximum. Dimensions: 0,15m d'épaisseur.

7. Calcestruzzo armato gettato in opera in solai di pavimenti o tetti 20N/mm^2: cemento di tipo Portland normale, inerti di 20mm. Dimensione 150mm spessore.

8. Reinforced insitu concrete in columns 20N/mm^2. Ordinary Portland Cement, 20mm coarse aggregate. Size 400 x 400mm.

8. Béton armé pour poteau 20N/mm^2. Ciment de Portland ordinaire, agrégats de 20mm dimension maximum. Dimensions: 0,40m x 0,40m.

8. Calcestruzzo armato gettato in opera in pilastri 20N/mm^2: cemento tipo Portland normale, inerti di 20mm. Dimensioni 400 x 400mm.

9. Reinforced insitu concrete in isolated beams 20N/mm^2. Ordinary Portland Cement, 20mm coarse aggregate. Size 400 x 600mm deep.

9. Béton armé pour poutre isolée 20N/mm^2. Ciment de Portland ordinaire, agrégats de 20mm dimension maximum. Dimension: 0,40m de largeur x 0,60m de hauteur.

9. Calcestruzzo armato gettato in opera in travi isolate 20N/mm^2: cemento tipo Portland normale, inerti di 20mm. Dimensioni 400 x 600mm di profondità.

10. Precast concrete slab (including reinforcement as necessary). Contractor designed for total loading of 3N/mm^2 5.00m span.

10. Plancher préfabriqué en béton (y compris les armatures). Portée maximum de 5m, acceptant une charge maximum de 3N/mm^2.

10. Calcestruzzo in soletta prefabbricata (compreso il ferro di armatura necessario), progettata dall'impresa per un carico totale di 3N/mm^2, luce 5,00m.

Formwork
(Assume a simple repetitive design which allows 3 uses of formwork)

Coffrage
(Coffrage ordinaire permettant trois utilisations)

Casseforme
(Considerare casseratura semplice ripetitiva che permetta il riutilizzo delle casseforme tre volte)

11. Softwood or metal formwork to concrete walls. Basic finish. (one side only).

11. Coffrage en bois ou métal pour mur en béton. Face lisse (Une face)

11. Casseforme in legno o acciaio per muri in calcestruzzo, faccia liscia. (Un lato soltanto).

12. Softwood or metal formwork to concrete columns. Basic finish. Columns 1600m girth.

12. Coffrage en bois ou métal pour poteau en béton. Face lisse. Développé: 1,60m.

12. Casseforme in legno o acciaio per pilastri in calcestruzzo, facce lisce. Pilastri 1,60m di contorno.

7. Ort-Stahlbeton für Deckenplatten oder Dächer, 20N/mm². Normaler Portlandzement, Korngröße 20mm. Dicke 150mm.

7. Hormigón para armar de cemento portland y áridos de 20mm en losas de 150mm de espesor.

7. Reinforced insitu concrete in suspended floor or roof slabs 20N/mm². Ordinary Portland Cement, 20mm coarse aggregate. 150mm thick.

8. Ort-Stahlbeton für Stützen, 20N/mm². Normaler Portlandzement, Korngröße 20mm. Dimensionen 400 x 400mm.

8. Hormigón para armar de cemento portland y áridos de 20mm en pilares de 400 x 400mm.

8. Reinforced insitu concrete in columns 20N/mm². Ordinary Portland Cement, 20mm coarse aggregate. Size 400 x 400mm.

9. Ort-Stahlbeton für isolierte Balkenträger, 20N/mm². Normaler Portlandzement, Korngröße 20mm. Breite 400mm, Tiefe 600mm.

9. Hormigón para armar de 20N/mm² de cemento portland y áridos de 20mm en vigas de 400 x 600mm.

9. Reinforced insitu concrete in isolated beams 20N/mm². Ordinary Portland Cement, 20mm coarse aggregate. Size 400 x 600mm deep.

10. Fertigbetonplatte (einschließlich erforderlicher Bewehrung). Für eine Gesamtbelastung von 3N/mm² vom Unternehmer hergestellt. Spannweite 5.00m.

10. Losas de hormigón pretensada (incluyendo la armadura necesaria), para una carga total de 3N/mm² de 5,00m de luz.

10. Precast concrete slab (including reinforcement as necessary). Contractor designed for total loading of 3N/mm² 5.00m span.

Schalungen
(Anzunehmen ist, daß eine einfache Konstruktion eine dreimalige Nutzung der Schalungen ermöglicht)

Encofrados
(Se supone una forma simple y repetitiva que permite 3 usos del encofrado)

Formwork
(Assume a simple repetitive design which allows 3 uses of formwork)

11. Schalungen aus Weichholz oder Metall für Betonwände. Glatte Fläche (nur eine Seite).

11. Encofrado metálico a una cara, en muros, acabado para revestir.

11. Softwood or metal formwork to concrete walls. Basic finish. (one side only).

12. Schalungen aus Weichholz oder Metall für Betonstützen. Glatte Fläche. Umfang der Stützen 1600mm.

12. Encofrado metálico en pilares de 1600mm de perímetro, acabado para revestir.

12. Softwood or metal formwork to concrete columns. Basic finish. Columns 1600m girth.

13. Softwood or metal formwork to horizontal soffits of slabs. Basic finish. Slab 150mm thick, not exceeding 3.50m high.

13. Coffrage horizontal en bois ou en métal pour plancher en béton. Face lisse. Plancher de 0,15m d'épaisseur d'une portée maximum de 3,50m.

13. Casseforme in legno o acciaio per solai orizzontali, faccia liscia, solai 150mm di spessore, non oltre 3,50m dal piano di appoggio.

Reinforcement

Armature en acier

Armature

14. Reinforcement in concrete walls. Hot rolled high tensile bars cut, bent and laid, 16mm diameter.

14. Armature en acier à haute adhérence, y compris la coupe, le pliage, l'assemblage et la pose. Diamètre de 160mm. Pour mur en béton.

14. Armatura in muri di calcestruzzo: barre di acciaio ad aderenza migliorata, tagliate, piegate e posate, 16mm di diametro.

15. Reinforcement in suspended concrete slabs. Hot rolled high tensile bars cut, bent and laid, 25mm diameter.

15. Armature en acier à haute adhérence, y compris la coupe, le pliage, l'assemblage et la pose. Diamètre de 250mm. Pour plancher en béton.

15. Armatura in solai di calcestruzzo: barre di acciaio ad aderenza migliorata, tagliate, piegate e posate, 25mm di diametro.

16. Fabric (mat) reinforcement in concrete beds (measured separately). Weight approximately 3.0 kg/m^2. Laid in position with 150mm side and end laps.

16. Armature en treillis soudé pour plancher en béton. Poids approximatif: 3 kg/m² posé avec un recouvrement de 0,15m.

16. Rete in acciaio elettrosaldata in solette di calcestruzzo, peso approssimato 3,0 Kg/m^2, posata in opera con risvolti ai lati di 150mm.

Steelwork

Charpente métallique

Carpenteria metallica

17. Fabricate, supply and erect steel framed structure. Including painting all steel with one coat primer.

17. Fabrication, fourniture et mise en oeuvre d'une structure métallique y compris l'application d'une couche primaire antirouille.

17. Fabbricazione, fornitura e montaggio di struttura metallica compresa l'applicazione di una mano di minio.

Brickwork and blockwork
(Assume a notional thickness of 100mm for bricks and blocks. Rates should be for the nearest standard size to 100mm)

Maçonnerie
(Les prix indiqués sont ceux de maçonneries de l'épaisseur standard locale se rapprochant le plus de 10cm)

Opera di muratura
(Prezzi per mattoni e blocchi di dimensioni di normale produzione, il più vicino possibile a 100mm di spessore)

18. Precast lightweight aggregate hollow concrete block walls. Gauged mortar. 100mm thick.

18. Mur en élévation en parpaing creux hourdés au mortier de ciment. Epaisseur 10cm.

18. Blocchi cavi di calcestruzzo alleggeriti, posati con l'impiego di malta bastarda, in muri spessore 100mm.

13. Schalungen aus Weichholz oder Metall für horizontale Unterseiten von Deckenplatten. Glatte Fläche. Dicke der Deckenplatten 150mm, Höhe der Decke maximal 3.50m.

13. Encofrado de madera en losas de 150mm, acabado para revestir a menos de 3,50m de altura.

13. Softwood or metal formwork to horizontal soffits of slabs. Basic finish. Slab 150mm thick, not exceeding 3.50m high.

Bewehrung

Armaduras

Reinforcement

14. Bewehrung für Betonwände. Warmgewalzte hochzugfeste Stäbe, schneiden biegen und verlegen, Durchmesser 16mm.

14. Armadura en muros de hormigón con acero de alta resistencia de 16mm de diámetro, colocada.

14. Reinforcement in concrete walls. Hot rolled high tensile bars cut, bent and laid, 16mm diameter.

15. Bewehrung für Deckenplatten. Warmgewalzte hochzugfeste Stäbe, liefen schneiden biegen und verlegen, Durchmesser 25mm.

15. Armadura en losas de hormigón con acero de alta resistencia de 25mm de diámetro, colocada.

15. Reinforcement in suspended concrete slabs. Hot rolled high tensile bars cut, bent and laid, 25mm diameter.

16. Baustahlmatten für Bodenplatten (separat gemessen). Gewicht ungefähr 3.0 kg/m^2. Mit 150mm Überlappung an Seiten und Enden verlegt.

16. Mallazo en pavimentos de hormigón (medido aparte) con un peso aproximado de 3.0 kg/m^2.

16. Fabric (mat) reinforcement in concrete beds (measured separately). Weight approximately 3.0 kg/m^2. Laid in position with 150mm side and end laps.

Stahlbauarbeiten

Estructura metálica

Steelwork

17. Fertigung, Lieferung und Montage eines Konstruktions-Stahlrahmens. Mit Grundanstrich für alle Stahlflächen.

17. Acero en estructura metálica, colocada, incluyendo una capa de pintura de imprimación.

17. Fabricate, supply and erect steel framed structure. Including painting all steel with one coat primer.

Mauererarbeiten
(Die Kostenkennwerte sollen für Mauersteine, Dicke 100mm, oder nächstliegendes örtliches Standardformat gelten)

Paredes de ladrillo y bloques de hormigón
(Se supone un grueso de 100mm para ladrillos y bloques. Los datos tendrían que referirse a medidas standard lo más proximas posibles a 100mm)

Brickwork and blockwork
(Assume a notional thickness of 100mm for bricks and blocks. Rates should be for the nearest standard size to 100mm)

18. Wände aus Betonhohlblocksteinen. Zementkalkmörtel. Wanddicke 100mm.

18. Pared de bloque de hormigón hueco, tomado con mortero, de 100mm de espesor.

18. Precast lightweight aggregate hollow concrete block walls. Gauged mortar. 100mm thick.

19. Solid (perforated) clay or concrete common bricks (priced at .... per m² delivered to site). Gauged mortar. 100mm thick walls.

19. Mur en élévation en briques pleines en terre cuite, de fabrication courante, hourdées au mortier de ciment. Epaisseur 10cm.

19. Mattoni pieni di laterizio o di cemento (al prezzo di ...... lire/m² franco cantiere), posati con l'impiego di malta bastarda, in muri spessore 100mm.

20. Solid (perforated) sand lime bricks (priced at .... per m² delivered to site). Gauged mortar. 100mm thick walls.

20. Mur en élévation en briques pleines de chaux et sable, de fabrication courante, hourdées au mortier de ciment. Epaisseur 10cm.

20. Mattoni pieni di calce e sabbia (al prezzo di ...... lire/m² franco cantiere), posati con l'impiego di malta bastarda, in muri spessore 100mm.

21. Facing bricks (priced at .... per m² delivered to site). Gauged mortar, flush pointed as work proceeds. Half brick thick walls.

21. Mur en briques pleines à parement, joints de 10mm, faces alignées. Epaisseur 10cm.

21. Mattoni "a cortina" (al prezzo di ...... lire/m² franco cantiere), posati con l'impiego di malta bastarda con giunti stilati, in muratura ad una testa.

Roofing

Toiture

Tetti

22. Concrete interlocking roof tiles 430 x 380mm (or nearest equivalent). On and including battens and underfelt. Laid to 355mm gauge with 75mm laps (excluding eaves fittings or ridge tiles).

22. Couverture en tuiles à emboîtement en béton de 430 x 380mm (ou dimensions standards locales les plus proches). Y compris les liteaux et le film étanche. Pureau de 35,5cm, recouvrement de 7,5cm.(Sont exclus les faîtages, arêtiers et rives)

22. Tegole di cemento ad incastro 430 x 380mm (o di simili dimensioni), posate a 8 tegole/m², su listelli di legno e cartonfeltro bituminato compresi nel prezzo. Sono esclusi colmi ed altri pezzi speciali.

23. Plain clay roof tiles 260 x 160mm (or nearest equivalent). On and including battens and underfelt. Laid to 100mm lap (excluding eaves fittings or ridge tiles).

23. Couverture en tuiles plates de terre cuite, 260 x 160mm (ou dimensions standards locales les plus proches). Recouvrement de 10cm. Y compris les liteaux et le film étanche. (Sont exclus les faîtages, arêtiers et rives)

23. Tegole piane in laterizio 260 x 160mm (o di simili dimensioni), posate a 60 tegole/m², su listelli di legno e cartonfeltro bituminato compresi nel prezzo. Sono esclusi colmi ed altri pezzi speciali.

24. Fibre cement roof slates 600 x 300mm (or nearest equivalent). On and including battens and underfelt. Laid flat or to fall as coverings for roofs.

24. Couverture en plaques de fibre ciment de 600mm x 300mm (ou dimensions standards locales les plus proches). Y compris les liteaux et le film étanche. Pose en couverture.

24. Lastre di fibro-cemento 600 x 300mm (o di simili dimensioni), posate in piano o in pendenza su listelli di legno e cartonfeltro bituminato compresi nel prezzo.

19. Normale Vollziegel (bzw. Hochlochziegel) oder Betonsteine (Preis ..... pro m² mit Lieferung). Zementkalkmörtel. Wanddicke 100mm.

19. Pared de ladrillo perforado (PVP de 650 pesetas por m², puesto en pie de obra) tomado con mortero, de 100mm de espesor.

19. Solid (perforated) clay or concrete common bricks (priced at .... per m² delivered to site). Gauged mortar. 100mm thick walls.

20. Kalksandvollstein (bzw. -Lochstein) (Preis ..... pro m² mit Lieferung). Zementkalkmörtel. Wanddicke 100mm.

20. Pared de ladrillo calcáreo (PVP de 960 pesetas por m², puesto en pie de obra) tomado con mortero de 100mm de espesor.

20. Solid (perforated) sand lime bricks (priced at .... per m² delivered to site). Gauged mortar. 100mm thick walls.

21. Verblendsteine (Preis ... pro m² mit Lieferung). Zementkalkmörtel, Vollfugen auszuführen bei der Hochmauerung. Halbsteindicke Wände.

21. Pared de ladrillo en cara vista (PVP de 1100 pesetas por m², puesto a pie de obra), con junta enrasada, de medio pie de espesor.

21. Facing bricks (priced at ...... per m² delivered to site). Gauged mortar, flush pointed as work proceeds. Half brick thick walls.

Bedachungsarbeiten

Cubiertas

Roofing

22. Betonflachdachsteine, 430 x 380mm (oder nächstliegendes Standardformat). Mit Lattung und Dachpappenunterlage. Lattenweite 355mm, Überdeckung 75mm. (Ohne Traufenabschluß und Firststeine).

22. M² de cubierta con tejas de hormigón de 430 x 380mm incluyendo piezas especiales con longitud vista de 355m y un solape de 75mm (no se incluyen uniones con los canalones o las piezas de remate).

22. Concrete interlocking roof tiles 430 x 380mm (or nearest equivalent). On and including battens and underfelt. Laid to 355mm gauge with 75mm laps (excluding eaves fittings or ridge tiles).

23. Biberschwannzziegel 260 x 160mm (oder nächstliegendes Standardformat). Mit Lattung und Dachpappenunterlage. Überdeckung 100mm. (Ohne Traufenabschluß und Firststeine).

23. M² de cubiertas de tejas planas cerámicas de 260 x 160mm incluyendo piezas especiales con un solape de 100mm y excluyendo las entregas a canalones o piezas de remate.

23. Plain clay roof tiles 260 x 160mm (or nearest equivalent). On and including battens and underfelt. Laid to 100mm lap (excluding eaves fittings or ridge tiles).

24. Faserzementplatten für Dach 600 x 300mm (oder nächstliegendes Standardformat). Mit Lattung und Dachpappenunterlage. Flach oder im Gefälle als Dachhaut verlegt.

24. M² cubierta con fibrocemento en piezas de 600 x 300mm incluyendo las juntas y un fieltro inferior.

24. Fibre cement roof slates 600 x 300mm (or nearest equivalent). On and including battens and underfelt. Laid flat or to fall as coverings for roofs.

25. Sawn softwood roof boarding, preservative treated 25mm thick. Laid flat or to fall.

25. Volige à planches jointives, en sapin de pays traité. 25mm d'épaisseur. Pose en toiture horizontale ou en pente.

25. Tavolame in abete impregnato con protettivo, 25mm di spessore, posato in piano o in pendenza.

26. Particle board roof coverings with tongued and grooved joints 25mm thick. Laid flat or to fall.

26. Panneaux en particules de bois avec languettes et rainures. 25mm d'épaisseur. Pose en toiture horizontale ou en pente.

26. Lastre di truciolare 25mm di spessore, con giunti ad incastro a linguetta, posati in piano o in pendenza.

27. 3 layers glass-fibre based bitumen felt roof covering. Finished with limestone chippings in hot bitumen. To flat roofs.

27. Revêtement d'étanchéité tri-couche en feutre bituminé, y compris finition par une couche d'émulsion bitumineuse et granulats minéraux fins. Pour toiture terrasse.

27. Manto impermeabile prefabbricato, composto da 3 strati di feltro bituminoso a base di fibre di vetro, coperto con emulsione bituminosa ed inerti reflettenti, su coperture piane.

28. Bitumen based mastic asphalt roof covering in 2 layers. On and including sheathing felt underlay, with white chippings finish. To flat roofs.

28. Revêtement d'étanchéité bi-couche en asphalte coulé, y compris finition par granulats minéraux fins. Pour toiture terrasse.

28. Due strati di asfalto con manto di feltro sottostante e finitura di inerti reflettenti, per coperture piane.

29. Glass-fibre mat roof insulation 160mm thick. Laid flat between ceiling joists.

29. Isolation de toiture par laine minérale en rouleaux, pose à plat entre les chevrons. Epaisseur 160mm.

29. Coibentazione del tetto con materassino di lana di vetro, spessore 160mm, posato piatto tra morali.

30. Rigid sheet resin-bonded loadbearing glass-fibre roof insulation 75mm thick. Laid on flat roofs

30. Isolation de toiture terrasse par laine minérale en panneaux rigides. 75mm d'épaisseur.

30. Coibentazione di coperture piane composta di pannelli rigidi di lana di vetro, spessore 75mm.

31. 0.8mm troughed galvanized steel roof cladding in single spans of 3.00m with loading of 0.75 KN/m^2. Fixed to steel roof trusses with bolts. To pitched roofs.

31. Couverture par bacs en acier galvanisé, d'une portée de 3,00m et de 0,80mm d'épaisseur, acceptant un poids maximum de 0,75 KN/m^2. Y compris fixation à la structure acier. Pour toiture en pente.

31. Copertura di tetti a pendenza con lamiera grecata zincata spessore 0,8mm, luce di 3,00m, con sovraccarico di 0,75 KN/m^2, imbullonata a capriate metalliche.

25. Weichholzdachschalung mit Schutzbehandlung. Dicke 25mm. Flach oder im Gefälle verlegt.

25. Remate de madera tratada previamente de 25mm de espesor, colocado plano o inclinado.

25. Sawn softwood roof boarding, preservative treated 25mm thick. Laid flat or to fall.

26. Spanplattendachdeckung, Stöße mit Nut und Feiler. Dicke 25mm. Flach oder im Gefälle verlegt.

26. Cubierta con tablero de partículas, incluídas las juntas, de 25mm de espesor, colocado plano o inclinado.

26. Particle board roof coverings with tongued and grooved joints 25mm thick. Laid flat or to fall.

27. Dreilagige Dachhaut aus Glasfaserbitumendach-dichtungsbahn. Überdeckt mit Kalksteinsplitt, in Heißbitumen eingebettet. Für Flachdächer.

27. Impermeabilización con tela bituminosa de tres capas de fibra de vidrio, acabada con árido visto sobre emulsión asfáltica en caliente en cubiertas.

27. 3 layers glass-fibre based bitumen felt roof covering. Finished with limestone chippings in hot bitumen. To flat roofs.

28. Zweilagiger Bitumen-Mastixdachbelag. Mit Dachpappenunterlage, überdeckt mit Weißsplitt. Für Flachdächer.

28. Impermeabilización de cubiertas con dos capas de tela asfáltica incluyendo el fieltro inferior, acabado con árido visto.

28. Bitumen based mastic asphalt roof covering in 2 layers. On and including sheathing felt underlay, with white chippings finish. To flat roofs.

29. Wärmedämmung aus Glasfasermatten, Dicke 160mm. Zwischen Unterzügen verlegt.

29. Aislamiento con manto de fibra de vidrio de 160mm de espesor colocado entre las viguetas del techo.

29. Glass-fibre mat roof insulation 160mm thick. Laid flat between ceiling joists.

30. Wärmedämmung aus tragenden harzverleimten Glasfaserplatten, Dicke 75mm. Auf Flachedächern verlegt.

30. Placa autoportante de resina rígida con aislamiento de fibra de vidrio de 75mm de espesor, colocada en techos planos.

30. Rigid sheet resin-bonded loadbearing glass-fibre roof insulation 75mm thick. Laid on flat roofs

31. Trapezprofildachdeckung aus verzinktem Stahl, Dicke 0.8mm, Spannweiten maximal 3.00m, Belastung 0.75KN/m². Mit Bolzen an Stahlbinder befestigt. Für Schrägdächer.

31. Cubierta de acero galvanizado de 0,8mm de espesor colocada en vanos de 3m, con carga de 0,75KN/m². Fijado a soportes metálicos de techo con tornillos.

31. 0.8mm troughed galvanized steel roof cladding in single spans of 3.00m with loading of 0.75 KN/m². Fixed to steel roof trusses with bolts. To pitched roofs.

Woodwork and metalwork
(Hardwood should be assumed to be of reasonable exterior quality)

Menuiserie et métallerie
(Les bois durs considérés étant d'une qualité raisonnable pour l'extérieur)

Opere in legno e in ferro
(Il legno duro sarà di buona qualità, adatto per l'uso esterno)

32. Preservative treated sawn softwood. Size 50 x 100mm. Framed in partitions.

32. Sapin de pays traité. Dimensions 5cm x 10cm. Pour ossature de cloison.

32. Legno segato impregnato con protettivo, dimensioni 50 x 100mm, per ossatura di tramezzi.

33. Preservative treated sawn softwood. Size 50 x 150mm. Pitched roof members.

33. Sapin de pays traité. Dimensions 5cm x 15cm. Pour charpente de toiture en pente.

33. Legno segato impregnato con protettivo, dimensioni 50 x 150mm, per struttura di tetto a pendenza.

34. Single glazed casement window in (............) hardwood including hardwood frame and sill. Including steel butts and anodised aluminium espagnolette bolt. Size approx. 650 x 900mm with 38 x 100mm frame and 75 x 125mm sill.

34. Fenêtre à un vantail en bois dur. Y compris châssis et appui en bois dur. Y compris charnières en acier et fermeture de type espagnolette en aluminium anodisé. Dimensions approximatives : 650 x 900mm. Sections du châssis de 38 x 100mm et de l'appui de 75 x 155mm.

34. Finestra ad anta semplice in legno duro del tipo ...., compreso telaio (38 x 100mm), soglia (75 x 125mm), cerniere e chiusura in alluminio. Dimensioni approssimate 650 x 900mm.

35. Two panel door with panels open for glass in (............) hardwood including hardwood frame and sill. Including glazing with 6mm wired polished plate security glass fixed with hardwood beads and including steel butts, anodised handles and push plates and security locks. Size approximately 850 x 2000mm with 38 x 100mm frame and 38 x 150mm sill.

35. Porte en bois dur à double vantaux vitrés, y compris huisserie et seuil en bois dur, vitrage armé de 6mm d'épaisseur, charnières en acier, poignées, serrures de sûreté et plaques de propreté anodisées. Dimensions approximatives 850 x 2000mm. Sections de l'huisserie de 38 x 100mm et du seuil de 38 x 150mm.

35. Porta a vetri a due ante in legno duro tipo ...., compreso telaio (38 x 100mm), soglia (38 x 100mm), vetri retinati spessore 6mm fissati con bacchette in legno duro, cerniere in acciaio, maniglie e piastre in alluminio e serratura di sicurezza. Dimensioni 850 x 2000mm.

36. Solid core half hour fire resisting hardwood internal flush door lipped on all edges. Unpainted, including steel butts, anodised handles and push plates and mortice lock. Size approximately 800 x 2000mm.

36. Porte intérieure pleine à un vantail en bois dur, coupe-feu une demi heure. Y compris huisserie, charnières en acier, poignée, serrure à mortaiser et plaques de propreté anodisées. Dimensions approximatives 800 x 2000mm. Finition non comprise.

36. Porta interna del tipo antincendio REI 30' in legno duro non verniciato, compreso cerniere in acciaio, maniglie e piastre in alluminio e serratura ad incasso. Dimensioni approssimate 800 x 2000mm.

Holz und Metallarbeiten
(Anzunehmen ist, daß das
Hartholz von orderntlicher
äußerer Qualität ist)

32. Schnittweichholz, 50mm x
100mm, mit
Schutzbehandlung. Für
Trennwandgerippe.

33. Schnittweichholz, 50mm x
150mm, mit
Schutzbehandlung. Bauteile
für Schrägdach.

34. Flügelfenster, einfach
verglast, aus (.....) Hartholz,
mit Rahm und Fenstersbank
aus Hartholz. Mit
Stahleinstemmbändern und
Espagnolettenverschluß aus
adonisiertem Aluminium.
Ungefähre Dimensionen 650
x 900mm, Rahmen 38 x
100mm, Fensterbank 75 x
125mm.

35. Zweifüllungstür mit
Sicherheitsverglasung, aus
(....) Hartholz, mit Rahmen
und Fensterbank aus
Hartholz. Zwei Füllungen aus
Spiegeldrahtglas, Dicke
6mm, Hartholzglasleisten
befestigt. Mit
Stahleinstemmbändern,
adonisierten Griffen und
Stoßschildern, und
Sicherheitsschlössern.
Ungefähre Dimensionen 800
x 2000mm, Rahmen 38 x
100mm, Schwelle 38 x
150mm.

36. Innen-Flächenvolltür aus
Hartholz, mit halbstündiger
Feuerwiderstandszeit, alle
Ränder mit Umleimern.
Unbemalt, mit
Stahleinstemmbändern,
adonisierten Griffen und
Stoßschildern, und
Einsteckschloß. Ungefähre
Dimensionen 800 x 2000mm.

Carpintería y cerrajería
(Se considera la madera de
una calidad exterior
razonable)

32. M de montante de
madera de 50 x 100mm,
acabado natural con
tratamiento previo.

33. Travesaño de 50 x
150mm con tratamiento
previo, acabado natural
colocado en cubiertas.

34. Ventana de una hoja de
pino melis incluyendo
marco y hoja, incluso
bisagras de acero y
españoletas de aluminio
anodizado. Medidas
aproximadas de 650 x
900mm con huecohojas
de 38 x 100mm, marcos
de 75 x 125mm.

35. Puerta vidriera de dos
hojas en pino de melis
incluyendo marco y
hojas, incluso vidrio de
seguridad armado pulido
de 6mm, con bisagras de
acero, y herrajes de
aluminio anodizado, así
como cerradura de
seguridad. Medidas
aproximadas de 850 x
2000mm con armazón de
38 x 100mm y marco de
38 x 150mm.

36. Puerta anti-incendios RF-
30 de madera, anclada en
todo su perímetro, sin
pintar, incluyendo
bisagras de acero,
herrajes de aluminio
anodizado y cerradura de
muesca. Medidas
aproximadas de 800 x
2000mm.

Woodwork and metalwork
(Hardwood should be
assumed to be of reasonable
exterior quality)

32. Preservative treated sawn
softwood. Size 50 x
100mm. Framed in
partitions.

33. Preservative treated sawn
softwood. Size 50 x
150mm. Pitched roof
members.

34. Single glazed casement
window in (............)
hardwood including
hardwood frame and sill.
Including steel butts and
anodised aluminium
espagnolette bolt. Size
approx. 650 x 900mm
with 38 x 100mm frame
and 75 x 125mm sill.

35. Two panel door with
panels open for glass in
(............) hardwood
including hardwood
frame and sill. Including
glazing with 6mm wired
polished plate security
glass fixed with
hardwood beads and
including steel butts,
anodised handles and
push plates and security
locks. Size approximately
850 x 2000mm with 38 x
100mm frame and 38 x
150mm sill.

36. Solid core half hour fire
resisting hardwood
internal flush door lipped
on all edges. Unpainted,
including steel butts,
anodised handles and
push plates and mortice
lock. Size approximately
800 x 2000mm.

37. Aluminium double glazed window and hardwood sub-frame. Standard anodised horizontally sliding double glazed in (.............) hardwood sub-frame and sill. Including double glazing with 4mm glass, including all ironmongery. Size approximately 1200 x 1200mm with 38 x 100mm sub-frame and 75 x 125mm sill.

37. Fenêtre coulissante en aluminium anodisé avec double vitrage 4/6/4, châssis et appui en bois dur. Y compris toute quincaillerie et ferrage. Dimensions approximatives 1200 x 1200mm avec un châssis de 38 x 100mm et appui de 75 x 125mm.

37. Finestra in alluminio di normale produzione con ante scorrevoli e controtelaio in legno duro del tipo ...., vetro camera con lastre di vetro di 4mm, compresa tutta la ferramenteria. Dimensioni approssimate 1200 x 1200mm, contro telaio di 38 x 100mm e soglia 75 x 125mm.

38. Aluminium double glazed door set and hardwood sub-frame. Standard anodised aluminium, double glazed in (.............) hardwood sub-frame and sill. Including double glazing with 4mm glass, including all ironmongery. Size approximately 850 x 2100mm with 38 x 100mm subframe and 75 x 125mm sill.

38. Porte à un vantail en aluminium anodisé avec double vitrage 4/6/4, châssis et seuil en bois dur, compris toute quincaillerie et ferrage. Dimensions approximatives 850 x 2100mm avec un châssis de 38 x 100mm et un seuil de 75 x 125mm.

38. Porta finestra in alluminio di normale produzione con contro telaio in legno duro del tipo ...., vetro camera con lastre di vetro di 4mm, compresa tutta la ferramenteria. Dimensioni approssimate 850 x 2100mm, controtelaio 38 x 100mm e soglia 75 x 125mm.

39. Hardwood skirtings. Wrought (..............) hardwood. Fixed on softwood grounds. Size 20 x 100mm.

39. Plinthe en bois dur fixée sur un tasseau de bois tendre. Dimensions 2cm x 10cm.

39. Zoccolino battiscopa in legno duro piallato del tipo ...., fissato con tasselli di legno. Dimensioni 20 x 100mm.

40. Framed structural steelwork in universal joist sections. Bolted or welded connections, including erecting on site and painting one coat at works.

40. Ouvrage assemblé de métallerie en profilés du commerce, y compris les perçages, boulons, rivets ou soudures nécessaires à l'exécution des ouvrages. Peinture de protection réalisée en l'atelier.

40. Profilati di ferro a doppio T intelaiati, compresi i giunti saldati o bullonati, il montaggio in cantiere e una mano di minio protettivo.

41. Structural steelwork lattice roof trusses. Bolted or welded connections, including erecting on site and painting one coat at works.

41. Ouvrage assemblé de charpente métallique en profilés du commerce, y compris les perçages, boulons, rivets ou soudures nécessaires à l'exécution des ouvrages. Peinture de protection réalisée sur le chantier.

41. Capriate in ferro, compresi i giunti saldati o bullonati, il montaggio in cantiere e uno mano di minio protettivo.

37. Horizontalschiebefenster aus anodisiertem Aluminium mit Doppelverglasung, Glasdicke 4mm. Innenprofil und Fensterbank aus (..............) Hartholz. Mit allen Beschlägen. Ungefähre Dimensionen 1200 x 1200mm, Fensterprofil 38 x 100mm, Fensterbank, 75 x 125mm.

37. Ventana corredera de aluminio anodizado standard con doble vidrio y premarco de flandes, incluyendo doble vidrio de 4mm y toda la cerrajería. Medidas aproximadas de 1200 x 1200mm, con sección de premarco de 38 x 100mm y 75 x 125mm de sección de alfeizar.

37. Aluminium double glazed window and hardwood sub-frame. Standard anodised horizontally sliding double glazed in (..............) hardwood sub-frame and sill. Including double glazing with 4mm glass, including all ironmongery. Size approximately 1200 x 1200mm with 38 x 100mm sub-frame and 75 x 125mm sill.

38. Türanlage aus anodisiertem Aluminium mit Doppelverglasung, Glasdicke 4mm. Innenprofil und Schwelle aus (....) Hartholz. Mit allen Beschlägen. Ungefähre Dimensionen 850 x 2100mm, Innenprofil 38 x 100mm, Schwelle 75 x 125mm.

38. Balconera de aluminio con anodizado standard, con doble vidrio y premarcos de madera en flandes, incluyendo doble vidrio de 4mm de espesor, y toda la cerrajería. Medidas aproximadas de 850 x 100mm y 75 x 125mm de sección de alfeizar.

38. Aluminium double glazed door set and hardwood sub-frame. Standard anodised aluminium, double glazed in (..............) hardwood sub-frame and sill. Including double glazing with 4mm glass, including all iron-mongery. Size approxi-mately 850 x 2100mm with 38 x 100mm sub-frame and 75 x 125mm sill.

39. Sockel aus gehobeltem (.....) Hartholz. An Weichholzstreifen befestigt. Dimensionen 20 x 100mm.

39. Zócalo de madera de melis colocado sobre pavimentos de madera de medidas de 20 x 100mm.

39. Hardwood skirtings. Wrought (..............) hard-wood. Fixed on softwood grounds. Size 20 x 100mm.

40. Rahmentragwerk aus Doppel-T-Stahl. Bolzen oder Schweißverbindungen, mit Montage vor Ort und einem Anstrich vor Lieferung.

40. Acero estructural en elementos principales, en perfiles universales, atornillado o soldado, colocado, incluso una capa de pintura de imprimación.

40. Framed structural steelwork in universal joist sections. Bolted or welded connections, including erecting on site and painting one coat at works.

41. Sägedachbinder aus Stahl. Bolzen oder Schweiß-Verbindungen, mit Montage vor Ort und einem Anstrich vor Lieferung.

41. Acero estructural para estructuras espaciales, atornillado o soldado, colocado, incluso una capa de pintura de imprimación.

41. Structural steelwork lattice roof trusses. Bolted or welded connections, including erecting on site and painting one coat at works.

Plumbing
(Sizes of sanitary installations
and pipes are indicative)

Plomberie
(Les dimensions des installations
sanitaires et des canalisations
sont données à titre indicatif)

Impianto idrico sanitario
(Le dimensioni degli impianti
e delle tubazioni sono
puramente indicative)

42. UPVC half round eaves gutter. Screwed to softwood at 1.00m centres. 110mm external diameter (excluding bends, outlets etc.).

42. Gouttière demi-ronde en PVC posée sur crochets fixés à la charpente. Fixation tous les mètres. Diamètre extérieur de 110mm. Coudes, talons et naissances non compris.

42. Canale di gronda di sezione ad U in UPVC, di diametro esterno 110mm, con un supporto ad ogni metro di lunghezza, fissato con viti. Sono esclusi angoli, uscite ecc.

43. UPVC rainwater pipes with pushfit joints. Screwed to brickwork at 1.50m centres. 100mm external diameter (excluding bends, outlets etc.).

43. Descentes d'eau pluviale en PVC d'un diamètre de 100mm, y compris fixation sur maçonnerie tous les 1,50m. Dauphin et naissance non compris.

43. Pluviali in UPVC di diametro esterno 100mm, con giunti ad incastro, fissati al muro ogni 1,5m. Sono esclusi gommiti, uscite ecc.

44. Light gauge copper cold water tubing with compression or capillary fittings. Screwed to brickwork horizontally at 1.00m centres. 15mm external diameter.

44. Tuyau en cuivre pour distribution d'eau froide, diamètre extérieur de 15mm (ou dimension du commerce la plus proche), avec tous raccords, joints, etc. Y compris fixation sur maçonnerie tous les 1,50m.

44. Tubazione leggera di rame per acqua fredda, 15mm diametro, con giunti saldati o a compressione, fissata orizzontalmente ogni metro di lunghezza.

45. High pressure polypropylene, polythene or UPVC (as appropriate) pipes for cold water supply. Fixed horizontally to brick walls at 1.00m centres. 15mm external diameter, complete with fittings.

45. Tuyau à haute pression en polyéthylène pour alimentation en eau froide, diamètre extérieur de 15mm (ou dimension du commerce la plus proche), avec tous raccords, joints, etc. Y compris fixation sur maçonnerie tous les 1,50m.

45. Tubazione di UPVC, polipropilene o polietilene ad alta pressione per distribuzione di acqua fredda, fissata orizzontalmente alla muratura ogni metro di lunghezza, completa di gomiti, raccordi ecc. Dimensione 15mm diametro esterno.

46. Low pressure polypropylene, polythene or UPVC (as appropriate) pipes for cold water distribution. With plastic compression fittings 20mm external diameter, laid in trenches.

46. Tuyau d'alimentation en polyéthylène, posé en tranchée. Diamètre extérieur de 20mm (ou dimension du commerce la plus proche).

46. Tubazione di UPVC, polipropilene o polietilene a bassa pressione per distribuzione di acqua fredda, posata in trincea, completa di giunti in plastica a compressione, manicotti, raccordi ecc. Dimensione 20mm diametro esterno.

Installationen
(Die Dimensionen der
Sanitäranlagen und Leitungen
gelten als Richtlinien)

Instalaciones sanitarias
(Las medidas de las
instalaciones sanitarias y
cañerías son indicativas)

Plumbing
(Sizes of sanitary installations
and pipes are indicative)

42. Halbrunde Dachrinne aus
Hart-PVC. In Entfernungen
von 1.00m an Weichholz
geschraubt.
Außendurchmesser 110mm.
(Ohne Bögen, Auslauf usw.).

42. Canalón semicircular de
UPVC, atornillado a
madera cada 1m,
diámetro exterior de
110mm (excluyendo
curvas salidas).

42. UPVC half round eaves
gutter. Screwed to
softwood at 1.00m
centres. 110mm external
diameter (excluding
bends, outlets etc.).

43. Regenfallrohr aus Hart-PVC,
mit Einsteckverbindung. In
Entfernungen von 1.50m an
Mauerwerk geschraubt.
Außendurchmesser 100mm.
(Ohne Bögen, Auslauf usw.).

43. Canalones de UPVC de
recogida de lluvias con
juntas. Atornillado a
muro de ladrillo cada
1,50m. Diámetro exterior
de 100mm (excluyendo
curvas y salidas).

43. UPVC rainwater pipes
with pushfit joints.
Screwed to brickwork at
1.50m centres. 100mm
external diameter
(excluding bends, outlets
etc.).

44. Leichtkupferrohre für
Kaltwasser, mit
Verbindungsstücken. In
Entfernungen von 1.00m
horizontal an Mauerwerk
geschraubt.
Außendurchmesser 15mm.

44. Tubos de cobre para
agua fría con juntas de
unión a compresión.
Atornillados a muro de
ladrillo horizontalmente
cada 1m, diámetro
exterior de 15mm.

44. Light gauge copper cold
water tubing with com-
pression or capillary
fittings. Screwed to
brickwork horizontally at
1.00m centres. 15mm
external diameter.

45. Hochdruckrohre aus
Polypropylene, Polyäthylen
oder Hart-PVC (wie
angemessen) für Kaltwasser-
versorgung. In Entfernungen
von 1.00m horizontal an
Mauerwerk befestigt.
Außendurchmesser 15mm.
Mit Formstücken.

45. Conducto de
polipropileno de alta
presión, polietileno o
PVC, para suministro de
agua fría, fijado
horizontalmente en
paredes de ladrillo cada
1m, de 15mm de
diámetro exterior,
acabado incluso
complementos.

45. High pressure
polypropylene,
polythene or UPVC (as
appropriate) pipes for
cold water supply. Fixed
horizontally to brick walls
at 1.00m centres. 15mm
external diameter,
complete with fittings.

46. Niederdruckrohre aus
Polypropylen, Polyäthylen,
oder Hart-PVC (wie benötigt)
für Kaltwasserverteilung. Mit
Kunststoffverbindungs-
stücken. Außendurchmesser
20mm. In Gräben verlegt.

46. Conducto de
polipropileno de baja
presión, polietileno o
PVC, para distribución de
agua fría, 20mm de
diámetro exterior con
juntas de presión
plástica, colocado por
capas.

46. Low pressure
polypropylene,
polythene or UPVC (as
appropriate) pipes for
cold water distribution.
With plastic compression
fittings 20mm external
diameter, laid in
trenches.

47. UPVC soil and vent pipes with solvent welded or ring seal joints. Fixed vertically to brickwork with brackets at 1.50m centres. 100mm external diameter.

47. Tuyau en PVC pour canalisation d'eau usée, y compris joints et fixation à la maçonnerie par anneaux tous les 1,50m. Diamètre extérieur de 100mm (ou dimension du commerce la plus proche).

47. Tubi per fognatura in UPVC, con giunti saldati o con anelli di gomma, fissati verticalmente alla muratura ogni metro di lunghezza. Dimensione 100mm diametro esterno.

48. White vitreous china WC suite with black plastic seat and cover and plastic low level cistern, 9 litre capacity. Complete with ball valve and float and flush pipe to WC suite. Fixed to concrete.

48. WC en porcelaine vitrifiée de couleur blanche avec abattant et couvercle en PVC. Réservoir de chasse d'eau de 9 litres. Y compris mécanisme de chasse d'eau et robinet d'arrêt. Fixation à la maçonnerie.

48. Vaso igienico bianco con sedile in plastica nera con coperchio, casetta di 9 litri capacità, valvola, galleggiante, tubo di collegamento con rosone ecc, fissato al pavimento di cemento.

49. White vitreous china lavatory basin with 2 No. chrome plated taps (or medium quality chrome plated mixer taps). Including plug, over-flow and waste connections (excluding trap). Size approximately 560 x 400mm, fixed to brickwork with concealed brackets.

49. Lavabo en porcelaine vitrifiée de couleur blanche avec 2 robinets chromés (ou un robinet mélangeur standard) avec bonde et toutes fournitures nécessaires. Dimensions approximatives 560 x 400mm. Fixation à la maçonnerie.

49. Lavabo in porcellana dura vetrochina da 560 x 400mm, compresi due rubinetti cromati, (o miscellatore), piletta, tappo, catenella ed allacciamenti (escluso sifone), fissato alla muratura con mensole.

50. Glazed fireclay shower tray. Including overflow and waste (excluding trap). Size approximately 750 x 750 x 175mm, fixed to concrete.

50. Receveur de douche en porcelaine vitrifiée de couleur blanche. Robinetterie exclue. Dimensions approximatives 750 x 750 x 175mm. Fixation à la maçonnerie.

50. Piatto doccia in maiolica vetrificata bianca da 750 x 750 x 175mm, compreso piletta e griglia (escluso sifone), fissato al pavimento di cemento.

51. Stainless steel single bowl sink and double drainer (excluding taps). Including plug, overflow and connections (excluding trap). Size approximately 1500 x 600mm, fixed to softwood sink unit (excluding sink base).

51. Eviers en acier inoxydable comprenant une cuve, deux égouttoirs et le vidage. Robinetterie exclue. Dimensions approximatives 1500 x 600mm. Fixation sur meuble bas, ce dernier étant exclu du prix.

51. Lavello in acciaio inox con vasca singola e doppia scolatura, da 1500 x 600mm, da montare su base di legno. Compreso piletta, tappo ed allacciamenti, escluso rubinetti, sifone e base.

47. Lüftungs und Abwasserrohre aus Hart-PVC, mit Quellschweiß oder Dichtungsringvergindungen. In Entfernungen von 1.50m mit Rohrhaltern vertikal am Mauerwerk befestigt. Außendurchmesser 100mm.

47. Conducto de ventilación con soldadura disuelta o juntas de cierre de anillo. Fijados verticalmente a muro de ladrillos con soportes cada 1,5m, de diámetro exterior de 100mm.

47. UPVC soil and vent pipes with solvent welded or ring seal joints. Fixed vertically to brickwork with brackets at 1.50m centres. 100mm external diameter.

48. Klosettanlage aus weißem Kristallporzellan mit Brille und Deckel aus schwarzem Kunststoff und Tiefspülkasten mit neun Litern Kapazität. Mit Schwimmerventil und Spülrohr ausgestattet. An Betonunterlage befestigt.

48. WC de porcelana vitrificada blanca con asiento y tapa de plástico negro, cisterna de plástico de nivel bajo, (capacidad 9 litros) completado con válvula esférica, flotador y tubo para el suministro de agua al WC. Fijado a hormigón.

48. White vitreous china WC suite with black plastic seat and cover and plastic low level cistern, 9 litre capacity. Complete with ball valve and float and flush pipe to WC suite. Fixed to concrete.

49. Waschbecken für WC aus weißem Kristallporzellan mit zwei verchromten Wasserhähnen (bzw. einer verchromten Mischbatterie mittlerer Qualität). Mit Stöpsel, Überlaufloch und Anschluß an Abwasserleitung (ohne Siphon). Ungefähre Dimensionen 560 x 400mm, mit verdeckten Konsolen an Mauerwerk befestigt.

49. Lavabo de porcelana vitrificada blanca con 2 grifos cromados, calidad media. Incluyendo tapón, rebosadero y desagüe (excluyendo colector), de medidas aproximadas de 560 x 400mm fijado al ladrillo con soporte oculto.

49. White vitreous china lavatory basin with 2 No. chrome plated taps (or medium quality chrome plated mixer taps). Including plug, over-flow and waste connections (excluding trap). Size approximately 560 x 400mm, fixed to brickwork with concealed brackets.

50. Keramik Duschwanne. Mit Überlaufloch und Anschluß an Abwasserleitung (ohne Siphon). Ungefähre Dimensionen 750 x 750 x 175mm, an Betonunterlage befestigt.

50. Plato de ducha de cerámica vitrificada. Incluyendo rebosadero de agua y desagüe (excluyendo colector). Medidas aproximadas de 750 x 750 x 175mm fijado a hormigón.

50. Glazed fireclay shower tray. Including overflow and waste (excluding trap). Size approximately 750 x 750 x 175mm, fixed to concrete.

51. Spültisch aus Edelstahl mit einem Becken und zwei Ablaufplatten (ohne Wasserhähne). Mit Stöpsel, Überlaufloch und Anschlüssen (ohne Siphon). Ungefähre Dimensionen 1500 x 600mm, an Weicholzunterteil befestigt (Unterteil nicht enthalten).

51. Fregadero de acero inoxidable de un seno y dos escurridores. Incluyendo rebosadero de agua, tapón y desagüe (excluyendo colector). Medidas aproximadas de 1500 x 600mm, fijado a soporte de madera (excluyendo el soporte).

51. Stainless steel single bowl sink and double drainer (excluding taps). Including plug, overflow and connections (excluding trap). Size approximately 1500 x 600mm, fixed to softwood sink unit (excluding sink base).

Electrical work

52. PVC insulated and copper sheathed cable, 450/750 volt grade, twin core and ECC 6mm^2 cross section area. Fixed to timber with clips.

53. 13 amp, 2 gang flush mounted white, unswitched socket outlet. Including 6.0m of 2.5mm^2 concealed PVC insulated copper cable (excluding conduit). Flush mounted to brickwork including all fittings and fixing as necessary.

54. Flush mounted 20 amp, 2 gang, 1 way white light switch. Including 6.0m of 1.5mm^2 concealed mineral insulated copper cable (excluding conduit). Flush mounted to brickwork including all fittings and fixings as necessary.

Electricité

52. Câble flexible sous gaine PVC et renforcée, capacité de 450 à 750 Volts. Section 3 x 6mm². Fixation par cavaliers sur structure en bois. Conduit exclu.

53. Double prise de courant de 13 ampères, pose encastrée y compris 6m de câble en cuivre de 2,50mm² sous gaine PVC. Fixation sur maçonnerie, y compris tous systèmes de fixation. Conduit exclu.

54. Foyer lumineux simple de 20 ampères avec un interrupteur (installation encastrée). Y compris 6m de câble en cuivre de 1,5mm² sous gaine. Fixation sur maçonnerie, y compris tous systèmes de fixation. Conduit exclu.

Impianto ellettrico

52. Cavo bipolare in rame isolato con PVC, di tipo 450/750 volt con conduttore di terra, 6mm^2 sezione, fissato su legno con graffette.

53. Doppio presa di corrente ad incasso 13 amp, montata al muro. Sono compresi 6m cavo di rame 2,5mm^2 di sezione isolato in PVC, scatole, plache, piastre ecc. Esclusa canalizazzione.

54. Doppio interrutore ad incasso ad azione singola, montata al muro. Sono compresi 6m di cavo di rame 1,5mm^2 sezione con isolamento minerale, scatola e tutti gli accessori necessari. Esclusa canalizazzione.

Finishings

55. 2 coats gypsum based plaster on brick walls 13mm thick. Floated finish.

56. White glazed tiles on plaster walls size 100 x 100 x 4mm. Fixed with adhesive and grouted between tiles.

57. Red clay quarry tiles on concrete floors size 150 x 150 x 16mm. Bedded and jointed in mortar.

Revêtements

55. Enduit en plâtre de 13mm en deux passes, sur maçonnerie de brique. Finition lisse.

56. Faïence murale de couleur blanche posée à la colle sur un support en plâtre lisse. Joints remplis. Dimensions 100 x 100 x 4mm.

57. Carreaux en terre cuite rouge posés à bain de mortier sur sol en béton. Joints remplis. Dimensions 150 x 150 x 16mm .

Finiture

55. Due strati di intonaco a base di gesso, applicato sul muro di mattoni, 13mm di spessore, finito a fratazzo.

56. Piastrelle di maiolica bianche, 100 x 100 x 4mm, fissate con colla alle superfici di muri intonacati, compresa la sigillatura dei giunti con cemento bianco.

57. Piastrelle di gres rosso, 150 x 150 x 16mm, fissate e sigillate con malta di cemento al pavimento.

Elektroanlagen

52. PVC-isoliertes Kupferkabel, Klasse 450/750V, zweiadrig mit Erdschutzleiter, Querschnitt 6mm². Mit Klemmen an Holz befestigt.

53. Zweifache weiße Steckdose ohne Schalter, 13A. Mit 6.0m verdecktem PVC-isoliertem Kupferkabel, 2.5mm² (ohne Kabelrohr). Putzbündig an Mauerwerk versetzt, mit allen anfallenden Nebenarbeiten.

54. Weißer Lichtschalter mit zwei Serienschaltern, 20A. Mit 6.0m verdecktem mineralisoliertem Kupferkabel, 1.5mm², (ohne Kabelrohr). Putzbündig an Mauerwerk versetzt, mit allen anfallenden Nebenarbeiten.

Verkleidungen

55. Zwei Lagen Gipsputz auf Ziegelmauerwerk, Dicke 13mm, abgeglättet.

56. Weiße Glasurplatten auf Gipswänden, Dimensionen 100 x 100 x 4mm, mit Kleber befestigt. Mit Fugenausfüllung.

57. Rote Keramikplatten auf Betonböden, Dimensionen 150 x 150 x 16mm. In Mörtel eingebettet, mit Fugenausfüllung.

Instalaciones electricas

52. Cable de 450/750v., con dos conductos y 6mm² de sección y tubo de PVC fijado a la pared con grapas.

53. Batería de dos cajas de derivación de 13 amp con toma de corriente desconectada, de color blanco, incluyendo 6m de cable de cobre de 2,5mm², protegido con PVC (excluyendo el tubo protector). Colocado sobre pared, incluidos los accesorios y elementos de fijación necesarios.

54. Batería de dos interruptores empotrados de 20 amperios de color blanco de un solo encendido. Incluyendo 6m de cable de cobre de 1,5mm² protegido con tubo (se excluye este). Colocado sobre pared incluídos todos los accesorios y elementos de fijación necesarios.

Acabados

55. Revoque y enlucido con yeso en paredes de ladrillo, acabado a "buena vista".

56. Alicatado en paredes con cerámica vidriada de 100 x 100 x 4mm. Tomado con cemento cola, incluso la borada de las juntas.

57. Solado con tobas de 150 x 150 x 16mm sobre suelo de hormigón, tomado con mortero de c.p., incluso borada.

Electrical work

52. PVC insulated and copper sheathed cable, 450/750 volt grade, twin core and ECC 6mm² cross section area. Fixed to timber with clips.

53. 13 amp, 2 gang flush mounted white, unswitched socket outlet. Including 6.0m of 2.5mm² concealed PVC insulated copper cable (excluding conduit). Flush mounted to brickwork including all fittings and fixing as necessary.

54. Flush mounted 20 amp, 2 gang, 1 way white light switch. Including 6.0m of 1.5mm² concealed mineral insulated copper cable (excluding conduit). Flush mounted to brickwork including all fittings and fixings as necessary.

Finishings

55. 2 coats gypsum based plaster on brick walls 13mm thick. Floated finish.

56. White glazed tiles on plaster walls size 100 x 100 x 4mm. Fixed with adhesive and grouted between tiles.

57. Red clay quarry tiles on concrete floors size 150 x 150 x 16mm. Bedded and jointed in mortar.

58. Floor screed. Cement and sand screed to concrete floors 1:3 mix. 50mm thick. Floated finish.

58. Chape en mortier de sable et ciment (dosage 500 kg). Epaisseur 5cm et finition lissée.

58. Massetto di cemento e sabbia (1:3), 50mm di spessore, gettato sul pavimento di calcestruzzo e finito liscio.

59. Thermoplastic floor tiles on screed 2.5mm thick. Fixed with adhesive.

59. Dalles thermoplastiques de 2,5mm d'épaisseur pose collée sur chape.

59. Pavimentazione in piastrelle termoplastiche di spessore 2,5mm, fissate con colla.

60. Suspended ceiling system. Fissured mineral fibre tiles size 300 x 300 x 15mm. On galvanized steel concealed suspension system. Fixed to concrete soffits with 500mm drop (excluding lamp fittings).

60. Faux plafond en fibre minérale, dalles rainurées de 300 x 300mm, épaisseur 15mm. Aspect fissuré. Y compris ossature en acier galvanisé. Retombée par rapport au niveau du plafond brut : 50cm. Luminaires exclus.

60. Controsoffitto composto da un telaio in acciaio galvanizzato, sospeso 500mm dal soffitto di cemento, e piastrelle di fibra minerale ad incasso, 300 x 300 x 15mm, (escluse le plafoniere).

Glazing

Vitrerie

Opere in vetro

61. Glazing to wood. Ordinary quality 4mm glass. Softwood beads.

61. Verre simple de 4mm d'épaisseur pose sous parclose en bois sur un châssis en bois.

61. Lastre di vetro semplice 4mm di spessore, fissate a finestre di legno con listelli.

Painting

Peinture

Opere da pittore

62. Emulsion on plaster walls. One coat diluted sealer coat and 2 coats full vinyl emulsion paint.

62. Peinture murale, sur enduit en plâtre, composée d'une couche d'impression diluée et de deux couches de finition en émulsion vinylique.

62. Tinteggiatura lavabile (idropittura) con emulsione vinilica, la prima mano diluita con acqua e due mani successive non diluite.

63. Oil paint on timber. One coat primer and 2 coats oil based paint.

63. Peinture à l'huile, sur bois, composée d'une couche primaire et deux couches de finition à base d'huile.

63. Una mano di primer e due mani di smalto a base d'olio su superfici in legno.

58. Bodenestrich aus Zement und Sand (1:3) für Betonböden. Dicke 50mm, abgeglättet.

58. Recrecido de suelos con 50mm de mortero de c.p., de dosificación 1:3, acabado enlucido.

58. Floor screed. Cement and sand screed to concrete floors 1:3 mix. 50mm thick. Floated finish.

59. Thermoplastische Bodenplatten auf Estrich verklebt. Dicke 2.5mm.

59. Pavimento termoplástico en piezas de 2,5mm de espesor, fijado con adhesivo sobre embaldosado o capa de hormigón.

59. Thermoplastic floor tiles on screed 2.5mm thick. Fixed with adhesive.

60. Hängedecke aus Mineralfaserplatten, Dimensionen 300 x 300 x 15mm. 500mm Abhänghöhe, verdeckte Aufhängevorrichtung aus verzinktem Stahl, an Betondecke befestigt. (Ohne Leuchten).

60. Falso techo formado por piezas de fibra mineral de 300 x 300 x 15mm sobre estructura de acero galvanizado fijado al techo con varillas de 500mm (no se incluyen los accesorios de iluminación).

60. Suspended ceiling system. Fissured mineral fibre tiles size 300 x 300 x 15mm. On galvanized steel concealed suspension system. Fixed to concrete soffits with 500mm drop (excluding lamp fittings).

Verglasung

Vidrio cristanina

Glazing

61. Verglasung für Holzelemente. Glas normaler Qualität, Dicke 4mm. Weichholzglasleisten.

61. Vidrio cristanina de 4mm colocado sobre madera.

61. Glazing to wood. Ordinary quality 4mm glass. Softwood beads.

Malerarbeiten

Pintura

Painting

62. Emulsionsfarbe auf Putzwänden. Ein verdünnter Grundanstrich und zwei Anstriche Voll.

62. Pintura de paredes enyesadas, con una capa de imprimación y dos de pintura vinílica.

62. Emulsion on plaster walls. One coat diluted sealer coat and 2 coats full vinyl emulsion paint.

63. Ölfarbe auf Holz. Ein Grundanstrich und zwei Anstriche Ölfarbe.

63. Pintura de aceite sobre madera con una capa de imprimación y dos de aceite.

63. Oil paint on timber. One coat primer and 2 coats oil based paint.

Appendix four: Statistical notes

Statistical notes

Introduction

Statistics are the results of attempts to present diverse activities of the real world in a numerical format, which purports to represent accurately the totality of various aspects of the economy. As it is not possible to add up varied units of physical production in a meaningful way, these physical units have to be expressed in money terms. The process involves the solving of a number of problems and results in an approximation of the real situation. This is the case with even the most sophisticated statistical methods. When the data are inadequate and the methodology poor, the statistics become subject to very wide margins of error.

The problem is compounded in attempting to produce international comparisons because it is necessary to convert domestic data into some common unit of measurement – normally using a rate of exchange – which is often inadequate for the purpose.

Thus virtually every figure in this volume is in some way an estimate. The figures for certain countries are more dubious than others, notably those for Eastern and Central Europe, and Greece and Turkey. Caveats on the reliability of the data have not usually been included in the text because it would be repetitive to do so, but the likely margins of error must be borne in mind.

Nevertheless, the authors believe the data presented in this volume is the best available. The sources and methods used to arrive at them are discussed below.

Gross domestic product, exchange rates and PPP methodology

For most countries, data are given for gross domestic product (GDP) and are generally quoted at market prices. Gross domestic product refers to the total output of goods and services produced in the country. There are now more sources of data for Eastern and Central Europe than hitherto. There are estimates based on money values in local currencies converted at exchange rates; others are based on purchasing power parity (PPP) methodology. The global sources do not always tally with country sources. In general, global sources have been used for greater comparability.

Conversion of GDP figures at the exchange rate in use often produces unrealistic levels of output. Exchange rates are determined by international trade, by movements of money and by levels of confidence in the economy and expectations for the future. Construction is not normally traded. The problem is especially serious where the currency is not convertible for all transactions so that the exchange rate becomes particularly unreal This is the case in states of the CIS. Trade between them and, to some extent, with other countries is still small and payment is sometimes made in kind rather than money. Attempts have been made in Part One and in the country sections to deal with this problem by using data on GDP on a purchasing power parity basis from the CIA *World Factbook*. This has the effect of making the data more or less comparable across Europe.

For Western European countries the difference between money and PPP estimates is not very great. However, for countries which have had very high inflation and a partially non-convertible currency the differences can be very great, especially for the countries of Eastern and Central Europe.

Construction output, exchange rates and PPP methodology

Construction output is the gross value of output.

As in the case of GDP, an attempt has been made to reduce the difficulty that the exchange rate is not satisfactory in converting output from one currency to a common currency. In table 1.3 construction output is quoted on PPP basis as well as on an exchange rate basis. The PPP value is based on all goods and services not just on construction output. The ratio of the exchange rate basis to the PPP basis is the same as that used for GDP. In fact, however, the rates of the purchasing power of a US dollar compared to the exchange rate calculation for all goods and services may be very different from that for construction output and the difference will vary for each country. The use of a PPP basis based on GDP calculations for construction is of limited value but it has been included because it is probably a nearer representation of 'truth' than exchange rates.

Definitions of construction output

The broad definition of the construction industry is similar for most Western European countries but in terms of detail, it can vary considerably. The data that are available and the work which is included are certainly far from uniform. The data on value of construction output normally refer only to work undertaken by the construction industry and the figure of GDP is usually based only on construction industry output.

Because of the harmonization of the European System of National Accounts, some of the data for EU countries for 1998 are estimates only. For Eastern and Central Europe countries, there is often greater uncertainty as to what is included because it is difficult to establish how far they have changed from the usual procedures of the FSU.

Installation of plant and equipment

For Eastern and Central Europe it is often unclear whether installation of plant and equipment in new industrial buildings is included in construction. In the FSU they were included and still are in Russia. Other countries may be moving towards the Western system of keeping buildings and industrial plant and equipment separate.

Renovation, rehabilitation, repair and maintenance

These terms are used without any clear definitions. In the UK statistics, new work includes extensions, major alterations and improvements. No clarification on these matters has been found in the EC *General Industrial Classification* (NACE) except for housing where work done on improvements, extensions and alterations and house/flat conversions is included under repair and maintenance. Statistical tables produced by Euroconstruct, the forecasting group, refer to renovation but generally include repair and maintenance. In this volume, renovation, repair and maintenance are used loosely to refer to all the types of work above, including refurbishment, where these data are separately available.

Most countries collect data on new work and repair and maintenance but in some cases, for example the USA, no statistics are available for repair and maintenance. In Germany regular repair and maintenance is excluded but rehabilitation is included. In others there are statistics of 'construction output' for example in Turkey, but it is not known whether or not they includes an allowance for repair and maintenance. This last problem generally arises in countries where the data are not very firmly based for other reasons. Where the information is available, comment on what is included is given in the country sections.

Construction by other sectors, the black market and DIY

Although the data on value of construction output normally refer only to work undertaken by the construction industry, there are some exceptions. In particular in Finland all construction is covered in 'construction output including work by other sectors and an estimate of DIY'. No split is possible to separate them out. Euroconstruct generally provides estimates of the construction work carried out, which is not included in the official output of the sector. Estimates must by the nature of the work be approximate. The work by the black economy covers work which is undertaken illegally in some way and therefore will be deliberately hidden. Construction work by other sectors is rarely separated in national statistics and DIY, expenditure if known, is difficult to translate to value of construction output. The estimates of these types of output given vary from 11% to 34% of official output of the industry in Western Europe and from 22%

to 54% in Central Europe. These figures are not inconsistent with estimates of the total black underground economy in various countries made by Professor Friedrich Schneider for *The Economist.*

The information in the country sections is as specific as possible on the magnitude of 'unrecorded' construction work.

Military and defence work

Construction expenditure on defence is generally not included in the statistics, though it may be substantial. Where it is included (if at all) it may distort the figures because it may be included in whole or in part in some, but not all, countries.

Taxes

The value of output data in the country sections for Western European countries excludes taxes unless otherwise specified. For Greece, Turkey and Cyprus it is not known whether taxes are included or not.

Data collection

As important as homogeneity in definitions is the efficiency of the data collection process. Because the construction industry in Western Europe has a very large number of small firms and a large proportion of its output is in small projects spread geographically, it is very difficult to collect comprehensive data on the output of the industry, no matter whether it is collected from clients, from approvals, or from contractors. In Eastern and Central Europe the system for the collection of statistics from state organizations, which was probably reasonably efficient, is breaking down with the increasing growth of smaller firms and with privatization. Most of these countries have no mechanism for collecting data from very small firms and no power to enforce completion of returns.

Western Europe

Statistics for countries in the European Union, with the exception of Greece, and for the other countries of Western Europe except Cyprus and Turkey are based mainly on Euroconstruct data. The advantage of Euroconstruct data is that they are up-to-date and are provided by organizations within each country which have good access to the available data on that country.

Statistics for Greece, Cyprus, and Turkey are estimated by the authors using a number of sources including Economist Intelligence Unit Reports and national statistics data. All the data suffer from the problems of definition and coverage referred to above.

Eastern and Central Europe

There are now more sources of data for Eastern and Central Europe but there is very little agreement among them. The estimates of GDP per capita, for example, include those published by the International Monetary Fund, in *International Financial Statistics;* those by the Economist Intelligence Unit in their *Country Reports*; and those by Euroconstruct for a few countries and some national statistics. Some of these estimates are based on PPPs in US dollars, others on exchange rates, some on a mixture of PPPs and exchange rates and some on adjusted exchange rates. The results are so diverse as to make the estimates of doubtful value. Different sources even yield a different order of ranking of countries by value of output. There are very few published estimates of construction output for Eastern and Central Europe. An exception is Euroconstruct for Poland, Hungary, the Czech Republic and the Slovak Republic. The Economist Intelligence Unit gives data on the percentage of net value-added construction output of GDP for a few countries and also gross domestic fixed capital formation.

Presentation and rounding of figures

There are some comments which refer to all tables but have not been included in footnotes to avoid unnecessary repetition. Because the figures are not accurate beyond about three digits and in many cases not even that, figures have been rounded. Generally totals and percentages are based on unrounded figures. Therefore items do not always sum exactly to the total and percentages may be marginally different from the percentages of the rounded numbers.

Currencies and exchange rates

In the country sections values are given in local currencies. In each case the value of construction output is converted to ECUs at mid 1998 values based on *The Financial Times* or other published sources. The ECU came into force with the European Monetary System (EMS) on 13 March 1979 as the unit of account for the European Union. It is a 'basket' of currencies of the Member States, each currency contributing a fixed percentage of the ECU's value.

The Euro superseded the ECU on the 1ˢᵗ January 1999. 1998 figures in this book are often expressed in ECUs as well as local currencies. The 1998 ECU is, for most countries, very similar to the value of the Euro in January 1999 but if for any country the difference is greater than 1%, that is stated.

The cost and price data for local construction activity have been given in the currencies of each country because, until the Euro replaces the individual currencies for domestic purposes that is the currency in which transactions will take place.

Different rates of internal inflation affect the relative values of currencies and, therefore, the rates of exchange between them. However, the reasons behind exchange rate fluctuations are complex and often political as much as economic; they include such factors as interest rates, balance of payments, trade figures and, of course, government intervention in the foreign exchange markets, and, for that matter, other action by governments.

Appendix five: Davis Langdon & Everest

Davis Langdon & Everest

Structure, resources and locations

Davis Langdon & Everest (DLE) is an independent firm of Quantity Surveyors and Construction Consultants, providing managed solutions for clients with construction and property requirement. DLE is the European component of Davis Langdon & Seah International (DLSI). DLSI operates worldwide, with offices throughout the UK, mainland Europe, Middle East, Australasia and Asia, Africa and America; with associated offices in Canada, Ireland and West Africa. The firm employs some 2,200 staff worldwide and is managed by over 100 Partners, located in 80 offices, in 25 countries around the world.

Europe
- United Kingdom   : Davis Langdon & Everest
                              Davis Langdon Consultancy
- Poland   : Davis Langdon Polska
- Spain   : Davis Langdon Edetco
- France   : Davis Langdon Economistes
- Czech Republic   : Davis Langdon Cesko*Slovensko

Middle East
- Bahrain   : Davis Langdon Arabian Gulf
- Qatar   : Davis Langdon Arabian Gulf
- United Arab Emirates   : Davis Langdon Arabian Gulf
- Lebanon   : Davis Langdon Lebanon

Asia :
- Singapore   : Davis Langdon & Seah Singapore
- Hong Kong   : Davis Langdon & Seah Hong Kong
- Malaysia   : Davis Langdon & Seah Malaysia
- Brunei   : Davis Langdon & Seah Brunei
- Philippines   : Davis Langdon & Seah Philippines
- Thailand   : Davis Langdon & Seah Thailand
- Vietnam   : Davis Langdon & Seah Vietnam
- China   : Davis Langdon & Seah China

Australasia:
- Australia   : Davis Langdon Australia
- New Zealand   : Davis Langdon Knapman Clark

America   : Davis Langdon Adamson

Africa   : Davis Langdon Africa

Professional Services and Market Sectors

DLE specialises in the financial management of construction projects, from inception to completion, and the firm's range of services includes:

- Feasibility Studies
- Cost Planning
- Construction Cost Management
- International Procurement
- Tender and Contract Documentation
- Project Management
- Claims Negotiation/ Resistance
- Development Economics and Appraisals
- Risk Analysis and Management
- Value Analysis and Management (whole life costings)
- Project Audit (due diligence & forensic)
- Construction Litigation Services
- Construction Insolvency Services
- Construction Industry Research
- Consultancy and Best Practice guidance

The firm's experience extends to managing a wide range of construction projects including:

- Airports and Airport Buildings
- Arts and Cultural Buildings
- Bank Buildings
- Business Park Developments
- Cinemas and Theatres
- Civic Buildings
- Civil Engineering Projects
- Distribution Centres
- Contaminated land remediation
- Educational Buildings
- Factories and Industrial Buildings
- Headquarters Buildings
- Health and Welfare Buildings
- Hotels
- Housing and Residential
- Manufacturing/Research Including Process, Food and Drink, Packaging
- Office Developments and Fit-outs
- Petro-chemical Projects
- Pharmaceutical Laboratories
- Power Stations
- Shopping Centres and Retail
- Road and Rail Terminals
- Sports and Leisure Centres
- Transportation
- Water and Waste Projects
- Zoological Gardens

Practice Statement

Davis Langdon & Everest (DLE) is an independent firm of Chartered Quantity Surveyors & Construction Consultants, which is committed to representing, protecting and ensuring client interests on construction projects, large or small, locally, nationally or internationally.

The strategic and integrated management of cost, time and quality – the client's 'risk' areas of a contract – are essential functions, which are necessary to ensure the satisfactory planning, procurement, implementation and operation of construction projects.

DLE specialises in the costs and time management of construction projects, their risk and value aspects, from project inception to completion.

The firm employs highly qualified and skilled professional staff, with specialist experience in public and private domains of the construction industry and specialist experience in all sectors, buiding types etc.

It operates a sophisticated information support system, based on the latest computer technology, enabling large-scale capture and retrieval of construction cost and relevant market data.

The highest operational standards are observed to ensure quality of product and the firm has Quality Assurance programmes in respect of its services in those countries where formal accreditation is available.

DLE draws upon its international network of offices, but works in manageable teams under direct Partner leadership and maintains personal client contact at all stages of a project.

The firm concentrates on:

- being positive and creative in its advice, rather than simply reactive;
- providing value for money via efficient management, rather than by superficial cost monitoring;
- giving advice that is matched to the Client's specific requirements, rather than presenting standard or traditional solutions;
- paying attention to the whole life cycle costs of constructing and occupying a building, rather than to the initial capital cost only.

The overall objective of the firm is to control cost, limit risk and add value for its clients.

Davis Langdon Consultancy

In addition to the financial management of construction projects, DLE, through Davis Langdon Consultancy (DLC), also undertakes varied construction industry research and consultancy assignments, worldwide, ensuring a broadly based and truly international knowledge-based service for their clients.

DLC contributes much to the firm's international experience, via research and information, which is distilled into the strategic advice and services provided to DLE's individual clients and which also provides essential data for publications, such as this current *European Construction Costs Handbook*.

DAVIS LANGDON & SEAH

Davis Langdon & Seah manages client requirements, controls risk, manages cost and maximises value for money, throughout the course of construction projects, always aiming to be – and to deliver – the best.

TYPICAL PROJECT STAGES, INTEGRATED SERVICES AND THEIR EFFECT

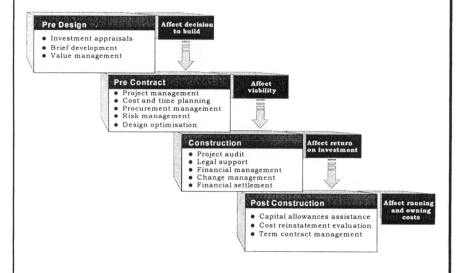

ASIA ⟷ AUSTRALASIA ⟷ EUROPE ⟷ AFRICA ⟷ AMERICA

GLOBAL REACH ⇨ *LOCAL DELIVERY*

DAVIS LANGDON & SEAH INTERNATIONAL
www.davislangdon.com

DAVIS LANGDON & EVEREST
DAVIS LANGDON CONSULTANCY
DAVIS LANGDON MANAGEMENT

Davis Langdon & Everest manages client requirements, controls risk, manages cost and maximises value for money, throughout the course of construction projects, always aiming to be – and to deliver – the best.

TYPICAL PROJECT STAGES, INTEGRATED SERVICES AND THEIR EFFECT

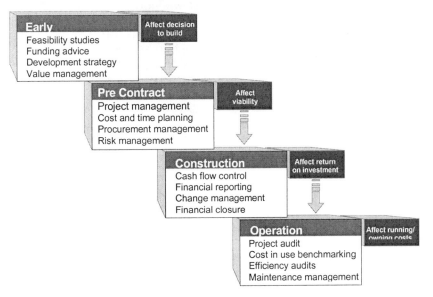

UK: London, Birmingham, Bristol, Cambridge, Cardiff, Edinburgh, Gateshead, Glasgow, Leeds, Liverpool, Manchester, Milton Keynes, Norwich, Oxford, Peterborough, Plymouth, Portsmouth, Southampton

France: Paris **Spain:** Girona, Barcelona **Czech Republic:** Prague

Bahrain: Manama **Qatar:** Doha **United Arab Emirates:** Dubai

Lebanon: Beirut **Saudi Arabia:** Jeddah

and in

ASIA ⇔ AUSTRALASIA ⇔ AFRICA ⇔ AMERICA

which together form

DAVIS LANGDON & SEAH INTERNATIONAL
www.davislangdon.com

GLOBAL REACH ⇨ ***LOCAL DELIVERY***

DAVIS LANGDON & EVEREST

www.davislangdon-uk.com

OFFICES AND CONTACTS

UNITED KINGDOM

Central Contacts: Paul Morrell or Alastair Collins - London

LONDON
Princes House
39 Kingsway
London WC2B 6TP
Tel: 0171-497 9000
Fax: 0171-497 8858
Contact: Rob Smith

BRISTOL
St Lawrence House
29/31 Broad Street
Bristol BS1 2HF
Tel: 0117-9277832
Fax: 0117-9251350
Contact: Alan Trolley

CAMBRIDGE
36 Storey's Way
Cambridge CB3 ODT
Tel: 01223-351258
Fax: 01223-321002
Contact: Stephen Bugg

CARDIFF
4 Pierhead Street
Capital Waterside
Cardiff CF1 5QP
Tel: 01222-497497
Fax: 01222-479111
Contact: Paul Edwards

EDINBURGH
74 Great King Street
Edinburgh EH3 6QU
Tel: 0131-557 5306
Fax: 0131-557 5704
Contact: Ian McAndie

GATESHEAD
11 Regent Terrace
Gateshead
Tyne and Wear NE8 1LU
Tel: 0191-477 3844
Fax: 0191-490 1742
Contact: Peter Millidge

GLASGOW
Cumbrae House
15 Carlton Court
Glasgow G5 9JP
Tel: 0141-429 6677
Fax: 0141-429 2255
Contact: Hugh Fisher

LEEDS
Duncan House
14 Duncan Street
Leeds LS1 6DL
Tel: 0113-2432481
Fax: 0113-2424601
Contact: Tony Brennan

DAVIS LANGDON & SEAH INTERNATIONAL
WWW.davislangdon.com

LIVERPOOL
Cunard Building
Water Street
Liverpool L3 1JR
Tel: 0151-236 1992
Fax: 0151-227 5401
Contact: John Davenport

MANCHESTER
Boulton House
Chorlton Street
Manchester M1 3HY
Tel: 0161-228 2011
Fax: 0161-228 6317
Contact: Stephen Frood

MILTON KEYNES
6 Bassett Court
Newport Pagnell
Buckinghamshire MK16 OJN
Tel: 01908-613777
Fax: 01908-210642
Contact: Mike Sharman

NORWICH
63A Thorpe Road
Norwich NR1 1UD
Tel: 01603-628194
Fax: 01603-615928
Contact: Mike Ladbrook

OXFORD
Avalon House
Marcham Road, Abingdon
Oxford OX14 1TZ
Tel: 01235-555025
Fax: 01235-554909
Contact: Paul Coomber

PLYMOUTH
3 Russell Court
St Andrews Street
Plymouth PL6 2AX
Tel: 01752-668372
Fax: 01752-221219
Contact: Gareth Steventon

PORTSMOUTH
Kings House, 4 Kings Road
Portsmouth
Hampshire PO5 3BQ
Tel: 01705-815218
Fax: 01705-827156
Contact: Chris Tremellin

SOUTHAMPTON
Clifford House, New Road
Southampton SO14 OAB
Tel: 01703-333438
Fax: 01703-226099
Contact: Richard Pitman

Davis Langdon Consultancy
Princes House, 39 Kingsway
London WC2B 6TP
Tel: 0171-379 3322
Fax: 0171-379 3030
Contact: Jim Meikle

DAVIS LANGDON & SEAH INTERNATIONAL
WWW.davislangdon.com

MAINLAND EUROPE

SPAIN
Central Contacts: Francesc Monells - Girona
Jon Blasby - London
DAVIS LANGDON EDETCO

GIRONA
C/Salt 10
Girona 17005 SPAIN
Tel: (00 34 97) 2238000
Fax: (00 34 97) 2242661
Contact: Francesc Monells

MADRID
C/Ferrer del Rio, 14
28028 – Madrid
Tel: (00 34 1) 3611805
Fax: (00 34 1) 3612951
Contact: Francesc Monells

BARCELONA
C/Muntaner 47912"
Bracelona – 08021
Tel: (00 34 3) 4186899
Fax: (00 34 3) 2110003
Contact: Francesc Monells

FRANCE
Central Contacts: Andrew Richardson- Paris
Jeremy Horner – London
DAVIS LANGDON ECONOMISTES

PARIS
1 Rue Edouard Colonne
Paris, 756001
France
Tel: (00 33 1) 53409480
Fax: (00 33 1) 53409481
Contact: Andrew Richardson

THE CZECH REPUBLIC
Central Contact: Alan Willis - London
DAVIS LANGDON CESKO*SLOVENSKO

PRAGUE
Londynska 28
120 00 Prague 2
Tel: (00 422) 254655
Fax: (00 422) 254590
Contact: Alan Willis

DAVIS LANGDON & SEAH INTERNATIONAL
WWW.davislangdon.com

MIDDLE EAST

Central Contacts: David Galbraith - Manama
Paul Morrell or Alistair Collins – London

DAVIS LANGDON ARABIAN GULF

BAHRAIN
PO Box 640
Manama
Tel: (00 973) 251755
Fax: (00 973) 232291
Contact: David Galbraith

UNITED ARAB EMIRATES
PO Box 7856, Diera
Dubai, UAE
Tel: (00 9714) 227424
Fax: (00 9714) 220069
Contact: Neil Taylor

QATAR
PO Box 3206
Doha
Tel: (00 974) 328440
Fax: (00 974) 437349
Contact: Geoffrey Thompson

SAUDI ARABIA

TO BE ADVISED

LEBANON
Central Contacts: Muhyiddin Itani - Beirut
Derek Johnson- London

DAVIS LANGDON LEBANON
PO Box 135422 - Shouran
Australia Street
Chatila Building
Beirut
Tel: (00 9611) 809045
Fax: (00 9611) 603104
Contact: Muhyiddin Itani

ASIA PACIFIC

Central Contacts: Seah Choo Meng - Singapore
Paul Morrell or Alastair Collins – London

DAVIS LANGDON & SEAH

BRUNEI
DAVIS LANGDON & SEAH

BANDAR SERI BEGAWAN
No. 1 First Floor
Block H
Abdul Razak Complex
Gadong
P O Box 313
Bandar Seri Begawan BS8670
Tel: (00 6732) 446888
Fax: (00 6732) 440893
Contact: Tony Teoh

KUALA BELAIT
Suite S6 Hong Kong Bank Chambers,
Jalan McKerron
PO Box 811
Kuala Belait KA1131
Tel: (00 6733) 330457
Fax: (00 6733) 335839
Contact: Adrianus Bieshaar

PETROJAYA SDN. BHD and
PETROKON UTAMA SDN. BHD
No. 3 First Floor
Block H
Abdul Razak Complex
Gadong
Bandar Seri Begawan
P O Box 1188
Bandar Seri Begawan 1911
Tel: (00 6732) 441384
Fax: (00 6732) 441382
Contact: Tony Teoh

DAVIS LANGDON & SEAH INTERNATIONAL
WWW.davislangdon.com

ASIA PACIFIC (continued)

CHINA
DAVIS LANGDON & SEAH CHINA LTD

SHANGHAI
Room 2409, 24FShartex Plaza
88 Zhun Yi Nan Road
Shanghai 200335
Tel: (00 8621) 219 1107
Fax: (00 8621) 219 3680
Contact: Joseph Lee

GUANGZHOU
Guangzhou Representative Office
Unit 04, 11/F New Century Plaza
Tel: (00 8620) 432 8565
Fax: (00 8620) 432 8567

BEIJING
Beijing Representative Office
Room 427, Swissotel Beijing
Hong Kong Macau Center
Dongsi Shitiao Lijiaoqiao
Beijing 100027
Tel: (00 861) 5013816
Fax: (00 861) 5013816

HONG KONG
DAVIS LANGDON & SEAH HONG KONG LTD
21st Floor, 2101 Leighton Centre
77 Leighton Road
Hong Kong
Tel: (00 852) 2576 3231
Fax: (00 852) 2576 0416
Contact: M L Ku

INDONESIA
DAVIS LANGDON & SEAH INDONESIA PT
Wisma Metropolitan 1, Level 13
Jalan Jendral Sudirman
Kav. 29, PO Box 3139/Jkt
Jakarta 10001
Tel: (00 6221) 5254745
Fax: (00 6221) 5254764
Contact: Ian Reynolds

MALAYSIA
DAVIS LANGDON & SEAH MALAYSIA

KUALA LUMPUR
124 Jalan Kasah
Damansara Heights
50490 Kuala Lumpur
Tel: (00 603) 254 3411
Fax: (00 603) 255 9660
Contact:

KOTA KINABALU
Suite 8A, 8th Floor
Wisma Pendidikan
Jalan Padang, PO Box 11598
88817 Kota Kinabalu
Tel: (00 6088) 223369
Fax: (00 6088) 216537
Contact: P H Hey

KUCHING
2nd Floor, Lot 142
Bangunan WSK
Jalan Abell
93100 Kuching
Sarawak
Tel: (00 6082) 417357
Fax: (00 6082) 426416
Contact: H O Wong

PENANG
No 29 (1ˢᵗ floor) Lebuh Pantai
Victoria
10300 Pulau Penang
Tel: (00 604) 2642071
Fax: (00 604) 2642068
Contact: David Yap

JOHOR BAHRU
49-01 Jalan Tun Adbul Razat
Susur 1/1 Medan Cahaya
80000 Johor Bahru
Tel: (00 607) 2236229
Fax: (00 607) 2235975
Contact: E Amer

DAVIS LANGDON & SEAH INTERNATIONAL
WWW.davislangdon.com

ASIA PACIFIC (continued)

PHILIPPINES
DAVIS LANGDON & SEAH PHILIPPINES 4th
Floor, Kings Court 1 Building
2129 Pasong Tamo
Makati City
Metro Manila
Tel: (00 632) 8112971
Fax: (00 632) 8112071
Contact: M W Anderson

SINGAPORE
DAVIS LANGDON & SEAH
1 Magazine Road
#05-01 Central Mall
059567 Singapore
Tel: (00 65) 2223888
Fax: (00 65) 5367132
Contact: Seah Choo Meng

THAILAND
DAVIS LANGDON & SEAH (THAILAND)
8th Floor
Kian Gwan Building
140 Wireless Road
Bangkok 10330
Tel: (00 662) 253 7390
Fax: (00 662) 253 4977
Contact: C P Leong

VIETNAM
DAVIS LANGDON & SEAH VIETNAM
#405 North Star Building
4 Da Tuong Street
Hoan Kiem District
Hanoi
Tel: (00 844) 240395/6
Fax: (00 844) 240394
Contact: James Milner

AUSTRALIA & NEW ZEALAND
*Central Contacts: Mark Beattie - Melbourne
Paul Morrell or Alastair Collins – London*
DAVIS LANGDON AUSTRALIA
Level 19, 350 Queen Street
Melbourne
Victoria 3000
Tel: (00 613) 99338800
Fax: (00 613) 99338807
Contact: Mark Beattie

Level 9, Suite 902
181 Miller Street, North Sydney
NSW 2060
Tel: (00 612) 956 8822
Fax: (00 612) 956 8848
Contact: Alec Horley

Level 10, 241 Adelaide Street
Brisbane, QLD 4000
Tel: (00 617) 3221 1788
Fax: (00 617) 3221 3417
Contact: Malcolm Butcher

53 Salamanca Place
Hobart, Tasmania 7000
Tel: (00 613) 62348788
Fax: (00 613) 62311429
Contact: Steve D Gay

Level 8, 251 Adelaide Terrace
Perth WA6000
Tel: (00 6189) 22218870
Fax: (00 6189) 22218871

CAIRNS
Level 2, Suite 8
101 Spencer Street
Cairns, Queensland 4870
Tel: (00 617) 40517511
Fax: (00617) 40517611

DAVIS LANGDON & SEAH INTERNATIONAL
WWW.davislangdon.com

ASIA PACIFIC (continued)

DAVIS LANGDON KNAPMAN CLACK
M.H.B. PROJECT SERVICES PTY LTD
2 Victoria Road, Hawthorn
Victoria 3122
Tel: (00 613) 882 7044
Fax: (00 613) 882 8612
Contact: J Troedel

USA OFFICES

DAVIS LANGDON ADAMSON OFFICES
Central Contacts:

SAN FRANCISCO
170 Columbus Avenue, Suite 301
San Francisco CA 94133, USA
Tel: (00 1415) 981 1004
Fax: (00 1415) 981 1419
Contact: Martin Gordon

_____

SANTA MONICA
301 Arizona Avenue
4th Floor, Santa Monica
Los Angeles, CA90401
Tel: (00 1310) 3939411
Fax: (00 1310) 3937493
Contact: Nicholas Butcher

_____

SACRAMENTO
331 "J" Street, Suite 200
Sacramento, CA 95814, USA
Tel: (00 1916) 444 5797
Fax: (00 1916) 444 5799
Contact: Peter Morris

_____

SEATTLE
1000 2nd Avenue, Suite 1770
Seattle, WA98104, USA
Tel: (00 1206) 343 8119
Fax: (00 1206) 343 8541
Contact: Nicholas Butcher

_____

SOUTH AFRICA OFFICES
DAVIS LANGDON FAARROW LAING NTENE
Central Contacts:

CAPETOWN
22 Riebeeck Street, 9th Floor
Safmarine House, Cape Town, SA
Tel: (00 2721) 254 016
Fax: (00 2721) 418 1865
Contact: Fanie du Plessis

DURBAN
Third Floor, Liberty Life House
269 Smith Street,
4001 Durban, SA
Tel: (00 27 31) 304 6441
Fax: (00 27 31) 304 6434
Contact: Roy Turner

JOHANNESBURG (PARKTOWN)
Second Floor, MPF House
32 Princess of Wales Terrace
Sunnyside Park, 2193 Parktown
Tel: (0027 11) 484 2330
Fax: (0027 11) 642 2289
Contact: Rob Black

KLERKSDORP
2nd Floor, 22 Boom Street
2570 Klerksdorp, SA
Tel: (00 27 18) 464 1641
Fax: (00 27 18) 464 1644
Contact: Bertus van Eeden

LESOTHO
2nd Floor, Ecumenical Centre
Cnr. Old School and Institution Road
Maseru, Lesotho. SA
Tel: (0027 9266) 312 241
Fax: (0027 9266) 310 123
Contact: Pusetso Makote

DAVIS LANGDON & SEAH INTERNATIONAL
WWW.davislangdon.com

SOUTH AFRICA (continued)

PRETORIA
Suite No. 3, Schoeman Street Forum
1157 Schoeman Street
0028 Hatfield, Pretoria, SA.
Tel: (00 27 12) 342 2733
Fax: (00 27 12) 43 4045
Contact: Johan Kemp

SANDTON
Sand House, 45 Wierda Road West
Wierda Valley, 2196 Sandton, SA
Tel: (00 27 11) 884 5559
Fax: (00 27 11) 884 6476
Contact: Chris du Toit

STELLENBOSCH
9 Herold Street
7600, Stellenbosch, SA
Tel: (00 27 21) 866 5987
Fax: (00 27 21) 886 5906
Contact: Sam Kelbrick

SWAZILAND
4th Floor, Mbandzeni House
Church Street,Mbabane
Swaziland, SA
Tel: (00 27 9268) 40 43 658
Fax: (00 27 9268) 40 43 658
Contact: Harris Kamanga

DAVIS LANGDON & SEAH INTERNATIONAL
WWW.davislangdon.com

Index